CRASH LANDING

BORN AGAIN FOR?

SHANE MANILAL
B.A, B.A Hons, P.G.C.E

Foreword by Dr. Brian Wilson

INDIA • SINGAPORE • MALAYSIA

Notion Press

Old No. 38, New No. 6
McNichols Road, Chetpet
Chennai - 600 031

First Published by Notion Press 2019
Copyright © Shane Manilal 2019
All Rights Reserved.

ISBN 978-1-64650-618-7

This book has been published with all efforts taken to make the material error-free after the consent of the author. However, the author and the publisher do not assume and hereby disclaim any liability to any party for any loss, damage, or disruption caused by errors or omissions, whether such errors or omissions result from negligence, accident, or any other cause.

While every effort has been made to avoid any mistake or omission, this publication is being sold on the condition and understanding that neither the author nor the publishers or printers would be liable in any manner to any person by reason of any mistake or omission in this publication or for any action taken or omitted to be taken or advice rendered or accepted on the basis of this work. For any defect in printing or binding the publishers will be liable only to replace the defective copy by another copy of this work then available.

Contents

Foreword 5
Preface 7

BOOK 1. FIRST HALF

Chapter 1: Early Beginnings 15
Chapter 2: Childhood 22
Chapter 3: Primary School Highlights 32
Chapter 4: Sporting Rivalry 45
Chapter 5: The Same, But Different? 50
Chapter 6: A Natural Love for Her 58
Chapter 7: High School 64
Chapter 8: Standard 7 71
Chapter 9: Remember the Time? 80
Chapter 10: Standard 9 – Soccer Distraction 89

Chapter 11:	Durban, Twice as Naughty?.................	110
Chapter 12:	Life Presentation	132
Chapter 13:	Matric, Finally!..........................	136

BOOK 2. SECOND HALF

Chapter 1:	R U Ready?..............................	159
Chapter 2:	City of Gold	180
Chapter 3:	City of Insanity?.........................	189
Chapter 4:	Crash Landing?..........................	205
Chapter 5:	Born Again	225
Chapter 6:	One Hand Job?..........................	241
Chapter 7:	Moonshine	256
Chapter 8:	A Working Holiday	267
Chapter 9:	Back In South Africa	279
Chapter 10:	The Book Begins	282
Chapter 11:	New Thinking?..........................	286
Chapter 12:	Planting Something Fishy?................	297

BOOK 3. EXTRA TIME

Chapter 1:	First Half – Short Stories.................	319
Chapter 2:	Second Half – Etherealism	428

Afterword 487
Glossary 493
Republic of South Africa 497

Foreword

I met Shane Manilal recently, about 4 months ago, (January 2017) at the Grey Goose Game Lodge, outside Newcastle, KwaZulu-Natal, South Africa. I spent some time speaking to him and have also read this book.

It is clear that we share a number of views on life, especially on the importance of the care and preservation of nature. He is an interesting man who is anxious to save the world and humanity from its various and burgeoning crises.

His honesty, concern and passion for life and humanity are very clearly evident in his writing.

A very serious accident, with a near-death experience, multiple major surgeries and permanent physical handicaps changed Shane's attitude forever and set him on his present course, which still needs to fully materialise and come to fruition.

He is indeed a brave soul. I read this book with great interest, especially as Shane grew up in Newcastle, in South Africa, against a backdrop of great political change and turmoil.

Colonialism and apartheid are collapsing everywhere, and the birth of Nelson Mandela's Rainbow Nation offers hope to all, a vision that is clearly portrayed here.

A hope for unity in humanity, to shift focus to our planet's existence—that should be everyone's main focus that Shane's life story embodies.

Think 'out the box', and you will enjoy this book as much as I did.

Dr. Brian Wilson

Newcastle

Preface

(This has to be read first, to help you understand my different work)

There is a conception about fate and destiny—things happen how and when they are supposed to happen—Universal Timing. This book also follows that premise because it changed three times.

Personally, my dreams and aspirations are hard for me to crush because I do know that I have a word to spread. Even the actual typing of my word is, for want of a better word, challenging.

I say this because I am touch-typing with just the index finger of my now main left hand. I'm now a mild epileptic, and my right hand is disabled, as well as my skills, training and talent—if disability equates to non-usage. I was once so much more, but I am now, honestly, wasting my life away.

Please excuse my totally unintended vanity. I am honestly a simple, humble person. You may even take my life story to be a bit too egotistical, as a fabrication, but it is a true reflection of my life's incidents—what happened to me.

We do tell stories. Yes, mine may not be exact 100% replications of the past, but the main facts are adhered to. I have given a few pseudonyms, yes, but these incidents have happened. I seem to now have a strong memory of my past.

I keep getting told by others that my life is a different story. Yes, we all have unique lives, so we are all different and special. I'm documenting mine in this book.

As pointed out in the sub-heading, I have to explain its difference to place you in context. We all know that writing is creativity; artistry in a sense. I will, therefore, classify this book as abstract art.

It is different as it has my story in normal font, with me (Shane) as the lead character, *(but I also join you in bracketed italics with a lot of my opinions, diarised incidents and visions. I even joke and Laugh Out Loud (LOL ☺) with you. I also ask 'neh', which is slang for affirmation like a 'yeah' and Afrikaans' 'Ja', which means 'yes' and is pronounced 'yah'.)*

That is my status quo, my trademark as a writer. I join you as you read. *(Like this. Why not? I'm being me. There are also a few stories that make the bigger story. Plus, a glossary of the way I speak, with a few slang words as well as my cultural terms etc. and a map at the end.)*

This book's difference is reflective of my simple theory in life. If you are given two separated points, let's call them point A and point B, and are asked to go from A to B, most people would hastily follow a straight line and go directly.

But what if you went your own way—up and down or perhaps backwards or in another direction altogether? As long as you reach B eventually, there is no time limit to your personal journey, is there?

I mean, you are achieving the objective of getting there, in your own time and way, because there was no specification as to the fastest way or easiest way. You were just told to join the points together.

So too with my writing. This book, being true to me, is eccentric, sporadic and abstract.

There are also a lot of characters, some with real names and some pseudonyms. Maybe too many characters, but this is my journey. I have been around the block and nobody will mind being mentioned at all.

I also love poetry, which I think is the actual reflection of written creative freedom because they are true to emotional depictions.

Creativity cannot be truly and routinely followed, can it? My main point here in the preface is that we are all unique in our individuality. Be true to yourself, decide on your own route to your life's destination. Be true to you. I am while writing.

As the saying goes, when life gives us lemons, we must just make lemonade. But what if you get too many lemons and you have no juicer or even honey, what do you do then?

Must we cover our heads and hope adversity goes away? Great sentiment, but we have to make a choice to lift our heads, take it, and work it to our advantage.

How we deal with challenges is what defines us. My challenge was the reduction and reworking of this book, because of its own planned destiny.

Now, sticking to my love of sport as a figure of speech, Book 1 is the 'First Half', focusing on my school years. In my day, there was Class 1 and Class 2. Then, Standard 1 to 10. So then, Class 1 is also Grade 1 and Standard 10 (Matric) is Grade 12.

Then the highlights from the rest of my life are in Book Two, 'Second Half', until when I began my new thinking.

Book Three is 'Extra Time', the 'First Half' of which is relevant short stories that led to the 'Second Half' – 'Ether**real**ism', my new thinking, that ends this match with the actual answer to my posing question in the title.

I have been told that I am a visionary in a sense, with a gifted hand that tells my story. Again, please don't think that this book is vain and see it not as me just tooting my own horn but as my reality, my destiny, my life.

My Life

I am so lost, with my inner lust,
that sometimes I wonder,
will travelling on dust,
get me yonder?

Where I am headed
I do not know?
If only it was embedded
in something, tangible to sow?

I seem to be wild and free,
floating in fantasy,
settled in an abyss,
of pure untapped bliss.

Somehow, there is inner fulfilment
and no detriment,
to anything I desire,
but this life isn't burning like a fire!

Writing, I find happiness.
It puts an end to the painful inner stress.
So, where exactly this book leads me,
I cannot wait to see.

In my own mind, my destination seems to be a secret,
maybe it will be revealed,
by a reader's patient wit?
Then and only then, will it be sealed!

Tap into your patient wit and join me. I am wishing and hoping that I inspire you as you sail your life's dream.

My dream to help Mother Earth primarily, and then humanity, will be met as I vow to donate proceeds from this book. *Okay, enough now, Shane, get back to the book.* (Oh, I speak to myself as well! *Crazy?* Why not? *LOL* ☺)

I always approach life with a smile and laughter in a fun way. Step out of your comfort zone and join me to sing with the Seven Dwarfs, "Hi ho, hi ho, it's off to work I go, tarra rum pum pumm!"

My work is to write and yours, to read. Off we go!

Love & Light,

Shane ☺

BOOK 1
FIRST HALF

Chapter 1

Early Beginnings

It is early morning on 12 March 1977, at the house behind the shop in a small residential district called Lennoxton, which is an Indian suburb in Newcastle, a small mining town in the Northern Natal province of apartheid segregated South Africa.

Inside the house, the sound of a bell being rung is heard. A four-year-old boy, Rickesh *(known as Ricky)*, rings the bell louder and chants, "Matha *(Hindi word for godmother)*, give me a baby brother. Matha, give me a baby brother!"

Ricky's two slightly older sisters, Sunitha *(Suntha)* and Reshma *(Resh)*, are very worried. They were told that a baby is coming.

Resh and Suntha are only five and six, but being little girls, they somehow know what their mother, Sheila, is going through. Besides, they can hear her screaming in the other room even though their father, Manilal *(Money-lahl)*, nicknamed 'Manny' *(Man-knee)*, is there.

Also in the room with Sheila is the children's grandmother, their 'Nani' *(the Hindi word for grandmother)*. Sheila's sister Leela 'Mosi' *(the Hindi word for mother's sister)* and Manny's sister Betty 'Pooah' *(father's sister)* are there as well.

When he is told of this story, Shane is still amazed by it. *(My special story!)* His entrance into this world as a South African Indian, given the day and times, was more than what anyone would consider challenging!

He is also told that it was a surprise entrance, as prior arrangements regarding hospital and doctor bookings had not been made.

Manny, given the trying times, also had no car to take Sheila to hospital. Shane was born at home, in a bedroom, with his Nani and aunts as the midwives.

Before Shane came, Manny, given the almost desperate state he was in, stayed calm as his other son, Ricky, was also born at home.

He had to eventually go for a doctor on this rainy night and walked all the way to the doctor's house, knocked on the door and explained his story to the doctor, who good-naturedly dropped everything and told Manny to get into his car.

They sped off but when they reached the house, Shane had already 'crashed' into this world. It can really be seen as a symbolic crash then, especially when one considers all of today's technology. There are specialist gynaecologists, induced labour, caesarean sections and medication to make the actual process a little easier.

It was different in those days though, especially because it was under South Africa's apartheid system's discrimination of Indian families.

Shane's parents came from and grew up in more than the clichéd 'trying times'. Their relationship is actually a very touching love story.

(Let me rather give it the highlight it deserves as a sub-story, rightfully giving it grace, because at the end of the day, had it not have happened, I would not have been here, right?)

Divine Love

In 1955, when Manny was eight years old, he met Sheila, who was five. His family, the Praags, was a big one. He was the second youngest of the five brothers and seven sisters.

They lived in a designated Indian smallholding called Mayoyo *(Ma-yo-yo)* in Dannhauser, a smaller town about 50 kilometres from Newcastle.

The reason for this Indian area designation is that, in that time of apartheid, there were segregated areas for the different races. The vacant plot next to the Praag family's was now being occupied by the Sivparsadhs, Sheila's family.

At that time, her family consisted of three brothers and four sisters, of which she was the third-born. With it being slightly younger than Manny's and also with them being new to the area, a kind of authority was given to Manny's family.

Even with this taken into account, it was a very communal way of life in those days. All the children became friends, and even though there were a lot of chores for them to do, they would still make time to socialise and play games.

These were truly very trying times, as there was no electricity or running water. They lived in clay houses, with smeared cow dung on the floors. There was a well for water, a coal stove for cooking, candles and gas lamps for light and an outside toilet with the bucket system.

The yards were quite big, so they kept cows for milk and a few chickens and other livestock. Children would have to wake early, do the chores, milk the cows, sweep the yard and help with other things before going to school.

Even though it may seem like tough times based on the lack of luxuries, the love within the family, respect and the close-knit bonds they formed remained.

The outside areas were massive, so the boys converted a small section into a soccer field for playing on when they had time.

Manny was called 'Thunder' because he could kick the ball with a cracking thunderous strike.

Their lives were lived to the full because they made the most of what they had. This was the lives they were destined with.

As neighbouring families, the Praags and Sivparsadhs formed close bonds. Even with this neighbourly, almost sibling closeness, when Manny was 15 and Sheila 12, their love bloomed and their teenage romance began.

There was no high school in Dannhauser, so Manny now had to take a train to Newcastle, a slightly bigger town about 50 kilometres away, very early in the morning to attend the Indian School, Saint Oswald's Secondary.

After school, there was a long train trip back home to Dannhauser as well. Then the walk from the station back home. He used to pass Sheila's house every night, whistling his love tune for her to hear.

Love was in the air for them, but in those days, asking for more than a private chat over the fence would be too much.

At the end of that year, Manny left school with a Standard 8 qualification and went to the big city, Johannesburg, in search of work.

He got a job but still returned home to Dannhauser whenever he could for his sweetheart. Their love endured and they married in 1968. Sheila was 17 and he was 21. On the honeymoon night, Manny displayed his love for music. In his beautiful voice, 'Romeo' sang the following Hindi love song for his Juliet, Sheila:

> 'Thumi mereh manzil, thumi mereh pooja, thumi devata
> hoh, thumi devata hoh'
>
> (You are my destiny, you are my prayer, you are my
> goddess, you are my goddess.)

As a newly married couple, they went to start a life together in Johannesburg, where they lived with Manny's older sister in Actonville, the Indian area in Benoni, a town very close to the main city of gold.

Manny and Sheila survived there for a whole year, but it was too tough. Finding it too challenging to live there, with hardly any steady work, they returned to Dannhauser to be with their families.

Manny had to go through all the effort of taking the train to Newcastle to look for day jobs and then return. Eventually, the travelling became too costly, so they decided to relocate to Newcastle to live with his sister, who the kids would jokingly call Farm *'Pooah'* because the smallholding she lived on, with other family members, was called 'Rose Cottage' and it was basically like a farm.

In those days, there was a kind of hierarchy. These were extremely trying times indeed for anyone. The women had to wake up at three every morning, light the fire on the coal stove, boil some water and have a bath. They would then change into a sari, a form of clothing worn by women, which is basically a long cloth that is wound around the body.

Draping a sari is quite a complicated task. Once this was done, the women would then enter the kitchen to cook breakfast for all the working men.

Looking lovely in her sari, Sheila would also make a pile of roti, a hand – made Indian bread, to pack into their lunch tins. What everyone respected the most about Sheila is the fact that she would just bear it all out.

(My acronym is GABI – Grin And Bear It).

Sheila used to *GABI*, no matter how challenging it was. She always said that it was fate, her destiny, and she used to deal with it in the best possible way. It was also a challenging time for Manny again, who was struggling along, trying different jobs. He had the initiative though. He even used

(Found this oldest picture of my Creators.)

to drive the breakdown truck to collect accident-damaged cars for his brother-in-law, George, who was running a scrapyard at Rose Cottage.

Wanting more from life though, as these were very trying times, he used to walk around the whole day searching for work. But, although he did the odd job and other things, he never had any steady work. This did not discourage them though, as this was their life and they were making the best of it.

They moved on, and in the space of just over three years, Sunitha, Reshma and then Rickesh were born. There was a mess up at the registration offices, so their last name was not Praag, but Manilal.

Manny realised that Rose Cottage was getting a bit too crowded. He now had steady work at a construction company, so they could afford the rent at the house behind the shop in Lennoxton, where Shane crashed his first landing.

They were, however, given a deadline by the owners to vacate the house after Shane's birth! Even though this may seem like a bad omen, destiny was playing its own tune.

A new housing settlement area, Suryaville *(Sue-ria-ville – Surya is actually the Indian Sun God)*, was being built. Manny had applied to buy one of the houses in a residential district called Asnika Crescent, and just as Shane was born, Manny got the bond from the bank approved.

Even though the house, their *'Mandir' (temple)*, was a small two-bedroom unit with one inside toilet, kitchen and lounge, it was a brand-new house in a completely new residential area.

Yes, it was small by today's standards but palatial for them after the birth of their, as they say, good luck prince!

With a princely crown on his head, Shalendra's (nicknamed 'Shane') story of his 'riddled' life began in his palatial *Mandir*.

A Life Riddled…

I am what I am.
But what I am,
I cannot be!

For what I am,
I am not 'supposed' to be.
Yet what I am,
is all I can be!

So, what I am will never be!
Because, what I am,
is what I was,
and what I was,
is no more what I am!

What am I?
I am what I am.
BUT???

Chapter 2

Childhood

For those first few years as a toddler, with there being a four-year age gap between him and Ricky, five years between him and Resh and six years between him and Suntha, when they went to school, Shane was left alone at home with the maid, who he used to call 'Gogo' *(A Zulu name for an older woman, pronounced 'Goh-goh')*. This was because Sheila now had much-needed work as well. She would wait for Gogo's arrival, before leaving with Manny.

While Gogo performed the chores, Shane would spend most of the day out in the garden, in the sand, playing with bricks, sticks, stones and anything else he could find.

He also befriended the little girl from next door, Cookie. She was the same age as Shane, as she had been born just a day after him. They were best friends and would spend hours and hours together in the day, outside in the sand, in the garden.

Yes, they would both end up dirty, but Gogo would wash them before she fed them. With their tanks full, they would go back to the sand. They enjoyed themselves, played with bricks, which they would split to make cars, or slip into pretend mode and also played 'house, house', 'shop, shop' or 'doctor, doctor'. Childhood fun games!

At around 10h00, they would hear the horn in the distance. All excited, they would run to the end of the yard to wait for it to come up their street.

With a big smile on his face, Shane would put his hand into his pocket to take out the money that his mother bribed him with every morning to be quiet as they left. It was a whole 20 cent piece. It needed to be used, especially since he had a lady to treat!

Hand in hand, they waited anxiously at the gate for it. It could be heard. Their insides bubbled with anticipation.

They lived on the corner of Asnika Crescent. Because it was a circle, they'd only see it when it came slowly around the corner. The fruit and vegetable truck that actually used to tour through the district, rather like an ice cream truck, used to stop for Shane, who would stand on his toes to reach the counter and put his whole 20 cents down.

Abaahs, the owner of the truck, would smile and wink at him. Abaahs knew exactly what Shane wanted and would bend down and get them two 10 cent guava juices, one for his customer and one for his lady friend.

They would then go back into the yard, stand by the gate and wave Abaahs off as he drove on to the next stop. Because their juices were sealed and they were not yet strong enough to twist off the cap, they used to take them to Gogo to open.

As a kind of thank you reward, Shane always shared half with Gogo. Cookie also learnt Shane's trick and poured for Gogo some of her juice as well. Once their engines had been filled up, they would roar back to the garden for some more exuberant childhood fun.

Then later in the day, Ricky, Suntha and Resh would come home from school and play together with the other neighbourhood children from the district.

Shane and Cookie were still too young to join them fully, but they watched and idolised them with the typical childhood thought: 'One day, when I'm big, I'll join them.'

A jolly toddler childhood passed, and Shane grew big enough to join them but had to attend school first, even though he did not attend pre-school.

Suryaville Primary School was conveniently just a five-minute walk away from his home, and even though his parents objected to it, Shane wanted to walk to school with his brother and sister on his very first day. *(Let's go back to the past and join him in the present tense.)*

School Time

January 19th, 1983, Shane was finally in his uniform, bag in hand, getting ready to walk to school with Ricky and Resh. His parents had agreed to his tantrum to go all by himself.

Suntha, his eldest sister, was going to the big school that year, because, in those days, primary school went up to Standard 4, so she was now going to Standard 5.

At the front gate, Shane waved goodbye to his mother, who, out of concern, shouted to Resh to look after him as she waved at him. He smiled and set off, his excitement evident.

Cookie, like Shane, also hadn't attended pre-school as it was not a necessity, so he stopped outside Cookie's yard and called for her. Her mum came out and said that they would be taking her on her first day, as most other parents were.

Shane's feeling of being special grew even stronger; he felt big. He had manipulated his parents into not escorting him. It was a fun walk, stopping and waiting for a few other kids, like a team, a crew.

When they reached the school, Ricky ran off with his friends to play, while Resh took Shane to the office and asked the secretary where to take him. She was told that all the 'Class Ones' were going to be put in one of the bigger classrooms for the time being, and then, when they were divided, they would have their own class.

Resh took Shane to the class she was directed to, the bigger 'multi-purpose' room, as it was called. They could hear the commotion coming from the corridor before they even reached it.

Outside the room, there was a buzz of noise and sounds of howling, with lots of children crying and clutching tightly onto their parents.

At the door, Resh wished Shane luck, and after making sure he was fine, she left him outside and walked off to her class.

Shane said goodbye and walked in. He was so amazed by what he saw that he just stood silently and watched the fuss and commotion going on. Some children were comfortable, especially the ones who had attended pre-school.

Other first-time scholars were still outside, clutching onto their parents, crying and having tantrums. Shane, who honestly couldn't see what all the fuss was about, thought, 'Is this it? Where is the excitement? Where is the thrill? Why are they crying? They are now at school, all grown up.'

Pondering over these questions, he walked to the front and stood next to the chalkboard. There were no desks and chairs in the room, which was filled with about 60 children. The children had to sit on the floor in little groups. He scanned the scene, searching for Cookie, who wasn't inside yet.

Looking outside, he finally saw her. She was clutching her mother's hand and had a scared look on her face. But when she saw him, she let go of her mother's hand, told her that she would be fine and walked in.

Her mum, with soggy eyes, looked at her now grown baby stepping into her first classroom. Most of the parents still hung around, proudly looking in from the window.

Shane and Cookie sat down on the floor together and waited. Those who had attended pre-school and were already friends were in their groups. Shane noticed in a way that he and Cookie were the only mixed match in terms of gender.

The boys were in one group and the girls in another, even the ones who were there for the first time. Shane noticed the boy's circle looking at them and giggling in a mocking way. Most of the girls had tears in their eyes, so they didn't even notice.

Nonetheless, Shane and Cookie still sat together and carried on speaking to each other, waiting. After some time had passed and the children were finally settled, the headmaster welcomed them and explained what was going to happen. Their names were to be called out, separating them into different classes. All the children were scared, nervously holding onto their friends and hoping that they would still be together in the same class.

After quite a wait, Shane's name was called to go to a group. He had to leave Cookie, but she said that it would be fine. Shortly afterwards, her name was called.

He was hoping that she would be in his group, but unfortunately, she was in another one. His group had about 10 boys and 10 girls. When all the groups were complete, they were introduced to their form teachers. His was Mrs. Tooray, their 'Mam'. She smiled, said hello to them and told them to follow her to their classroom.

They were each given a place, with the desks in groups. Just as they were seated, the bell rang for the lunch break. Mam told them to take their lunches and go into the grounds, eat and play, and when the bell rang again, to assemble outside in their lines.

Shane took his lunch tin out of his bag, stood up and walked outside, ready to make new friends as well as see Cookie again. Before he could find her though, he was, in a sense, excluded.

All the boys were already in groups and had, in a way, excluded him by closing their circles. When he approached a group, a boy, Shannon, mockingly asked him whether he was looking for his girlfriend.

At that, they all started laughing and sniggering at Shane. Not realising that Shannon was a bully who had been to pre-school and already had a group, of which he was the boss, Shane asked him, "Why? You jealous?"

On hearing that, Shannon rushed at Shane with his head down. Shannon didn't realise that Shane was also tough, as he grew up with a bigger brother who used to knock him around silly.

Shane dropped his lunch tin as Shannon reached him, pushing head-first into his stomach, knocking him over and falling down with him.

He squirmed, tussled and twisted Shannon on the floor until he was on top. With Shannon on his back, Shane sat on him, jamming his hands between his knees—a solid lockdown.

With authority, Shane asked him why he had started with him and then warned him not to do it again. He then climbed off him, just before the prefects arrived to see what was going on.

When they asked what had happened, Shane reached his hand out to Shannon and said, "He fell down and I'm picking him up."

Shannon grabbed Shane's hand and pulled himself up.

The prefects could see from the faces of the crowd that had gathered that it was a lie, so when they responded, "Oh, okay!" and winked at him, he was stunned. He did not realise at the time that the prefects knew that he was the deputy head girl prefect's brother.

They smiled at him, winked and walked away. Shane was now seen in a different light. There was a new buzz in the air.

A few boys followed him as he walked around, still looking for Cookie. He found her in a small group of girls, friends that she had made in her class. From the way she looked at him, he could see what was obvious.

At school, in those early days, the norm was that boys and girls didn't play together. It was Cookie's turn to follow this set pattern. With his reading into this, Shane just asked her if she was okay, and she nodded. He nodded back and walked away, still with a few boys behind him.

Then, as he sat down to eat, he realised that his lunch tin was still on the floor where the fight had taken place. He was about to get up when Shannon approached him again.

'Oh no, not again,' thought Shane.

Shannon looked at him and said, "I think you left this behind?" He handed him the lunch tin. Shane was amazed, smiled and told Shannon to sit next to him. Smiling, Shannon did just that. They were to become the best of friends for those first few years.

A few other boys sat down with them as well, forming a new crew— the 'in' crowd. Like most of the other kids, after they had eaten, they played. It was mainly chasing, in big groups, or 'stuck in the mud'.

There was always one who no one else could catch. Shannon could run and run and run and not get tired like every other boy.

The bell rang and everyone rushed to their classrooms. Breathing heavily, they stood in their lines, girls against the wall, boys next to them. Prefects actually came along to make sure the lines were straight and led the children in when the teachers arrived.

Back in the classroom, Mam carried on from where they had left off and explained how things would work in her class. The very first thing for them to do was to introduce themselves to her and the rest of the class. She told them that she would give them each a turn to stand up and say who they were, and then the class had to greet them by name.

She then told them that they also had to say how old they were and tell the class a little bit about themselves, like what they liked and didn't like and what they wanted to be when they grew up. She then gave them about five minutes to prepare.

The class was quiet as everyone thought about what they would say. Their nervousness was quite apparent. After five minutes, Mam pointed at Shane's desk and stated, "Okay, let's start at that table, that boy with the green eyes."

Shane nervously stood up and said, "Hello, everybody. My name is Shane."

When the class said hello to him, he felt relieved, smiled and carried on, "I will turn six this year. I am so excited to be here. One day I want to be a doctor. I also want to be everyone's friend. Thank you."

Mam thanked Shane and called for the next pupil. All of the children had their turns, and all wanted to be doctors, lawyers or teachers. After they were all done, it was almost the end of the day and quite a few parents were already outside, waiting to collect their kids.

At 12h30, the block prefect started ringing the bell outside. This could be heard throughout the school as it was the bigger children's second lunch break as well. The noise made by the cheers when that bell sounded was understood by everyone because it was home time for the smaller children. This was because even though school was loved by everyone in those old days, especially the youngsters, more fun could be had on the way back home and later at home.

Many of the children ran into their mothers' arms. Some of them were leaving alone, mainly the ones who had come from pre-school. As Shane was about to join them, Resh came to check up on him.

"Are you okay, Shane?"

As the *Mandir* was just down the road and watching Shannon and the others almost leaving, he nodded his head and responded, "Yes, Didi jee," *(a Hindi respectful name for sister, pronounced 'Dee-dee-gee')* and ran off to them. Didi jee smiled and waved him goodbye.

Shane was already out of the gates and on his way home. He had always heard his brother and sisters and older friends talking about Gahaan, a street vendor, when they left school. He asked Shannon, who told him where Gahaan was.

Gahaan would have his car boot packed with Indian children's goodies like 'bor' *(a spicy Indian snack, like pickled dried fruity seeds)*. There was also figs, sour seeds, mango with masala, dried mangoes and other things.

There is a kind of Indian slang word as to what your taste buds do when you shudder in a way after you eat something sour. It's called 'june-june-knees' *(when your taste buds tingle and you start salivating)*. Also, because some of the delicacies were pungent, the children of Hindi origin would say 'thitha', while those of Tamil origin would say 'karoh'.

(What follows is probably the teacher in me that has to point out that Indians are not all the same. We have different cultures and traditions based on different languages that approach Hinduism traditionally. This is kind of similar to the difference between the majority of Caucasians. Some are of English descent and some, the 'Afrikaner', of Dutch descent. They are the same race and Christians but with slight differences.

There is a similar grouping in Hinduism. Within the religion, you mainly have Hindi, Tamil and Gujarati, based on the language they speak.

Then, looking very similar to them physically, there are also Muslims, who follow a totally different religion as well as culture. Let's get back to the story. I feel like a documentary disclosure now. School is finished.)

In the summertime, there were also 'aunties', the name given to all the female retailers who used to sell ice blocks at their homes. It was Kool-Aid frozen in a cup with a piece of wood for the ice cream stick.

Kids would have to go to these aunties' houses and stand in lines, while the aunties went to their freezers to get the frozen Kool-Aid blocks out of the cups. After acquiring one, they would then suck it as they walked home, ice melting and dripping all over their hands and sometimes messing their shirts.

Walking in groups, they would escort their friends to their homes, say goodbye and carry on, like a procession going home for their lunch. They would definitely meet up again in a short while though, as it was playtime on the streets after lunch.

When Shane finally reached home, he would have lunch with his parents, who had just come home for their break as well. The reason for this is that Newcastle is a small town, and it takes only about 10 minutes by car from the residential to the business district. When they had eaten, his parents would rush back to work.

After lunch, Shane would go next door, check for Cookie and they would play together till the older kids, who finished later, got home. Then, after the older group had eaten, they would all play together.

In the 'good old days', there were games like three tins *(stacking three tins and aiming at them with a tennis ball)* and hopscotch *(marking the ground in squares and throwing a tile into the blocks and then jumping into the blocks without touching the lines)*. Some girls also had hula hoops and skipping ropes. There were many other physical games too. It was a glorious time.

Apart from these games, sport and other fun activities also took place after school. There was a park in the centre of the two circular residential areas; it was big enough for the older boys to form teams and challenge the other district.

The sport depended on the season, with the main ones being soccer in winter and cricket in summer. The winners would get oranges or green mangoes, which would be cut up and eaten with salt and chilli powder. Then, as they grew older, they would use their pocket money to buy a bottle of Coke, like a trophy to be claimed.

When he was old enough to join them, it would either be Shane's favourite one, soccer, or cricket with a tennis ball or baseball with a broomstick for the bat.

Chapter 3

Primary School Highlights

Shane loved being in school and learning. He made friends and progressed. It was fun. At the end of those first two years, there were no tests, but in Standard 1, that situation changed.

Miss R. R. Singh, his new 'Mam', kind of changed his life forever, as her approach to education was balanced and not just didactic. Because it was her first year of teaching, she had lots of new ideas to try out. Shane loved how creativity balanced with studies as tests were now given.

He was a top performer, coming first in the class. His intellect flourished as did his creative side.

(Funny fond memory for me: I used to make a playdough helicopter like the one in 'Air Wolf', the cult television programme that all the children were hooked on at the time. Because Miss Singh had no objection to it, I placed it in my classroom cubicle. This became the 'in thing' and all my friends made one. In the mornings and on breaks, there was Air War. Oh, the glory of childhood memories.

Why don't you close this book for a bit, think back to your favourite, most happy one?

Welcome back. Smiling inside? That is what I love about good memories!)

This Standard 1 year was to be Shane's fondest memory of primary school. Also, his current knack and passion for public speaking came under the spotlight when he was chosen by Miss Singh to represent the school in the English-Speaking Board examinations in spoken English. He excelled and was awarded a certificate.

> **ENGLISH SPEAKING BOARD**
> (INTERNATIONAL)
>
> THIS IS TO CERTIFY THAT
>
> _____
>
> was awarded a _____ in
>
> **JUNIOR GRADE 1**
>
> OF THE BOARD'S EXAMINATIONS IN SPOKEN ENGLISH
>
> 6 DEC 1985

The end of that year also heralded the start of things to come. *(Not showing off but I was the top student and it was that way for the next few years.)*

With everything going so well on the academic side, Shane's mind was left with more room for balancing with other interests.

1986 was when the programme 'Shaka Zulu', about the Zulu tribe's king when the European settlers first got to the area, was featured on TV. His first performance at the school's variety concert was also a dance traditionally performed by the Shaka Zulu tribe.

With his friends, Shane also used to imitate the actors in the TV show and actually have – play wars across the river from their district. This playing of games at the river with his friends also developed his love for the outdoors and freedom. They would make rafts, swim and frolic in the river.

On weekends, Manny used to take his family to the dams at Normandien, the farming area that borders Newcastle, for freshwater fishing and picnicking.

(I found a surreal image of peace. Here is a picture of my father with his fishing rod, which was taken in the eighties. He is gazing into the distance, with his hands behind his back, which I love doing as well.)

Being out in the open just increased Shane's love for nature. A clear memory for him was how the van used to get bogged down in the mud on a wet road. They would then have to push it out, jamming logs under the wheel and so forth.

It was always worth the effort to be at the water's edge, fishing away and resting the mind. Shane's mind, however, never rests.

One day, while fishing, he wondered if fish ever ate chilli. Deciding to experiment, he broke up some of the brown bread they used for bait and kneaded it with some chilli. After casting his fishing rod into the water and setting it on the rod rest stand, he went off up-field to urinate.

Not realising that there was still some chilli juice on his fingers, he started urinating, holding his penis with his chilli flavoured hand. It took a few seconds before he felt the most agonising burning sensation on it. Screaming with the pain, he ran down the bank and into the dam in a frantic attempt to douse the flames. The cool water helped to alleviate it, and he stayed in there for quite a while.

Even with pains like that, it was actually more than a great childhood for Shane; it was well spent enjoying the outdoors and everything he did. He was having fun, playing sport at school and also after school by challenging the team in the other district. Socialising and making solid friendships went along with the fun.

With his playful side flourishing, Shane still somehow remained first in his grade. He was also captain of the soccer team and cricket team. School was a breeze for him. As well as after school. The boys played marbles for keeps, meaning the winner gets to keep the loser's marbles.

One day, Shane was beaten and ran out of marbles. He went back home sadly but then remembered that some small white balls came with his compendium of games. Taking them out, he saw that even though they were plastic, they were roughly the same size as marbles. He scooped them up and took them to the playground, where he told the boys that his father had bought him special marbles from China. The excited boys called them 'Chinas' and decided that they were each worth two marbles.

Shane was back in the game. He and the reigning marbles champion, Anesh, became the best of friends. They called themselves dirty 'brooders', their unique slang for brothers, and made the typical blood promise: cut your finger and exchange blood vow. This made them brothers for life. They shared the same passions, like marbles, bicycles, skateboards and sport, and specialised in their favourite game, soccer.

Anesh's dad was a soccer coach for the adult club, Stellas, and Shane and Anesh would spend the entire Sunday at the soccer grounds, playing on the side-lines and watching the other league matches. They also went off to the shop to buy bread and tinned fish for lunch.

When playing, they really worked well together, as Anesh provided Shane with beautiful through passes, splitting the defences. Shane would then finish clinically. The game thrilled them.

They even played barefoot in their district playground. It added to the fun, hardening the soles of their feet and even making their toes bleed because the toenails would come loose from playing on the tarred roads. Strangely enough, even the pain was fun.

There were no shin guards or any other form of protection. These things were not important, just the ball was. All the boys from the district would join in. Because Shane's father always spoilt him by buying the ball for him, he would walk around the district circle, bouncing the ball.

The mere sound of it brought all the boys out into the open field in the centre of the circle to play the game that they loved. They had a really fun time and would play till it got dark and they could not see the ball any more.

(An incident from the playground that the other boys and I will never forget had to do with us naming an uncle.)

One afternoon, while playing, because of the park being in the centre with houses all around, the ball went into Uncle Veda's yard. Being in his fifties, he was fed up with this. He was also a bit tipsy, as he used to drink. He came out with a big knife and shouted at the boys. But, instead of saying that he would poke the ball, he said, "I'll poke the knife!"

The boys collapsed with laughter. That is why Uncle Veda got the name 'Poke the Knife'!

Even when they played cricket or baseball with a tennis ball, it still went into his yard. The boys would cautiously put a guard in the front of the yard and then scamper in to pick it up. When they had the ball, they would run out as fast as they could, before 'Poke the Knife' saw them.

It was truly a jolly childhood for Shane with the children in his district. This took precedence in his life, having fun, as he could so easily do it and still produce top marks.

Shane was unique. *(I had a nice cover-up. Not being vain again, but I relied on my intellectual competency too much.)*

Most of the other boys, who were a bit naughty but did not have the ability to conceal their delinquency, got into trouble. Shane was also naughty, but he covered his tracks nicely.

On the other side of the scale though, Anesh preferred to focus only on his passion, soccer. He excelled at it and therefore neglected his schoolwork. He often used to abscond from school.

After leaving home in the normal fashion, in full school uniform, Anesh would head off across the river. He would spend the day there and only return home when the siren sounded. It was the perfect truancy.

One day, in his head boy year, Shane was asked by the principal to go and search for Anesh. It was quite a task for Shane as he had an entire river bank to cover, which was about a kilometre long.

After a vigorous search, an idea came into his mind. He was alone, but he started shouting out fellow friends' names and giving them instructions as to where to look for his friend.

About five minutes of this passed and Anesh heard Shane when he came close to where he was hiding. Coming out from behind a big bush, he ran up to Shane and pleaded, "Call the others off, please, my dirty brooder. I give up!"

Shane burst out laughing. He then made a deal: he would say that he went to his house and found him sick in bed. This was if, and only if, Anesh would not abscond anymore.

The deal was made, and Shane duly returned and gave the story to the principal. It worked very well, as the next morning, Anesh was back at school. *(Another incident regarding him comes to mind now.)*

Also, a favourite Indian boys' pastime that was not mentioned earlier is playing 'ghooli dunda'. *(Here's a description of ghooli dunda for interested parties to start playing.)*

It is basically a broomstick that is cut into two pieces, one big and one small. The big one is to use as a kind of bat, the 'dunda', and the smaller one as the 'ghooli'. This ghooli is sharpened on both sides, like a pencil. It is a solo game that is played against all the others.

A groove is dug in the ground, and a square box shape is marked around it. Then the dunda is placed on the backline of the box, with the ghooli across the grooved slash. Another, thinner 'lifter' stick is positioned in the groove, under the ghooli.

The player (batsman) goes to the box and then, flicking the lifter as hard as he can, the ghooli is launched as far as it can go. The other players (fielders) all stand around. If the ghooli is caught, the player is out. If not, it is aimed at the box from wherever it lands. The fielder has one throw to get the player out. This is either by throwing it to end up in the box or to hit the dunda. If either option works, the player is out.

But if not, the player goes to where it lands. He is allowed three chances to get it as far from the box as he can. With the dunda in hand, the player taps the sharpened end of the ghooli to make it rise into the air. When it is in the air, the player then tries to whack it as far as he can with the dunda. He does this three times. His score is determined by how many dunda measures it takes to cover the distance back to the launch box. Each person has their turns, and the highest score between them for that round is the winner.

Shane and his friends would often play a good few rounds of ghooli dunda because they loved it.

At the end of one of their ghooli dunda games, Anesh left on his bicycle. Up the road, Shane was already sitting on the pavement with the other boys, and as Anesh rode past, his naughty streak got the better of him.

He thought that if he threw the dunda between the spokes of the wheels, they would jam up and just make a screeching noise. As the bike passed him, he hurled the dunda between the spokes to make the bicycle screech to a halt.

But, unfortunately, he threw it into the spokes of the front wheel, and the bike, a 'Western Flyer', went flying, throwing Anesh into the air. He tried to break his fall, but he landed on his arm and screamed out in agony.

Shane really felt guilty and got Manny, who was home, to take Anesh to the hospital. Anesh's arm had been broken, and he ended up with a plaster cast. The break healed, but his arm is still slightly bent. That incident, however, was overlooked and laughed off, as their bond was, and always will be, strong.

(Hand breaking? Hmmm... I was younger, around eight years old, I think.)

Shane and Ricky would train for their sports day at home. On the grass of their yard, they would put up vertical bamboo poles which they would cut from the nearby river. They would use clothes pegs to clamp the upright pole, as the holder of the crossbar. The crossbar was a bamboo pole as well, as straight as they could find. They would then run towards it and jump in what's known as a 'scissors' style. This is basically leaping up and uncrossing and crossing one's legs over the crossbar. The movement looks like a pair of scissors, hence its name.

Like normal high jump, the one who jumps the highest was the winner. Because there was no protective mattress to land on, they stayed on their feet.

Shane jumped first. After he landed, he fell to the grass and stayed lying down. Then Ricky jumped and came flying through the air, clearing the bar, but his right foot landed on Shane's hand, which was in his way.

It really hurt and Shane screamed with pain but was afraid to tell their mum about the incident. Assuming it would heal as time passed, Shane listened to Ricky and went to bed with the pain.

In the middle of the night though, the pain and swelling became so unbearable that he was not able to sleep. In pain, he went into his parents' room and woke them to tell them what had happened.

They were shocked, and Sheila made a rule that her children should always tell her what happened. Shane was rushed to the hospital, where his broken hand was put in a plaster cast.

After the three-month healing period, he returned to the doctor to have his cast removed. While this was being done, the doctor grazed his forearm with the surgical instrument he was using. It was a nasty gash!

Shane then saw the trend in breaking bones and therefore did it again a few times? This was as he was trying to leap off a swing but ended up falling on the same hand. Because it had not yet healed completely, it broke again.

(Picture proof – Joint birthday party with Ricky, in the garage. Zoom in to the smallest guy, 2nd on the left, with a plaster cast on his hand.)

As if that was not accident-prone enough, he also fell off their veranda wall, hitting his head on a ceramic pot plant and slightly fracturing his skull. This left another scar that is still apparent on his forehead.

He also loved cooking, and still does, and saw no danger there. One afternoon, funnily enough, while removing a big heavy bowl from the microwave, he had yet another accident. The bowl dragged the glass turntable off the base of the oven, and it fell out right onto his bare toe, splitting it. It was stitched back together but took a long time to heal though.

As a naughty boy, Shane grew up rough and rugged, with broken bones, split toes and scars to prove it. Typically, there were also a few fistfights with the boys from the other district.

They also grouped together though, as they all lived close to the river where they would swim in it, build rafts, hike, fish and even attempt to shoot doves with their slingshots. The birds were safe though. Very few were actually shot with a slingshot, as it is a challenging weapon to aim at something so small. It worked for bottles and cans though.

These naughty boys were very adventurous and also had fun with a farmer on the other side of the river bank. They would borrow his horses, without him knowing, for the day and ride them bareback, and then return them. When the farmer found out about this, he hid behind a bush with his shotgun and waited to catch them.

Excitedly, the boys ran up to the horses, but when they were close enough, the farmer jumped out from behind the bushes and shouted, "Ja, julle wou my perde steel? Ek sal julle wys!" *(Yes, you wanted to steal my horses? I will show you!)* He pointed and fired.

The others were saved, but Shane was hit on the back of his thigh. As he had shorts on, it went into his flesh. He carried on running from shock and fear, even though his leg ached. He got away, crossed the river and made it home, where he lay on the grass, bawling!

When Resh came outside and found him crying, she asked him what had happened. He told her and showed her where the bullet had gone in. The wound was bleeding, and even though Resh has a phobia of blood, she said, "I'll have to get the bullet out!"

She ran into the kitchen and came back with some cotton wool, methylated spirits and a small spoon. She wiped the injury with the spirits and then dug in, using the spoon handle. She felt it and pulled it out. It was a very dangerous modern bullet, yellow in colour with a unique shape and tip. It was a... popcorn seed!

The farmer had removed the metal bullets from the shell and replaced them with popcorn seeds to just sting them. Even in his pain, Shane smiled!

Kite season was another popular, and safer, outdoor pastime in the open field next to the river bank. They would build the kites themselves.

(This is how:) Their secret recipe for building kites would firstly entail breaking off some 'black-jack' thorn tree sticks. *(Black-jacks are those cluster thorns that are black and stick to you.)* They would use these because the sticks are straight and light, but any straight, light stick of about half a metre long would do.

One stick would have to be longer than the other one. They would then take a big black plastic bin bag and cut it into a diamond shape. The longer stick would then be taped onto the plastic diamond vertically *(from the top to the bottom)* and the shorter stick would be taped onto the plastic diamond horizontally *(from side to side)*.

To balance in the air, the kite needs a tailpiece, which can be made with the rest of the plastic by simply knotting the plastic pieces together to make a string that is attached to the bottom of the diamond. They would sometimes tie a stone at the bottom of the tail to give the kite some weight in strong winds. Then a piece of gut line from their fishing reels is attached to the long stick, from top to bottom, making a sort of 'V'. This 'V' is tied to the rest of the line, and the kite is now ready to launch.

To launch, one person held the kite and the person flying the kite stepped away. When the wind picked up, the one holding the kite would release it and the kite would take off. The one flying the kite then released the line as it ascended.

Ricky was the champion, flying his kites till the end of his line. When they could not go any further, he would put his rod and reel in a metal pole and tie it down. If the wind remained constant, the kite soared.

The only problem, though, was that sometimes the wind blew quite strongly, then the line would snap and the kite would blow away. They would run joyfully after it, tracking it, and maybe find it attached to a tree or telephone wire high up.

If it was a tree, they would climb up to get it back and fly it again. Sometimes the tree was a mulberry tree though. If this was the case, the kite lost their attention because all the boys would climb up to have fresh mulberries.

Special trips would be made to the mulberry tree on the other side of the sewage pipe, which took skill to walk across. It was like a balancing act for true men, a testosterone boost for those who succeeded.

They even crossed the pipe on their short cut route to the public swimming pool on the other side. Most of the time Shane spent at the pool was with his close younger cousin Sudhir (aka Moona, aka Moonshine, aka Moonikes).

(Moona's 8th birthday. His hand is around me, his bigger brother.)

Even though Moona's father, Roshan 'Mama' (*Shane's mother's brother*), and mother, Anila 'Mami' *(Mama's wife)*, were in Ladysmith, he lived in Newcastle with Shane like a true younger brother.

Shane and Moona loved to swim and sometimes spent the entire day at the pool. They would not even go home for lunch. This was because a shop opposite the pool had their staple diet.

Like at the stadium, while playing soccer, it was a tin of fish and bread. Also, they were not really worried about getting black *(a tan)*, which their elders, cousins and friends had a problem with.

(On this subject, why? Is being darker bad? Especially amongst Indians. In my opinion, the origin of this thought in our country was in the apartheid days, where it was bad to be black.

In fact, this seems to be a global thought pattern. I think this is based on the Europeans, the Caucasians, aka the 'whites', communising the world.

According to my studies, Homo sapiens all originated in Africa. We are proven descendants who evolved into different races based on the geographical climate conditions. A lesson for those who do not know this factual information follows.

In layman's terms, before the world was colonised, the different races evolved for the climate they lived in. Will provide the two major examples only.

Hot Africa – pigment in skin as protection from the Sun, so 'black'. Also, short curly hair, broad noses for the heat and muscular buttocks because of the need for running away from predators.

Then eventual migration into Europe's colder climate. Over thousands of years, the body adapts. Lack of Vitamin D from the sun's rays, loss of pigment – lighter skin – 'white'. Colder, so longer hair, smaller noses, plus not that much running, so buttocks decrease.

This is proven. Every 'different' human is the same under the skin. We even share blood types, meaning that we are the same species, with different races, on the same planet. FULL STOP!

BUT we still fight each other even though we are the same? ☹ I will continue with this just now, as a story of the first time I realised that we are the same but different. It is Chapter Five.

Chapter 4

Sporting Rivalry

In his Standard 5 year, 1989, Shane was made the head boy of the prefects, the small group of students who helped the teachers run the school. He was still the captain of all the sports teams and was also part of the debating team, as he was very eloquent.

Being now 12 years old and pubescent, he met the crush of his life that year, when their debating team challenged another school. Suraya was the most beautiful, shy and intellectual girl he had ever met. She had the most amazing eyes Shane had ever seen. They made eye contact and his heart dropped to the floor. It was the clichéd 'love at first sight'.

(You will get more of my heart's 'flutter' later. Regarding the debate, I was actually nervous when I delivered my speech and we lost. Maybe I chose to let Suraya win? A case of cold feet but warm-hearted defeat?

Okay, let me be like a typical, clichéd man who cannot share too much of his feelings and change the topic now. Let's try to burn the crush away with some sporting rivalry.)

As the head boy, Shane was the proverbial 'king of the castle'. A few weeks into the year though, another Shalendra, also nicknamed

Shane, came to the school. He was a transfer student from Howick, another small town in the province of KwaZulu-Natal.

The other Shane was given the nickname 'Skeech' because his surname was Seechoonparsad. Aside from the coincidental names, their similarity went further.

Skeech was also top of his class at his old school and the captain of both the soccer and cricket teams. Oh, and he also had 'unique' green eyes just like Shane.

A difference, though, was that Skeech was shy and softly spoken—an introvert—and the total opposite of Shane, the extrovert. Also, Skeech had very neat handwriting.

Totally opposite, Shane, however, had untidy writing. His numbers were not clearly written either, and he used to get scolded by the strict mathematics teacher. This was mainly because they looked like other numbers. Shane's number '9', hastily written down, looked more like a '4'. For the same reason, after an incident one day, he was punished and had to stand on top of the desk as he had apparently written the wrong answer to a simple question.

As he climbed up on the desk, the matter went further as the class laughed at him. When standing there upright, he arrogantly told the teacher that his answer was correct and that he needed to re-check it, maybe with his glasses on.

The teacher then instructed Skeech to check it, and when Skeech merely stated that it did not look right, that the '9' looked like a '4', Shane lost it and jumped down from his desk.

Agitatedly, he went to Skeech's desk and pointed out how the '4' was actually a '9', but because he wrote so quickly, he did not have time to complete it properly. Because Skeech had spent extra time on his writing, he merely said to Shane that he must take more time and present neater work.

A kind of subtle argument developed, and the teacher screamed at them to stop. He also scolded Shane for jumping off his desk and brought out his cane. In those days, corporal punishment by teachers was legal. Shane cowered and took the blow.

When the teacher, in his anger, was about to strike Skeech, Shane grabbed the cane and held it back, saying that it was not right to hit him as well.

The teacher then lost it further and screamed at Shane for holding back his cane. In his defiance, Shane was too scared to let it go.

With all the noise going on, the Afrikaans language teacher from next door came into the class to ask what was going on. When she saw that Shane was involved, she said, "Oh, our head boy again. He is so naughty in my class as well. He takes advantage. He doesn't do his homework, relies on his brain. Let's de-badge him and take away his head boy status, so he will learn."

On hearing this, Shane said it was fine if they wanted to. The history teacher then entered and said to take the matter to the office because the rest of the classes were being disturbed. There, Shane explained to the principal what he had done and said that he was merely protecting Skeech, who had done nothing wrong, and that it had all been his fault.

Shane also added that he had deserved the maths teacher's anger and apologised graciously. He had a way with words and was merely given a warning that if he were ever to get in trouble again, he really would be de-badged.

Skeech then offered to help him do his mathematics properly the next time. They shook hands.

(I seem to be giving too much 'outside' detail away, hey? Warned you of my blabbering. Let's get to the story and start fresh.)

'Sporting Rivalry'

Because there were four different sporting 'houses', like groups, Alpha, Beta, Delta and Omega, in the school, there were inter-house tournaments. Shane's house, Alpha, had to play the soccer tournament finals against Skeech's house, Delta.

With the whistle blown, the game started. It was a tough game. At the end of the first half, the score was nil-nil. It stayed that way, but late in the second half, Shane got the ball just outside the box. He dribbled past the defender and was facing the goalkeeper.

Then, before he could shoot, he was challenged from the back by the defender he had just passed and brought down. The whistle went—a penalty.

As he stood up and walked to the penalty spot, he boastingly said to Skeech, "Watch, this is how it needs to be done!"

Skeech wished him well and said, "This is not the only game though, you know!"

This statement was mainly to remind Shane of his cricket prowess, which he preferred. Shane smiled. The ball was already on the penalty spot, but he would always go to pick it up and place it against his forehead.

Then, with his eyes closed, he asked the Creator for blessings, put the ball back on the penalty spot, turned around and took three big steps away. He turned around and waited for the whistle.

After it blew, he looked up and ran his three steps to the ball. When he reached it, he didn't even touch the ball with his right leg but deftly feigned a kick over it, thus sending the goalie to the right of the goal.

He followed this up by merely tapping the ball to the left with his left leg. His trick worked; the ball went in. Goal! His team was ecstatic, screaming and celebrating and jumping around him.

Skeech merely said, "Nice, but I will show you tricks at cricket. Watch out, my boy," giggled and tapped Shane on his shoulder.

The game carried on, but it stayed one-nil till the end, so Shane won the first part of their challenge.

Then the cricket season came, and everyone was excited. Shane was anxious because they had hardly played with the hard cricket ball outside of school. In fact, he was scared of it. With his friends, at home, they used a tennis ball, so there were no cricket pads on their legs.

Here was the scary game. His team won the toss, and he chose to bat first. He was the opening batsman. With his full protection kit on, Shane waddled to the crease, feeling like a knight in armour. He was not used to this. Skeech was going to open the bowling and had a gleam in his eye as he had a point to prove.

Shane assumed his position in the crease, the area marked off for the batsman, tapped the bat to the ground and waited. When the umpire signalled that the game could proceed, Skeech, on the other side, started to run in to bowl, with a smile on his shy face.

He did leapt in the air, landed as his feet and his hand turned around fast, sending a thunderous ball to Shane. It was perfect. But Shane was prepared. He raised his bat and tried to slog it to prove a point. As the fast ball approached him in his crease, he swung the bat. There was a cracking sound.

But Shane did not even feel the contact. When he looked back, he saw his stumps down. The bails had flown into the air, and the off stump actually broke. This was a 'royal duck', the term used when the batsman goes out on his first ball. Skeech had his revenge.

As Shane walked back with his head down, Skeech said, "Sorry, bru, but I had to do that to get my revenge!"

"Nice ball, you totally beat me, my friend."

After that, they became best friends, helping each other, playing sports after school, visiting each other. Kind of usual best friends?

(The end, but before I go, I have to tell you about the most memorable thing I did with my best friend at school. For a science project, we managed to turn a skeleton from a chicken into a fossilised dinosaur in a glass tank.

After suffocating it, gutting it, then boiling it whole in a big pot, we then shredded the remaining meat and left the complete skeleton to dry. After about a week, we bought white spray paint and painted it.

We then got a fish tank and put it in. Skeech had researched the names of the bones, and we stuck identification tags on them. We received the highest mark in the class for our team effort. Teamwork pays! Luckily, the chicken never turned into a 'royal' duck, like when you get out on the first ball bowled to you in cricket! LOL ☺)

Chapter 5

The Same, But Different?

In 1989, Shane and the head girl from his school were chosen as the school's representatives for the Rotary Club, an organised grouping of youngsters from different races.

With apartheid in full swing, this club's goal was to get the youth to interact with the different races in a controlled situation. Making up the group, the head prefects from most of the primary schools in Newcastle were chosen to meet and form the club. It was an equal mix of 20 youngsters from the four different races.

Essentially, the aim was to expose them to different cultures. This was because the government's apartheid system had the 'Group Areas Act' in place that separated and kept the same races on their own.

At the very first meeting in the clubhouse, before the formalities proceeded, they were merely told to mingle. Aside from Shane and his head girl, there were other Indians as well, but Shane was used to Indian company.

For almost the very first time, he spoke directly to and mixed with Whites, Blacks and 'Coloureds' (a combination of black and white) of his age. Meeting them, especially the exclusive Whites, gave Shane a totally different perspective on life.

He still had Gogo at home, met a few Coloured boys at the grounds for soccer, but never had direct contact with Whites.

Manny now worked as an accountant and office hand under White ownership for an automotive engineering company, so Shane had met the bosses and their children because they visited to watch them celebrate the Hindu religious festival of Diwali every year. This was just for them to witness, and they were treated like royalty. Shane had never been in the same boat as them, but now, for the very first time, he felt kind of equal?

With the general mingling done and new friendships formed, they were told to prepare a speech about themselves, their culture and their beliefs to deliver at the second meeting the week after next.

Those two weeks took the clichéd 'forever' for Shane, and when it finally passed, they gathered at the clubhouse for their meeting. Being overzealous, Shane went first. For his introductory speech, he firstly thanked them for being willing to share information about themselves and their lives with everyone there.

He proceeded to give them insights into his life and Indian lifestyles as a whole. He told them about rituals like Diwali, their religious holiday, with its dazzling displays of fireworks, and described the various games and sports he and his friends played and their communal way of life.

Close to the end of it, he made a joke about how a rainbow is made up of different colours and that maybe, if they all mixed with one another, they WOULD find the pot of gold.

There was a big round of applause and laughter. Then his final statement was that he always knew that they were all the same, human beings. Everyone nodded in agreement as he returned to his seat.

The others all spoke, and when the informative, much-needed day was almost over, they were told that for the next meeting, again in two weeks' time, they would meet at the Rotary Clubhouse and then venture into different areas. They would go to the homes of some of the members from each race group to get a first-hand viewing of their lifestyles.

All excited as always, Shane raised his hand and volunteered for the Indians along with Renell for the Coloureds, Vusi for the Blacks and Wendy for the Whites. They were then dismissed.

When Shane got home, he told his parents; they could see the excitement on his face. He saw astonishment on theirs though. They had never experienced this in their younger days, so their secluded upbringing was obvious. Older people came from the heyday of apartheid and knew no different.

When the visiting day arrived, Shane was thrilled to see his new and interesting friends again. The group met in the parking lot outside the club, and there was an excited buzz in the air when they boarded the minibus. The visiting order was Shane first, followed by Renell, Vusi and Wendy last. This order was arranged according to the proximity of the houses, in the different racial areas, from the clubhouse. Suryaville, then Fairleigh, then Madadeni *(Ma-dah-dhenny)* and finally Arbor Park.

When they reached Shane's house, he led them indoors to introduce them to his mother. As the normal tradition in most Indian homes is to be hospitable to visitors and make them feel at home, Sheila had already prepared traditional Indian snacks for the guests.

It was a huge assortment, a whole table of goodies, and only the other Indian children knew what was in the assortment. The other club members did not even know what they were eating while they tasted and relished the sweet and savoury delicacies.

Shane then took them on a tour around his home. Manny had now extended the Mandir, adding another room, so the boys had one and the girls, one.

In the boys' room, Shane showed them his collection of stones on his bedside table. When they asked where he had got them from, he told them how he always collects when he is outdoors and one kind of 'calls' to him to pick up.

This was mainly in the nearby river bed. He asked the teachers in charge if he could show them the river. The teachers were also curious, so they agreed, and Shane excitedly took them for a walk to the river. The beauty amazed them all.

Shane took it a step further and told them to walk around and pick the stone that 'calls' to them to keep as a token, a souvenir, of their visit to his home. The teacher said it was fine and gave them all five minutes to explore before they went back to the bus.

They each found a stone and thanked Shane, not only for the trip but for the idea as well. Then it was back to the bus as they had a long day ahead with the other houses to visit.

As the Coloured area, Fairleigh, was not that far off, they stopped at Renell's place in just over five minutes. It was a normal, comfortably furnished, three-bedroom council house. The thing that amazed Shane the most about Renell was that she had no father, only her mother, who was blind.

They had adapted the house for her, and she did everything as if she could see. She ran the house with Renell's older brother and younger sister. Plus, she had made a few pots of tea for them.

(Renell's mum was one of my initial motivations to adapt and make the most of life when I was bedridden!)

When they were finished with a pleasant visit, they went on to Vusi's house in Madadeni, the Black area out of town. It was a surprisingly long distance for them, mainly because the Black townships were located on the outskirts of most cities.

The trip was going to be about 20 minutes long. Shane and Vusi clicked from before as they shared the same passion for soccer. Also, Shane could speak a bit of Zulu, Vusi's mother tongue, which he had learnt from Gogo. It was not the proper language; it was broken with slang and English, termed 'Fanagalo' *(Far-nah-gah-loh)*. But it worked.

They sat together and spoke in Fanagalo about soccer teams and other things. Then, when Shane asked Vusi about the match of his favourite team, Manchester United, which had been on TV the night before, he was shocked when Vusi replied that they had no electricity and therefore didn't have a TV. What was even more shocking was that this was the first time that Vusi had heard about Manchester United!

He only knew of the local teams in his area whose games he went to watch. Nationally, Kaizer Chiefs was his favourite team, but he had only heard the commentary of their games on their battery-powered radio.

(It is a shocking reality that this still exists, and it is even worse for some poor people in our Third World country, even now, in the twenty-first century. Count your lucky stars that you are not in this situation.

On that note, again, I am seriously vowing to donate most of the proceeds from my book to charity, to lift some of these disadvantaged people off the ground. That is my main goal.)

They finally reached Vusi's home. It was a tiny, one-bedroom council house without any amenities such as electricity or running water. Most of the learners stepped down with trepidation. Shane and a couple of the others were really excited.

Vusi had a big family. His mother, father, three brothers and two sisters, as well as his granny, were sitting outside. They had a fire burning to boil water for tea for their guests. They were very kind and offered the tea from a big pot. This was poured into enamel mugs for the guests. Most of them were full up already, but some took it.

Shane sipped the tea, which he found sweet and divine, and thanked them. Then Vusi showed them around. This didn't take long as it was tiny inside, a small kitchen, sitting room and bedroom. At night, the sitting room, on the floor, is where the males slept, with the females in the bedroom. Outside, they were shown the toilet. Noticing no running water, Wendy asked about it. Vusi said it was a bucket system that was collected and emptied.

"So, there are no taps. Where do you have a bath?" asked someone from the back of the group in a mocking way, and there was a stifled giggle.

Vusi dropped his head and said, "We go to the tap in the other street with our buckets and collect the water, come back here and boil it on the fire and then fill it in our bathtub over there."

The whole group looked at the tub on the side, with a make-shift curtain around it, and their hearts sank.

With sincerity, Shane stated, "It is a lot of work, yes, Vusi, but you have the cleanest heart I have ever seen. You are our best friend, mine especially. Times are tough, but they will change. Well done!"

Vusi smiled as the group applauded. They boarded the bus again and went back into town to Wendy's house in Arbor Park, the exclusive White area.

What a difference! The streets were spotless, with lovely velvety green sidewalks and mansions all around.

The underprivileged learners were quite apprehensive. The electric gates at Wendy's home opened onto a long driveway lined with trees, rose bushes and other flowers. They even passed a fountain and a small fish pond.

The bus stopped in a parking space outside a huge open garage, which contained two Mercedes Benz cars, an Isuzu bakkie, *(pickup truck)* and a silver Suzuki Katana motorbike.

When they climbed out of the bus, everyone focused on the two fancy Mercedes. Shane, however, went to the bike, thinking that he would never be able to own one like that. Then they all grouped again, as Wendy's mother was already outside, waiting for them. She had a typical housewife's apron on.

After she greeted them, she took them into the house through the entrance foyer. There was a big staircase leading up, but they were first taken to the massive dining room. The huge oak table was laden with scones, cookies, doughnuts, cakes and a lot of saucers and little forks.

On the matching serving table, there were China tea sets and coffee pots with mugs. There was also a sugar basin and a whole collection of spoons.

Wendy's mother told them, "Please feel free to help yourselves. I hope it is okay?"

Nervously, most of the learners sat down at the table, which seated them all. To create more space for the youngsters, the teachers were taken into the kitchen by Wendy's older sister.

Everyone had tea and some goodies, and then they went on a lengthy tour. They went upstairs first. There were five bedrooms, as Wendy also had two older brothers. Apparently, the motorbike outside belonged to the oldest brother. All the enormous bedrooms had their own bathrooms, with double beds, bedroom sets, built-in cupboards, desks and television sets. They were amazed as they went around, walking on spotlessly clean, expensive carpets. Then they went downstairs.

In the guest lounge, there was a pricy genuine leather lounge suite and a fancy coffee table. There were also some fully-stocked bookcases and a wine rack with wine glasses.

The TV room had a few recliners that faced the biggest TV set Shane had ever seen, on an oak wall unit, which also held a hi-fi and video machine and a whole collection of records. There were two large floppy couches on either side of the wall unit.

Then they went into the fully-fitted kitchen, which contained a built-in hob, eye-level oven, double sink and a massive fridge. There was also a breakfast nook with six barstools and a scullery.

On their way out, they went into the entertainment room, which had a fully-stocked bar, small fridge, pool table, dartboard and a soccer machine.

Outside the house was a covered patio with a braai area, a huge pool with change rooms and showers. In addition, outside the massive yard with its beautiful gardens were the maids' quarters as well as the laundry and washing lines.

The two maids were already in the kitchen, washing up the dishes, when they went back into the house. It was a paradise, but Shane felt the cold atmosphere, and so did some of the others. It was not a 'home' but a house with fancy things, paintings, furniture and so on; it lacked a warm spirit.

When they finished off at Wendy's house, they went back to the clubhouse and had an open discussion about their visits with insights into different cultures. Most, for the first time, had seen how each other lived and were really shocked at the differences.

This opened up a lot of doors for them. When the meeting had ended, they were told that the next outing would be a weekend away in the Drakensberg Mountains in three weeks' time. They needed extra time to prepare for a camping expedition so that they would be able to appreciate the outdoors. Full of excitement, they left.

(My eyes being opened like this is the main reason for feeling the way I do now. Honestly, although this realisation was being affirmed at that time, inside I always saw people as people, regardless of different races or religions and backgrounds.

Essentially, we are all the same—Homo sapiens, human beings—people. Everybody IS the same inside. We just make and see differences that we are trained to see, especially by elders. There is a new focus in my mind now.

We are young and are in charge of our futures on this planet. On that trip, the next chapter, I fell more in love with our planet, the most important ingredient in our lives, our Mother Earth.)

Chapter 6

A Natural Love for Her

It was after school on a Friday afternoon. This was the climax, the final event for the Rotary Club of 1989. It was to be a trip to the Drakensberg Mountains, and even though Shane had been there to the Royal Natal National Park on a school excursion, he had not camped over as it had been a day trip.

You can imagine his excitement. Manny had bought him his tent, sleeping bag, hiking boots and a special backpack. He was more than ready for his trip. His things were all packed, and he was ready to go.

Everybody met at the club and excitedly jumped onto the bus. There was endless chattering and bantering and discussions about expectations on the long trip. They wound through the Natal Midlands Meander and went on into the mountains.

Their eyes reflected the inspiration they experienced at the mere sight and peace of Mother Nature around them. As they reached their rugged destination, it was twilight, almost dark. But they all got a shock because there was nothing there. It was merely an area cordoned off by wooden poles with a pathway going down into the valley.

The adults laughed and gave the instruction to get their packs and join the crew on a hike down. Most of them were excited, but the doubt on some of their faces was apparent.

At the bottom, after about half-an-hour, they could see that it was quite a well-organised campsite, with grassed areas for pitching their tents as well as pre-built braai *(barbeque)* stands. The four outside toilets, two for boys and two for girls, was an added bonus, plus there were two outdoor communal showers. It was totally liveable.

Because it was almost totally dark, the camp hands already had a few fires blazing. The group was then instructed to help each other to pitch their tents and then gather around the fire. After quite an effort, the tents were finally up, mainly because of sharing the task. Then there was a signal for gathering around the huge fire in the centre.

There was an introduction to their site for the weekend, and they were told that this was to be a celebratory night, with a braai and storytelling and singing around the campfire. Then early to bed as there was a long day ahead.

They were then told that they had to form two groups. In order for this to happen, they were each asked to choose a piece of paper that had either a one or a two written on it, and all those with like numbers were grouped together.

That split them into two braai groups as there were two braai stands. They were given the charcoal and firelighters as well as their portion of meat to braai. Shane and Vusi were together and because Vusi was an expert in lighting fires, he had theirs going quickly.

Vusi went to help the other group that was not doing too well. When the fires were ready, each group braaied their meat, chatting away. They put the cooked meat onto their paper plates and took salads from the main table.

Gathering around the campfire, they ate with their plates on their laps while chatting and telling a few jokes. After supper, they told stories and had typical campfire fun. It had to end though, as they were told to get some sleep as it would be an early morning.

The next morning, they all brushed their teeth at the outside tap and had cold showers. They were asked to wear shorts as it was hot. Everyone had cereal for breakfast and then formed into their groups for a hike.

Firstly, they went to a muddy pit and were asked to remove their shoes and walk in the mud to feel Mother Earth for real. It was a moving experience, feeling the earth between their toes.

They walked barefoot to a waterfall, where they had a refreshing shower again and washed the mud off their feet. The water was so clear that Shane drank some. The others also tasted a little and agreed that it was good.

In the pool, everyone swam for a bit, and afterwards, with damp clothes, the great hike began. As they followed the trail, they absorbed the natural beauty of their surroundings and made their way up towards the top of the valley.

They were actually bone dry when they reached there because it did take some time. On the other side of the trail, there was a cliff face, with ropes attached, for them to abseil down the mountain. Once they had all been brave enough to have a turn, they started the hike down the back.

It was early afternoon. Everyone was surprised when they were each allocated a spot, like a rock, and given some biscuits, cheese and juice as well as paper, a pencil and a torch. The task was to stay at their spot for as long as they wanted, and when they had had enough, to follow the path back to the site. They were asked not to shout out and talk to each other. They had to draw their surroundings and possibly write a poem or a passage about their experience.

Shane ended up on a high rock on the river bank. He noticed the river bank with the rocks and the never-ending continuous flow of water, the basic sustenance to life between the rocks. He felt so at peace and his love for nature intensified.

Totally inspired, he planned a combined task of drawing and writing. He drew the rocks of the bank on either side of the page. Just like a river, in the middle, as the 'water', he wrote...

Water, The Flow of Life

Between river banks,
we should always give thanks.
Because it has what we need in our lives,
on which our body thrives.

For its functioning,
as the body will refreshingly sing
when it quenches its thirst,
as refreshment comes first.

When the body is dried up,
it is had in a cup,
or like now,
it is even drunk from the river, like a cow!

Shane laughed at his silly last line that rhymed. He was at peace, loving the sounds of nature around him, the river, crickets and owls.

Challenging rules made by the 'system' was always a thrill for him, so he then pondered about how he could beat the no 'talk' system.

He had a loud whistle, so he put his fingers into his mouth and let go. Out of the dark distance, he received quite a few responses. The response was comforting as darkness was now descending.

When the hunger got too much to bear, Shane had his biscuits and cheese and drank his juice. The quiet surroundings, being alone in Her, really filled him with a love for this Mother. The whistles gradually died down as most of the other members started going back to camp.

Shane stuck it out though, as he loved it. He lay back and gazed at the stars, like he did when he would climb up the roof of his house, especially during blackouts. His searching mind was spellbound, thinking about his existence.

Being naturally competitive *(my excuse for my ego? LOL☺)*, he waited. There were no more whistles, so when he knew he was the last one, he climbed down. His torch beam fell on a small stone that said to him, 'Pick me up', so he did. He wrote the date on the back with his pencil and then left for the camp, carrying the stone.

As he entered, he received a round of applause and a few teasing comments like, "Finally, the boss arrives!" and 'He whistles in!"

With rhetoric, Shane laughed and stated, "Even good things must come to an end."

All of them were asked to gather, and when they were sitting around the fire, they were asked to pass their drawings around and recite their poems, if they had any.

The group was very creative. Some drawings were amazing as well as poems. Shane told the leaders that he had combined the two. He delivered his poem first and then passed it around for viewing. They then toasted marshmallows and sang a few campfire songs before going off to bed.

After breakfast the following morning, they were asked to pose for a special group photo in a natural setting. The photograph was taken with the latest Polaroid camera, which printed the image immediately.

Once they had all written special messages on the back of each other's, which would serve as a reminder of their trip, it was home time.

The hike up to the bus was a different, longer route, and again, they were told to walk alone, with a gapped spacing. This gave them time to reflect internally about their experience in solitude.

When all of them reached the bus, it was a three-hour-long trip back to Newcastle. Most had to be woken up as the bus entered the parking lot. Before they got off, they were told that for their final meeting, which was scheduled for the following Saturday, they had to give a talk on their Rotary Club experience. Nice and tired, they then said their goodbyes to their friends.

At the following week's Rotary meeting, Shane had no prepared speech, but he spoke from the heart and thanked everybody, especially the club, for showing him so much about life and Mother Nature.

More importantly, he said that tragically, because of apartheid, they were living apart with different lifestyles, so they would probably never meet physically again. He heightened that this is simply because on the outside they were different, but inside they were all the same.

He thanked the Rotary Club for helping him see and confirm that humans are humans and for exposing him to our most important ingredient for survival, our Mother Nature, like never before.

In about a month, Shane's final year at primary school was also ending, and even though he lost the top spot academically to Skeech, he was still the head boy and had to write a parting note in the school magazine.

HEAD BOY'S MESSAGE

This last message, not only as a Head Boy, but a standard 5 pupil of Suryaville Primary School is filled with sadness. To depart from the portals of Suryaville Primary in which I laid my foundation seven years ago fills me with sorrow because of the drama of parting from friends and teachers and most of all the institution which has become a home.

To Mr Patel and all the teachers who entrusted me with the opportunity of leadership I express my sincere and grateful appreciation. My enlightening experiences must all be attributed to the trust and faith placed upon me by the Principal and teachers. I would also like to thank all the prefects for their dedication, sacrifice and sincerity. The prefect body was always competent in executing its duties and to them I say "Well done for your fine performance".

I would like to wish my successor all the best; to the pupils of the Suryaville Primary School - "strive, preservere and succeed". I bid farewell to S.P.S. with tears in my eyes ...
"Sail in the sea of ambition and anchor in the ports of success."

Shalendra Manilal
Head Boy

Chapter 7

High School

January 1990 and Shane was more than ready for the first day at Lincoln Heights Secondary School, which he thought would be a breeze. It was the only Indian high school in Newcastle, and all the children from the three primary schools in the town went there.

On the first day, as always, it seemed that there would be an orientation where the older children would 'enslave' the young newcomers and make them do silly things for them, like polishing their shoes and so on.

Manny took him to school; Shane expected the worst. That year, Resh had left to go to Durban to further her studies, as she had matriculated with an exemption, leaving only the two brothers. Ricky was in matric and was not with Shane and their father in the car because he preferred to walk to school.

Dropped off outside the school gates, Shane nervously strolled in. The older pupils were waiting for their victims. He entered, expecting the worst. Nothing happened though. He was dumbstruck. Looking in, he saw Ricky leaning on the wall in front of the gate.

Because of his reputation of not being someone to mess with, Ricky had given the other boys specific instructions to lay off his younger brother, and this was adhered to.

Shane felt special. The Standard 6s gathered in the assembly area and were split into different classes, with the learners ranked according to their primary school results. Shane was in the top class, Standard 6A.

He looked around for his crush, Suraya, as she had excelled at her school and had been placed first. She was not there, however. Shane asked Zubair, his friend from Rotary, where she was and was told that she had gone to the Middle East for her religious pilgrimage, known in the Islamic world as Hajj.

Shane was disappointed but made friends with the other learners from the different schools. New bonds were being made. It was exciting because for those first two weeks, they did not have schoolwork as all the organisation was taking place.

That first day, they were all lost during the lunch break, so they went to the tuck shop to buy cool drinks. At that time, it was only 80 cents for a 300-millilitre glass bottle.

The older boys were collecting the bottles from the girls who were too lazy to walk back to the tuck shop to receive their 30 cents deposit for the bottle. Shane saw this and stepped in to 'earn' his share of the money.

He was seen as the cute, green-eyed boy and collected quite a few bottles. This pastime soon became a normal way of building up his funds. After school, on their walk back home, they would stop at the game shop.

Funds were now healthy. They would have a Coke or a 'Mello Yellow', a pine-nut flavoured cool drink. 'Down down' competitions were held as well, to see who would finish first. Then they would play each other at the soccer table.

It was whole-hearted fun, with teams of two developing and having competitions. Winners stayed in. There was also Pac-Man and other video games that they loved. It only cost 20 cents for a game.

After the funds were finished, they would go home, eat and go back to the field to play whatever sport was in season at the time. This was the routine that was established in that first week.

When Suraya returned to school the following week, she was placed in a different class. Shane never spoke to her, but when he saw her around, his feelings grew even stronger. He was falling so deeply; just a glance made his heart flutter.

For two weeks at the start of February, he saved up the money he got from the bottles by not spending it on games at the shop. On the day before Valentine's Day, he walked to town. After a long search, he could only afford a small teddy bear, a small bunch of fake roses and a card. Excitedly, he walked back home. He closed his bedroom door and wrote in his card how he felt about her and the flutter she caused in his chest.

The following day, during the lunch break, most of the boys were handing over their gifts. With his relevant goodies in his hand, Shane went to find the cause of his flutter.

Looking around, he saw her in the girls' playground. Taking a deep breath, as the flutter had now doubled and the testosterone was pulsating through his veins, he walked towards her.

He reached the group of Muslim girls, but their circle seemed to close up. One of her friends, Thasneem, came out and signalled for him to follow her, which he did.

As she turned around, Thasneem asked, "What do you want? Who do you want to see?"

"I want to make Suraya my Valentine."

Thasneem said that he could not speak to her, as it was not allowed for a Hindu boy to pursue a Muslim girl. She took the gift though, and as she turned around, she said that she would pass it on.

Shane's heart sank, but maintaining his pride, he said, "Okay, thank you."

He watched as Thasneem handed it over to the giggling group of girls. Suraya turned around to look at Shane, and their eyes met. With a deep red blush on her cheeks, she smiled and mouthed 'thank you'. Then she turned back. The flutter heightened as Shane now needed air to walk away.

At first, it was air needed to go ahead, and now it was to back away? He took a deep breath, relaxed his mind, turned around and stepped away.

The next day, he waited for some contact at least but received none. His heart sank. That afternoon, he wrote a letter to her, saying how he felt and that he actually wanted to talk to her and ask her out, to make her his girlfriend.

In his letter, Shane also asked for some response at least. During the break, he handed the letter to Thasneem to pass on to Suraya. When Suraya received it, she looked up, and again, her mere look at him made Shane quiver.

During the lunch break the following day, he went to receive his response. Thasneem was waiting for him, and she handed him a letter. Shane thanked her and again saw Suraya looking their way from a distance. He walked to the bench, sat down and nervously opened it.

"Dear Shane, before I go any further, aside from the religious difference, I am still too young to have a boyfriend. I want to focus on my schoolwork. You enjoy your life and all the best with the other girls. I know you will find someone else.

This is not meant to happen. Keep your spirits up.

All the best, Suraya."

A sinking heart brought tears to Shane's eyes. He was not asking for too much, was he? Merely to actually want to spend some time with her at least? He slumped on the bench, shutting himself off from the rest of the world.

Chivani, who Shane nicknamed 'Chiefy', approached him.

"Hey, Shane'o, the main 'ou'. Here is my bottle."

Snapping back into reality, Shane wiped his eyes.

"What's wrong, Shane'o? Your eyes are red. You don't look like the main 'ou'," she giggled. "What's up?"

Taking a deep breath, Shane stood up and held his head.

"I've just got a headache, Chiefy. Thanks for the bottle."

"That's okay, my bru," replied Chivani as the siren sounded. "See you in class, Shane'o."

Steadying himself, he took a deep breath and vowed to himself that he would attempt to pursue Suraya again when they were both older, but as she had advised, he would flirt with other girls to pass the time.

For the rest of that year, he passed his time by playing sport, mainly in the breaks. His sociable nature also helped him to develop strong bonds, not just with his fellow students but with his teachers as well.

He could adapt and fit into any group. This trend of Shane's worked better for females though. After all, Suraya had advised him to pursue other girls, so he would sit and talk and listen. The year went by, and unlike his male friends, who would rather kick a ball, Shane was in the girls' grounds, socialising.

One of the girls who fell for his charms was actually in matric. Merryl was beautiful, and Shane could not believe that she had asked him to take her to the matric ball. He was only in Standard 6.

Wholeheartedly, Shane agreed. On the big night, he dressed smartly in black pants, a black shirt, tie and a gold coat. He felt great, very important and honoured, as he was the only boy not from matric or Standard 9 who would be attending. The older boys had younger 'chicks', but he was the youngest 'lightie'.

It was Ricky's matric ball, so he went with his friends, while Manny drove Shane to pick Merryl up. With a corsage in his hand, he knocked on her door. Merryl's mum opened the door and greeted Shane.

"Hello, auntie," he replied.

She smiled as she called Merryl, who came out in a smart black gown. Shane was taken aback at how gorgeous she was. He took a deep breath, bowed and presented the corsage. She was thrilled and put it on. He then escorted her to the car and opened the back door for her to get in. When she was in, he sat next to her.

Manny drove them to the hall and said that he would be back at midnight to pick them up. Hand in hand, the dashing young couple entered the hall. They looked so good together that they caused quite a stir. The night progressed with them having fun on the dance floor and socialising.

In the middle of the night, the formalities took place. This was when the names of the king and queen of the night would be announced. The crowd was silent. The master of ceremonies was the head boy of that year, Mahesh.

The crowd was very anxious. Mahesh built up the suspense by waiting. He then announced, "And the queen for 1990 is... the most beautiful and glamorous woman for the night. That is... the one... and only... Merryl!"

There was a round of applause. Shane escorted her to the stage and left her just underneath. She went up, accepted the bouquet and stood on the stage. Looking ravishing, she took a few breaths away.

Mahesh then stated, "Even though he deserves to be the king, as he seems to be a true gentleman, Merryl's date, Shane, is not eligible as he is too young. He is not even in matric, but believe it or not, Standard 6. So, this has to go to a matriculant."

"Everyone will agree that it is his older version, Ricky, also a charmer, deserves it, right? He will be the king!"

Ricky was shocked. On his way to the stage, he tapped Shane on his shoulder and said, "Sharp, my lightie, your turn will come."

Ricky and Merryl were crowned king and queen of the ball for 1990. What an evening! Merryl came down from the stage and gave Shane a kiss. She was thrilled.

At midnight, the ball ended. Manny was waiting to take them home, and they walked to the car with huge smiles on their faces.

They reached Merryl's house, and Shane walked her to the door. He gave her a goodnight peck on the cheek, and she told him to visit her the next day for more as her parents would be at work. Shane blushed, then smiled, agreed and left.

He punched the air and jumped with joy on his way back to the car. Inside the car, Manny complimented him for knowing how to treat a lady, even though he was so young. That compliment made Shane's head swell even more.

The year was almost at an end and catching a 'jol' became an even more important pastime for Shane. He was wasting his life away so much that his schoolwork was suffering. That year, for the very first time, he received no book award at the end of the year and was also not even in the top five learners of the class.

But even that disappointment did not deter Shane from focusing on enjoying life. His light-hearted streak took over, and his only concern was to live life to the maximum, to have a good time, catch a 'jol' and distract his hormones.

Chapter 8

Standard 7

(I've decided to do a short story chapter highlighting each year at high school.)

In Standard 7, Shane's form class moved to the multi-purpose room. Their memorable form teacher, Mr. Naidoo, used to teach them English, but he was a drama teacher as well.

Like Shane's relationships with all his teachers, Mr. Naidoo became a close friend and even used to lend Shane most of his trendy shirts. This was because his wardrobe was very contemporary, the opposite of a clichéd didactic school teacher.

The multi-purpose room was bigger than other classrooms and was totally unique as well, just like him. The setup was different because it was used for drama as well, with a carpet in the centre and desks in a 'U' shape, along the walls, around the carpet to make space for the drama practicals.

As the school worked on a 'teacher-based' classroom method, it meant that the learners had to rotate between different classrooms for their various subjects during the day. A five-minute interval was given for the transition.

Shane's class, 7A, only had their English lessons in their form classroom and had to move to other classrooms for their other lessons. The boys were especially sad. The move to other normal classrooms meant that they could no longer look at the girls, who sat opposite and facing them.

(☺ What a view, especially when the girls crossed their legs because the school uniform meant that girls wore knee-length dresses. Male readers, get it? I know, I know girls are probably asking, "Is this 'italics' distraction just for that? Come on, Shane, get over it, why don't you?"

Do I know women or what? No offence again, my lovely female readers, see my 'silly' aka 'male' humour?)

On a normal day in the first term of that year, Mr. Naidoo also had to exit with the learners when their English lesson ended because he was called to the office. That meant that the room was free for a short while.

While leaving the classroom, Shane's naughty streak took over. That morning, he had brought a stink bomb to school with him, and his friends prompted him to let it off in the room for the next class to smell. He could not resist.

When the following matric drama class entered, they were hit by the stench, which was so bad that they could not even enter.

As Mr. Naidoo returned, they told him, and he went off to the office to report the incident. Mr. Ally, the deputy head principal of the school, went to inspect. He was known as 'The A-bomb' because he was very stern and also the strictest Afrikaans language teacher. He smelt it and summoned all the boys from 7A to his office.

They all gathered outside the office in a straight line. All of them knew it was Shane, but when asked by The A-bomb, they did not disclose his name, honouring their principle of sticking together, no matter what. The A-bomb's patience ran out, and he summoned all the boys to march behind him. They followed silently in fear. They did not know what was in store for them, as in those days, corporal punishment was legal.

Their procession stopped outside the boy's toilet. Everyone assumed that their punishment would just be to clean up the toilets but were shocked when The A-bomb told them that those who needed the toilet had to use it now. A few went inside, returned and then they continued.

When they reached the classroom, The A-bomb announced, "Okay, you boys, you thought you were being clever, hey? You will be detained in this classroom for the rest of the day. The door will be locked until the siren sounds for you to leave when school is over."

It was the middle of the day and they still had about two hours left. They squirmed as they entered. It stank. The door was locked. They ran straight to the windows, and opening them wide, they stuck their heads out to breathe the fresh air.

After a while, with open windows, the smell drifted away and it was bearable. They were locked in and bored. Skeech went into the smaller backroom that was used by the drama classes. Because there were mainly girls in the drama lessons, they used the room as a changing room to slip into the outfits they wore for drama. A few of their props for practicals were also there.

When he came across an empty orange sack, he had a bright idea. Why not stuff it with paper and make it into a soft ball for playing indoor soccer? This way, the windows would not break. He came out and told the boys.

The setup and size of the room were perfect. Also, because the floor was carpeted in the centre, it meant that there was a lot of protected space. Two desks were placed on either side so that they could use the legs as goalposts. There were no goalkeepers, but they divided themselves into two teams. The game was so much fun, and this was supposed to be punishment? When the bell rang for the second break, they were dog-tired and sat down, all hot and sweaty.

The girls from their form class came to see them. They laughed at first, as they obviously assumed that the effects of the stench had made the boys all hot and sweaty.

Shane pleaded, "Help us, please. We are dying here; our throats have gone so dry."

With their typical female sympathy, the girls gathered together their unclaimed deposits from their empty bottles. They could actually use the refund themselves now, as the money was normally collected by the boys. The girls returned the bottles themselves and bought the boys cool drinks, which they passed through the windows.

The boys thanked them profusely. When the siren sounded for the end of the break and the girls left, the boys jokingly thanked Shane for using his charm. They were now refreshed and ready to play again. There was only an hour more until the end of the day.

A decision was made that they would do this every morning, a bit like a league match where the winners would be awarded a cup. Only this time, the 'cup' was a cool drink. This was because the losing side would buy the winning side bottles of cool drinks in the break.

The end of day siren sounded, and the doors were finally opened. The next morning, they put their idea into practice. Their new punishment changed their whole day. Even boys from the other classes were so jealous of the 7A boys and used to gather around to watch from the windows.

One day, given the trend of the boys in his neighbourhood to compete on the soccer field, Shane approached the boys in his class and suggested that as the games were five a side, they should choose the top five and challenge the other classes.

His friends readily agreed. He approached Andy, the top soccer player from the other class, who was outside gazing in, and asked him if he was interested in the challenge for cool drinks for a whole week for the winning team.

Andy agreed. They discussed the rules of the game, which would have five players on each side. There was no referee, but they would play friendly with no fouls. They excitedly agreed to come early to school the next morning in order to start the match at 07h00 sharp.

They would play 15 minutes a half, with a five-minute break, and end the game at 07h35, giving them 10 minutes to freshen up for the assembly at 07h45. The captains shook hands on it.

The following day was a Friday. The word of the challenge had spread, and there were already pupils waiting at the windows to watch. Skeech, Shane, Rakesh, Vinay and Pravesh, the best players in 7A, were full of nervous excitement.

As soon as Andy arrived with his team—Amith, Ashin, Trevor and Bruce—the game started. It was tough and very aggressive, but goals were still scored. At half-time, 7A's experience dominated though; they had scored five goals to the others' two.

After Shane complimented his team at half-time, they were smiling. Andy, however, seemed very aggressive with his team, who were frowning. After all, he was the bully of the grade.

The game resumed. In their happy spirit, 7A scored twice. They were bubbling. The other team, however, became very aggressive, challenging hard for the ball.

Then Vinay, the smallest of the 7A boys, received the ball and headed towards the goal. Andy tackled him from behind, bringing him to the ground and taking the ball.

It was an obvious red card challenge and also went against the agreement made between the two captains. The 7A boys stopped playing and ran up to see how Vinay was.

Andy asked, "And now, why have we stopped?"

The boys never heard. Their concern was for Vinay. His ankle was swollen and twisted, and he was in tears. Shane went over to Andy.

"This is a game, man, not a war. Besides, we had a gentleman's agreement!"

"Well, if you all want to play like girls, it's like a house-house game for gentlemen!" scorned Andy mockingly and laughed loudly.

"We play soccer, Andy, not like your housework, to beat the dust out of the mattresses!"

"Well, that must be because you guys are not strong enough to play a man's game and beat a mattress," taunted Andy.

"Meaning what? Are you a mattress?" snorted Shane sarcastically.

"You want to see? Why don't you meet me after school and show me you are a man and can still beat a mattress?" stated Andy aggressively. His bullying attitude was obvious.

"Okay, I will see you at the field, just outside by the flats."

"Be frightened, boy. I'll show you," scoffed Andy as he left with his team.

The word spread, even reaching the ears of Miss Ramgoolam, the young first-time teacher who took 7A for computer literacy, the new subject at school and their last lesson for the day.

Out of white-faced concern, she asked him, "Are you sure about this, Shane? This Andy is the bully of the standard!"

"Yes, I am sure, Mam. What is meant to happen will happen." Out of pride, he added, "I can also fight. My big brother Ricky beats me up as well. I am well trained!"

"Okay, then. I hope you know that we can also settle this in the office if you want?"

"And then be labelled as a sissy boy after consultation with my mam?"

"I do understand, Shane. If this is what you want, all the best."

When the final siren sounded, Vinay limped up to him and asked, "Are you sure, Shane? I mean, this is Andy you're going to fight with. My leg will heal, leave it."

"It's okay, Vin. What needs to be done needs to be done, right? He needs to be put in his place! I'll be fine."

"Okay, if you're sure," said Vinay.

Nodding that he was, Shane proceeded. He had a whole entourage behind him. A group of pupils had already formed a circle at the spot. Fights, or 'speeches' as they were called, were kind of after school public events. They were something to watch in a boring town.

Andy stood in the centre of the crowded circle, waiting. Shane approached and the onlookers gave way to let him enter. Normally, fights would be mainly verbal, with swearing and mocking insults, followed by some pushing and shoving. It was quite physical but did not go too far; the 'hold me back, hold me back' saying was extremely popular. This was what was usually done.

Shane dropped his bag and entered the ring. He had his head down, like when taking his penalty kicks. He went towards Andy but didn't say anything. Instead, he looked up and half lifted his hands while raising his shoulders, symbolically asking, 'What?'

Andy responded, "You should know what! You are going to get beaten up!"

Usual speeches were normally a verbal battle at first—bantering and nit-picking each other. Shane, however, had a different approach.

As he reached Andy, he did not push but instead planted a right-hander on Andy's face that floored him. He then sat on him, like he did with Shannon in primary school, and lifted Andy up by his shirt.

He asked him, "This is a speech, neh? Is this what you want?"

Andy stubbornly responded, "This is nothing!" and tussled to get back on his feet.

Shane held him down with his right hand and then planted a left-hander on the other side of his face. Andy was out.

Not wanting more as his point was proven, Shane got up and walked away to a round of applause as the circle opened for him. As he made for home, the entourage behind him could not stop chatting.

"Yoh, did you see that?"

"Aye, Shane! You rock, man!"

"Don't mess with us, man!"

"We are the 'dallahs'! *(bosses!)*"

"Seven A all the way!"

Shane was quiet though. He did not really like it and actually felt bad that it had gone so far.

On Monday morning, he thought that it was over. As he entered the school gates, he saw Andy, with his head down, standing on the wall, waiting for him at the same place where Ricky used to stand the previous year.

Ricky had finished school and was now in a Technikon, studying to become an automotive engineer. Fixing cars was his passion, as was working with his father at the engineering shop. Shane was alone at Lincoln Heights this year.

As he walked in, Andy raised his head to check if it was him. Shane was shocked: Andy had two huge black eyes. He stared as Shane walked past him, but Shane put his head down and carried on walking towards his classroom.

When he got further up the path, he could hear footsteps behind him. He turned around just in time. Andy had a knife raised and was about to stab him from behind. Luckily, Shane grabbed his hand and kicked him off his feet.

Andy crashed to the ground, and the knife fell out of his hands as he tried to prevent himself from hitting the floor. Shane picked up the knife and carried on walking.

The word went around, and during that second lunch break, Mr. Naidoo summoned the two of them to meet him outside the Art Room. When the boys got there, their animosity was apparent.

"What is going on here, boys?" asked Mr. Naidoo.

Shane explained how it started from a mere soccer match and how he thought it had been settled. Andy responded by saying that it had not been settled and that he had a score to settle as his status had dropped.

"Come on, my friend, I thought it was settled and was over?"

"I still have a point to prove to you, Shane!" retorted Andy.

"Why, Andy, why?"

"Because I... am... embarrassed, okay!"

"Do you want more? I can't carry on!"

"You are right, Shane, this has become so childish!" agreed Mr. Naidoo.

"Come on, Andy. Why? We can move on. I am willing to not even hold it against you that you literally tried to stab me in the back!"

"Are you serious, Shane? It got that far? Knife and all?" demanded Mr. Naidoo.

"It never went that far, sir," replied Shane. He then lied, "Andy himself held back, sir; he threw the knife to the ground and then walked away. I picked it up."

Andy was lost in thought at what Shane had said. He was stunned. Then Shane bravely added, "Here, my friend. This is yours," and handed Andy his knife back, with his other hand out for a handshake.

Mr. Naidoo was shocked. Andy took the knife and saying thank you, put it in his pocket as Shane asked, "Peace?"

Andy replied, "Ja, my bru, peace!" and shook hands with Shane.

"Well, I am glad that's over! Now give me that knife, Andy. I'm confiscating it just to make sure!" sighed Mr. Naidoo.

Andy passed it over, and he and Shane then became friends. That year progressed with lots of soccer and 'playing' other games, like smoking in the toilets, bunking and being big boys—the cool guys.

At the end of that year, they received counselling lessons as to what profession to enter. This was because they would have to choose set subject courses of study for Standard 8.

Rakesh, Trevor, Vinay and Denise would be taking G278, a technical course consisting of English, Afrikaans, mathematics, technical drawing, physics and computer science.

It sounded interesting, so Shane chose it as well, because when the career counsellor mentioned civil engineering, it seemed good, as did drawing buildings with architecture as a perspective.

This was also the first time that computer science was being offered in Standard 8, and Shane followed the trend—the 'cool' thing, the 'in' thing.

Chapter 9

Remember the Time?

In Standard Eight, being in a mainly technologically based class with the other computer studies students, Shane's creativity felt a little stifled. When they were given a big project for their English class by their teacher, Miss Chetty, he was thrilled.

It was meant to be a group practical presentation about tourism and the different countries of the world. That same day after school, he met at the library with his group—Chivani, Sharon, Claire, Vinay and Trevor—to discuss the project and work on it. Because Chivani's sister was the librarian, she let them use the empty room at the back. Loving music in the background as always, Shane had brought his CD player along and put some music on. Everyone went in and sat on the floor in a circle to discuss the presentation.

"Okay, guys, we need to agree on which country to feature," said Chivani.

"Let's vye to Hawaii, Chiefy? You girls can wear bikinis!" joked Vinay, and the boys burst out laughing.

Not impressed, Sharon stated, "Not funny, Vin. Get serious, boys, come on!"

Trevor then smiled and added, "Yep, come on, Vin and Shane'o, let's get serious, not dangerous. These girls in bikinis? Come on!"

At this, all the boys started rolling on the floor with laughter. Very annoyed, Claire burst out, "Seriously, you boys are so immature!"

"Come on, girls, let's lighten up a bit. Work can be fun to start with as well. A light-hearted approach to begin with works as well. You guys, let's get serious now. It has to happen," said Shane calmly.

"Thanks, Shane," stated Chivani. "Now, again, where are we going? What will be interesting? Any suggestions?"

"Since we are all Indian, let's vye to India," suggested Vinay.

"But I'm Coloured, and you want me to wear a sari?' joked Sharon. "Besides, everyone will do that. Let's go somewhere else."

"That's right, Shaz. Almost every other group will do India, using their mother's saris and things to show," agreed Chivani.

Bringing in her passion for speech and drama, as it was one of her major subjects, Claire stated, "Well, then, if it does not have to be a show and tell, then let's make ours like a play."

"I would love that," responded Shane. "Let's be uniquely different. Mam just said that the topic is a country to represent. We can, therefore, represent it not by informing them about it but by performing as people of that country or something."

"Hold on there, Shane. Our school does not have a stage. How are we going to perform, huh?" put in Trevor.

"Well, Trevs, we can transform a bigger classroom, like the multi-purpose room where we played soccer last year. It is actually free in the last two lessons on Fridays when we have our double English period. We'll just tell Mam we'll do it then?" suggested Shane.

Everyone else agreed enthusiastically.

"Okay, guys, we have agreed on where to perform the play, but WHERE are we going to go?" asked Chivani again.

"Yep, we need to decide on that first," agreed Claire.

At precisely that moment, Michel Jackson's song 'Remember the time?' started playing on the CD player.

(I was about to type out the rest of that discussion in a form of narrative, but thought, instead, that I would build suspense and rather just present it as a form of written play shortly. In this way, you can follow the play when the curtain goes up.

It will be nice for me to personally explore playwriting, as I have never actually formally written a play before. It will be nicer for you to slip in and imagine it while you read it.

As mentioned before, I am sporadic in my chain of thought, and this is more proof that I don't follow normality in this book.

You must be used to that by now; storytelling, advice, education, opinions, comedy, plus poetry and now playwriting, all uniquely combined into one. As mentioned before, there is no restriction on creativity. Get from 'A' to 'B'. ☺)

After a lengthy discussion that afternoon, they finally reached a conclusion as to where they were going and how they were going to be uniquely different.

They met every day after school at the library for two weeks and excitedly prepared themselves, doing rehearsals together as well as making everything they would need to set up the show. Everyone's creativity and opinion were taken into account, and by Thursday afternoon, they were ready.

That Friday, during the second break, before English, they gathered in the multi-purpose room. They had already clarified with Mr. Naidoo that they could use the room for setting up the show. They had also invited him to join as well, as he was free during that period.

It was a lot of work for them, totally transforming the classroom into a performance space. The windows were even blocked out with black cardboard, not only to prevent anyone from looking in but also to create darkness. They used the full hour, and when they were finally ready in costume, just in time, they turned off the light and went into the small conjoining room in the back.

The actual door was open, but they sealed the doorway space with shoeboxes that had been filled with sand and painted to resemble bricks. They prepared to enthral the audience.

Skeech was told that when the siren for the break to end sounded, he was to gather the class outside the room. Then, when everyone was there, he had to tell the class to be quiet as they entered through the door, which he had left slightly open to allow them through one at a time.

They then had to feel their way around the room, which was almost completely dark, apart from the extremely dim light coming from the open door, and find a place at one of the desks, which kept their 'U' shape with space in the centre.

When everyone was seated, including Mr. Naidoo and Miss Chetty, Skeech shut the door with a slam, as requested, to signal to the cast that the audience was in.

Remember the Time?

Cast:
Girls: Chivani, Claire and Sharon

Boys: Vinay, Shane and Trevor

Setting:
There is darkness in the room; an enclosed space with the audience integrated into the performance space, sitting on the outskirts.

On the performance space, nobody is in sight. The cast is behind a wall, at the back, in the changing room. On the stage the, sound of crashing and banging is heard.

Sharon: Are you guys sure this is the way in?

Trevor: Yes, Sharon, we have researched the site and drawn maps and everything. This has to be the one.

Claire: I agree. We are headed in the right direction.

Chivani: I just hope so.

Vinay: Well, the path leads to this entrance. It is sealed; we will have to try to break it open. Hit it with your hammer, Shane.

Shane: Okay, guys, our destiny awaits. Brace yourselves. Here we go.

The shoe box 'bricks' are smashed through. Dust is created by the sand in the boxes, and the audience sees a gas light that is being held up by Vinay. Six archaeologists, dressed in typical safari suits with hard hats, enter the space, with their lights in their hands.

There is silence. Their jaws drop as they stand, lost in amazement. In the dim light, it is apparent that this is an Egyptian tomb. There are hieroglyphics on the wall and huge sealed caskets in the enclosed space, and a few vases and two mummies wrapped in bandages are on the top of a tomb.

Trevor: See, I told you guys that we were going to the right tomb. I am sure that this is Cleopatra's pyramid.

Sharon: Okay, Trevor, I may have been wrong, but is there still a possibility that this is some other Pharaoh's tomb?

Vinay: That is a possibility, but all the relics seem to be in place. Chivani can read the hieroglyphics. It seems that the story starts here.

He points to a wall inscribed with writing. Chivani makes her way to it and holds up her light.

Chivani: Well, from what I gather, this says that in this tomb lies the remains of a queen.

Trevor: What's her name? What's her name?

Chivani: It does not say, but it seems to say that there is a path of discovery that will lead one to reveal who she is. It begins here, with this arrow pointing to that casket with one stroke marked on it.

Shane: Well, then, off we go. Come on.

He battles with the lock of the casket. Then he beats at it with his hammer, but it does not open.

Claire: I have an idea. Why don't you just press that switch thingy on the side?

Everyone bursts out laughing.

Shane: What switch thingy? On which side? Oh, this?

He presses the knob she has indicated, and the casket opens. Inside it is a golden headdress, shaped like a snake. He takes it out.

Shane: Well, hello, this must be worth a fortune!

Chivani: It will be. It is 22-carat gold, I think?

Claire: Okay, mister expert opener. Get the other five.

Shane: Yes, Mam. But why don't we all open one? There are six of us. Whatever we find, we keep. Chivani can have that headdress.

Vinay: The caskets are on either side of the tomb though; perhaps that side with the male symbol is for the males and the other side for the females? So, guys, come with me. Girls, you stay.

Trevor: Hi ho, hi ho silver, away we go.

Laughing, the boys go over to the cases and open them.

Shane: And now?

Claire: Yay. Guys. Sharon and I got gold headdresses for us in our cases as well.

Sharon: What do you guys have?

Looking disappointed, the boys hold up the wreaths they found in theirs.

Vinay: This is a head thingy from Roman times, right?

Chivani: Yes, Vinay, looks like it.

They each try the wreaths on playfully. (While they do this, they improvise, making up their own dialogue.)

Chivani: That's enough comedy now, all of you. Hang on... look here. This huge tile has a story on it about Cleopatra and Julius Caesar.

Trevor: I told you – I am sure this is Cleopatra's tomb!

Sharon: Caesar and Cleopatra had a love story of note.

Vinay: What happened?

Trevor: I think that when the Romans invaded Egypt, Caesar fell in love with Cleopatra.

Shane: But because of their differences, they could not marry, I think. That is what I heard.

Vinay: Guys can't always get what they want. *(laughs)*

Claire: Forbidden love? You should know, Shane. *(She giggles and winks at Shane.)*

Shane: *(Sadly)* Tell me about it. I know the hurting part, not just unrequited love but incompatibility based on cultural differences.

Chivani: There is more. It says that this sealed casket underneath contains their love, which will last forever. It needs to be opened. Do the honours, please, Mr. Locksmith.

Shane: *(Stepping forward and looking at the box)* This one is different. There is no lock on the side. Only something that looks like a title. Come, Chiefy, come read.

Chivani: *(Steps forward)* Well, it looks like there are two sides. There are hands here.

Sharon: Let me see. Oh, I get it. There are six hands next to each other and six levers.

Vinay: A smaller girl's hand next to a bigger boy's hand, side by side, with levers on the outside. It says here, in order for this case to be opened, it needs three boys and three girls. In couples, they must stand side by side and join hands.

Shane: Side by side? Yeah, baby, yeah.

Chivani: Oh, grow up, Shane! Now, the couple has to hold hands and walk together to the box. Then each couple must put their inner hands on it and lift the lever with their outer hands, opening the doors of their lives together. We must all do this at the same time. Oh, the last part says that we must

	wear the headdresses we found earlier. Snakes for girls, and wreaths for boys.
Trevor:	Okay, who goes with whom? I got the hots for you, Claire. Come to me, baby! *(He laughs.)*
Sharon:	Okay, then I will go with Vin.
Chivani:	*(Almost jealously, she mutters under her breath)* You have a crush on him anyway...
Shane:	That leaves you and me, Chiefy.
Sharon:	*(Under her breath)* That is probably what you wanted, anyway!
Claire:	Come now, girls, get serious.

Holding hands, they walk to the casket. When they reach it, they place their one hand next to their partner's hand on the chest and count together.

All: One, two, three, lift!

As the doors of the chest open, a hissing sound is heard. Then smoke puffs, transforming each couple into 'Caesar and Cleopatra representatives.' At this point, the Michael Jackson song 'Do You Remember the Time?' blasts out.

The group performs the dance perfectly, mimicking Michael Jackson's music video.

At the end of the song, the couples hold each other's hands and raise them high as the house lights come on.

All: Together forever, in spirit at least.

(They take their bows)

The End

There was a standing ovation. Mr. Naidoo told them how impressed he was with their play, and Miss Chetty agreed. A job well done.

But the rest of the school year was not done well at all by Shane. He got into more trouble as the year passed on, distracting himself on what he thought to be 'cool' instead of schoolwork. Although he did pass at the end of the year, his marks dropped! The average 'A' student was now a 'B' student. But he was not really concerned. He was there to have a good time and enjoy his life.

Chapter 10

Standard 9 – Soccer Distraction

Well, it finally happened. With classes being shuffled, Shane and his five friends doing G278 course were combined with the accounting class in Standard 9A.

With the accounting course, 'S7', being the most difficult on offer, it was only for the most intelligent pupils. Whose class was Shane now put in?

The answer is easy because she was actually the head girl of her primary school. Yes, it was the extremely intellectual Suraya.

A bad situation just got worse for Shane! Those intense glances happened more frequently—in the morning, all day and at the end of the day. Even when he actively tried to resist the attraction with other distractions, it persisted. The pain was evident in Shane's eyes. What hurt him the most was that he could see the pain in her eyes as well, as it was obvious in a way that she was falling for him.

With this unattainable prospect becoming even more evident, Shane's personal decline continued. He had to have fun with something or the other to distract himself.

Luckily, his soccer skills didn't suffer though, as he was an active member of the local league and the top striker in his division.

In the middle of the season, Shane was selected for the under-16 provincial team. Shane was not shocked to see Vusi, his old friend from the Rotary Club, who was a part of the team as well. It was three years since they had last seen each other, and they were overjoyed. They had a quick catching up session and then it was back to official business.

After introductions, the team was told that they would have group training together for the next four weekends. They would have a month to prepare before going to the provincial tournament in the faraway city of East London, which would be held during their school break. They, therefore, had time to get into peak form.

Full of excitement, they trained together that night, learning about each member and becoming a team. During the following four weekends, everyone grew closer.

They were now a real team. The Indian coach, who they called Uncle Vasi *(Vah-see)*, was excellent. He had been a top player in his day and was now coaching on a voluntary basis, as there was very little funding for the team. They were on an extremely tight budget.

When the day arrived for departure, with the very tight budget, Uncle Vasi could only arrange an 18-seater Kombi *(minibus)*. With a squad of 15, two coaches and the driver plus baggage, it was packed full.

Early on a Saturday morning, they set off on their 14-hour trip to East London, all packed and ready to play. With the Kombi being packed to the limit, they had been warned beforehand to pack light. Some bags were placed under seats and others, on their laps.

In high spirits, they stuck out the lengthy trip, chatting away, playing games, telling jokes and listening to music, as there was a great sound system.

Every two hours or so, they stopped at a garage along the way to freshen up and fill up the tank and their tummies. This was the first time in most of the towns for all of them. The trip was down the south coast of KwaZulu-Natal to the Eastern Cape.

Finally, in East London, they went to their hotel, where they had all been allocated rooms to share. Shaun, a Coloured guy, Vusi and Shane were together. Being extremely tired, they unpacked and went to bed, where they tucked themselves in and spoke excitedly about the trip till they fell asleep.

The next morning, a beautiful Sunday, after a hearty breakfast of cereal and muffins, they set off to the field just outside the hotel for their training session. The humid sea air was totally different to the atmosphere they were used to.

It was a very long and tiring session, and when done, they went in for lunch, a fancy pasta dish. After eating, they jumped into the Kombi to go to the main stadium for the draw.

The teams came from every province, with a total of 16 altogether. *(Okay, sorry. My map at the back is the provinces of the new South Africa after democracy. Under the apartheid regime, there were only four – Cape, Transvaal, Orange Free State and Natal. I will use the same names here. So, don't refer to the map. Thank you. ☺)*

Wearing mix and match tracksuits, the Northern Natal team felt out of place because they had not been able to afford the team tracksuits that all the other provinces wore. They held their heads high though, as they were comforted by Uncle Vasi, who said, "So what? It is how you play, not what you wear!"

The four provinces were split into three groups: Northern on top, Midlands in the middle and South on the bottom. Because of its size, the Northern Transvaal had a fourth group, Far North.

Northern Natal was placed in Group B, together with Southern Transvaal, Eastern Cape Midlands and South Western Cape, making four split groups of four in total, Group A, B, C and D. They would play three group matches in a league system in matches on Tuesday, Wednesday and Thursday. Then the top two from each group would proceed to a quarter-final on Friday, the semi-final on Saturday and then the finals on Sunday.

It was almost as if the excitement and buzz of nervous anticipation in the air was tangible. It had been arranged that each group had one of the four different grounds. There would be two matches a day, which would alternate.

Play was to begin on Tuesday, which would enable everyone to squeeze in a training session at their allocated grounds the following day. Before the group meeting ended, they were also told that national selectors were watching each match.

The top 15 players would be selected for the under-16 country team. That meant that they had to perform at their best in all the matches. There was a loud cheer from everyone, "We'll show them!", "We'll rock!", "Watch me!"

After some strict instructions, the coaches finally managed to restore some order amid all the boasting banter. The teams all gathered back in their buses, with Northern Natal in front of their Kombi.

Shaun, a thrill seeker like Shane, asked the coach, "Uncle Vasi, where are we going? Back to the hotel?"

"Yep, that is the plan. Why, what do you guys want to do?" enquired Uncle Vasi, with a cynical smile.

"See East London, sir!" chorused everyone.

"Maybe I can arrange that," said Uncle Vasi.

"Please!" they all shouted.

"Okay, okay," smiled Uncle Vasi. "We will tour around and see this city from the Kombi."

That had been his plan all along. The players' happiness could be seen by the huge smiles on their faces.

While touring East London, they passed an aquarium.

"I know you don't want to stop anywhere, but please stop here, Uncle Vasi. I always wanted to go to one," begged Shane.

"Does everyone want to stop?" asked Uncle Vasi.

There was a roar of agreement.

"Okay, team, we will stop here. Go and see. And behave! You have about an hour. Meet back here at the Kombi."

Excitedly, they jumped out. For the first time, they would actually see sea life and the seals. They were amazed, especially Shane with his love for nature becoming heightened as this was a whole new realm compared to the one he was used to, and the hour seemed to rush past.

At the hotel, they had fish for dinner and then retired to their rooms. Feeling drained, they retired to bed.

After breakfast the following day, they did like they had done the day before and went to their training session. To get them accustomed to the climate, after lunch, Uncle Vasi surprised them with a trip to the beach. It was great fun as they frolicked in the sea.

That night, with them being so tired from swimming in the sea, they retired after their roast chicken dinner. Before they went to bed, Uncle Vasi told them to rest well as they would be starting the tournament the following day. All excited, they fell asleep.

After their shower on Tuesday, they could not stop talking about their prospective match against Southern Transvaal, with some of them even being too excited to finish their breakfast.

At the grounds, they kitted up and warmed up. They were very happy with their new sponsored black and white kits, which made them look like professionals and made up for the fact that they did not have any team tracksuits. Then the match started.

It was a challenging first half, with neither side scoring a goal. During the half-time break, Uncle Vasi was not happy, as even though they had gained most of the possession in the midfield, Shane and Vusi, who played as the other striker, were not being fed the ball.

This focusing on defending was probably from the nervous tension of their first match in the big league. Shaun, a more attacking midfielder, was then sent in. When the match resumed, they pushed forward, and early in the half, they got a corner.

Shaun took it, and the ball went high and soared into the air. Vusi, being as tall as he was, leapt high for the ball and headed it home. They scored. Finally!

They led for about 20 minutes, but then Southern Transvaal made a surging forward run, and their striker cracked an 18-yard drive that left the Northern Natal keeper standing. One goal each.

The match heated up but stayed locked, evenly matched. Then, in the dying minutes, Shaun threaded through and played a beautiful through ball to Shane, who passed the last defender and was faced with the goalkeeper.

As he entered the box, the goalie pushed forward towards him. Using one of his usual tricks, Shane skilfully mimicked a strike to the left by kicking over the ball. The keeper dived to the left, but the ball wasn't there. Shane merely tapped the ball into the goals on the right.

His teammates were ecstatic. Their team was ahead. After a few minutes, the whistle went. They were thrilled with their first win, and there was much rejoicing.

That day, they remained behind to watch the other match, Eastern Cape Midlands against South Western Cape. South Western Cape was the superior team by far. They were extremely good and won three-nil. Shane's Northern Natal team were a bit scared of them but looked past it, as they would play them last.

On Wednesday, they had an easy match against the Free State team. Shane scored in the box as well as Shaun, with a cracking midfield drive. They won two-nil. In the other match, South Western Cape pumped two against Southern Transvaal. With Northern Natal and South Western Cape both at the top of the group, their meeting would be the decider.

To ease the tension for the next day's match, Uncle Vasi took them to the games arcade, which also had a soccer table. They had their thrill and left for their hotel home.

The next morning, they anxiously headed to the field. Both the sides knew the potential of the opposition they were facing, and it was a very nervous start. It turned into an extremely tough match with the ball mainly in the middle of the field. At half-time, it was still nil-nil.

Uncle Vasi gave his by now normal team talk, "Come on, my boys. You are all fighting well. Push it now. Go for broke. You can do it, my boys!"

When play resumed, it was more challenging than the first half. An open attempt made by Vusi was the closest, with the ball striking the upright. Their goalkeeper made a few good saves for Northern Natal, and the game stayed deadlocked till the whistle blew.

As this was still the league stage, there was no extra time. They each received a point for the draw. They were the top two in Group B and had an equal seven points, with two wins and one draw each.

Because of a superior goal difference, South Western Cape led the group and Northern Natal was therefore second. They were still happy to go forward though.

In the other match, Eastern Cape Midlands and Northern Transvaal drew nil-nil, with neither team scoring. They were at the bottom of the group, which meant they were out. So, the top eight teams met at the main grounds the following day.

When the knockout draw was made, Northern Natal found that their next match for the quarter-finals was Southern Natal. Before the match the following day, Southern Natal was behaving in a very domineering way, bullying and threatening Northern Natal.

They laughed at them for coming from the 'farm', mockingly called them the 'farmers' or the 'skaapies' *(lambs)*. Northern Natal was not distracted at all though. They knew that playing the game would talk.

The next day, they ignored more scoffing and mocking before the match. At the toss, Southern Natal's captain mockingly said "moo", and Shane lost the toss.

"Farmers can't throw coins, maybe we need some cow dung?" jibed their rivals.

"Well, we'll rub the dung in your faces after we've thrashed you!" retorted Shane. "Bring it on!"

The match began. Early in the first half, Shane pounced on a square cross and headed it home. After the goal, he and his teammates started imitating the mooing sound Southern Natal's captain had made and chanting, "Yes, we are from the farm to provide the dung treatment!"

Then they burst out laughing. Southern Natal tried to stomach it as they centred, but just a short while later, Shane burst forward, dribbled past the last defender and skilfully lobbed the ball over the advancing goalkeeper. Two-nil! Moo! Chant again.

For a change, during their half-time break, Uncle Vasi didn't give his usual pep talk. Instead, he said, "Keep it up, my boys. Show them what you are made of! Rub the dung in. They can only boast, but you are all, how do you say, rocking them!"

The team laughed.

When the second half started, they were hungry to do some farming. A high pass from the centre of their defence came towards Shane. With his back facing the goal, he headed it down for a charging Vusi, who thundered it home.

Three-nil. Moo! Just before the final whistle, Northern Natal had a corner from the right. Shaun took the ball and whispered to Shane, "Far post."

He took the corner long, and the ball went sailing over everyone. Shane was waiting at the far post to thunder it home with a powerful volley. Four-nil, MOO!

The whistle blew. One of the selectors came up to Shane and said, "Nice hat-trick, young man! I have my eye on you," and walked off.

The boys were now even more excited; apart from thrashing Southern Natal, Shane was a possible candidate for the top 15 of the national team. They could not stop talking about it, even during supper.

Shane, however, noticed that Vusi was not really eating his fish and chips and asked, "What's wrong, mpinchi? *(friend)*. Yini ekuhluphayo? *(What's troubling you?)* I hope you are not jealous of me?"

"Me? Of you? Never, mpinchi! I just miss my phuthu *(ground maize meal)*! Angikuthanda lokudla *(I don't like this food)."*

Making his way to the kitchen, Shane smiled. He knew the chef, a Xhosa man by the name of Xholani, who he had already thanked for the chicken the night before. He asked Xholani for some phuthu for his friend, which Xholani readily gave him, as he had made some for himself.

As he approached the table to present it to Vusi, Shane held it behind his back. Then, in his loud voice, he announced, "Attention, attention, and the award for the best goal of the tournament goes to Vusi from Northern Natal for his cracking 18-yard drive against Southern Natal."

"What does he win?"

"This is something that he is dying for! A... trophy... full... of phuthu!"

The seated boys burst out laughing as they cheered him, and Vusi could not believe his eyes.

He smiled and said to Shane, "Ngiya bonga mpinchi *(Thank you, friend)."*

"My pleasure, my friend," responded Shane graciously, sitting down next to Vusi. He had not eaten yet, and when Vusi pushed his plate to him to join as well, he shared the phuthu.

(Eating together does bring people closer. Amazing to share from one plate. Some cultures put all the food in one big plate in the centre and sit around and eat. Suppose that this is rooted in the camaraderie of our primate origins? Apes do share their food willingly. Try it? ☺*)*

That night, they chatted till late and finally went to bed. The next day was the semi-final day. The final four teams drew randomly, with the chairperson choosing the teams to play.

Guess who Northern Natal was drawn against? A true challenge. Yep, remember their tough struggle in the group phases when they ended in a draw? South Western Cape again.

The first semi-final was Northern Transvaal versus Southern Cape. It was a tough match, but Northern Transvaal held onto their early one-nil lead and won. They were through to the finals the following day. Their opponents would be decided by the match that Northern Natal feared. It was time to face their opponents.

They won the toss and got off to an anxious start. This match would be just as challenging as their previous match. There were worthy attempts but no goals in the first half. After pushing them on, Uncle Vasi said that they needed to ease up and break the tension a bit. But after the second half started, the score remained nil-nil.

As this was a knockout match, it then went to extra time. This passed without a goal being scored though, with the score still nil-nil. The ultimate match decider would be a penalty shoot-out: the best of five. *(This is actually a tough decider as five penalties do not decide a match, but you do need a winner!)*

South Western Cape won the toss and chose to shoot first. Northern Natal decided on their striking order, and Vusi, Atish, Shaun and Mandla were chosen along with Shane, who was supposed to end with a bang.

The shoot-out started. Sashen braced himself, with his legs apart like the professional goalkeepers he had seen on T.V, but the other team scored first. Vusi equalised. Then the other side scored their second goal and now it was Atish's turn to equalise. Two goals each. Their third ball went in. Shaun equalised. It was now three each. Their fourth went in.

Nervously, Mandla placed his ball, ran towards it and sent it thundering towards the goal. Their keeper, however, guessed its direction and dived for it, but because of its force, he was not strong enough to keep it out. He merely fumbled it into the side netting. It was four all.

South Western Cape scored their fifth goal, making it five in a row. It was now Shane's turn. After his goal, it would go to 'sudden death', where the first one to miss in succession loses. Uncle Vasi had already started selecting the strikers for the sudden death.

Shane took the ball. As he walked up to place it on the spot, he could feel that all eyes were on him. He smiled as he followed his ritual. Blocking out the noise, the ball touched his forehead. He bent down to place it on the spot, then turned around and walked back to wait for the whistle.

The whistle blew. Shane raised his head and ran towards the ball. He did his normal trick of sending the keeper the wrong way by kicking over the ball with his strong right leg and tapping the ball the other way with his weaker left leg.

This time, however, the goalkeeper did not fall for it, as he had seen Shane doing it when they played Southern Transvaal. He deceptively started to move to the left but made a conscious effort to stop.

Obviously, as his trick shot was kicked with the weaker leg, Shane's left foot did not strike the ball with enough force. The keeper changed direction and dived after it. At full stretch, he reached it and pushed the ball, which hit the post and bounced out.

With his hands on his head, Shane was shocked. It was the first time he had ever missed a penalty. The shock then turned to embarrassment as this had been the most important kick of his life! For the first time ever, the crowd booed him.

He hung his head in shame as South Western Cape roared with cheers. His team was shocked; they had not expected this. As Uncle Vasi was tapping Shane on his shoulder to console him, the national team selector approached them.

"Why try to be fancy?" he asked Shane. "This is a simple game. That just cost you your place in the national team. You play very well, but it is not all about trickery and showing off. Cut yourself down to size! All the best for the future though."

Shane was dumbstruck. He only shook hands with the selector as he departed. For the very first time, he was speechless! He was so distraught that he did not speak at all on the way home.

(This sad incident made me realise that life is not all about trickery and showing off. There are serious times. I try to avoid them, as I

intentionally have a fun, playful way. But serious times will always be there. Even when clouded over, they are underneath. ☹

Whether we like it or not, these times can't be escaped! Come on now, Shane, put your chin up? Get serious. *I intentionally cloud all seriousness with humour! See?*

I even talk to myself, to console my mind. Why not, Shane, why not? Well, it helps for a bit, but these times will never be forgotten; they will always be there, even though underneath. ☹

To drown my sorrows as I write, I think I will document what happened that night with the 'heart' issue. I am catering for everyone, plus trying different things. So why not make it like a clichéd cheesy romantic story?)

A Night To Remember

It was the Saturday night after the soccer team had lost their semi-final match in a national tournament held in East London. Having let his team down, their top striker, Shane, needed to drown his sorrows. Even though they were only 16 years old, he made plans with his other thrill-seeking friend, Shaun, as Vusi, his other close friend, did not want to break the rules. They had to sneak out of their room to get to the bar on the ground level of the hotel they were staying in.

Climbing out of the window of their room onto the ledge, Shaun said, "Yoh, Shane, this is high!"

"Just don't look down, man. I am right behind you," Shane sneered.

"That will help a lot if I fall, you know," remarked Shaun.

"Just carry on, man. You are safe. It is fine," Shane reassured him.

They wiggled their way onto the narrow ledge to the main balcony, which had a low decorative wall around it. They climbed over the wall and found the doors to the hallway open. Their plan had worked.

On their way downstairs, Shane stated, "See, I told you it will be fine."

"But now there's a bigger problem," replied Shaun. "How do we get into the bar, Shane?"

"Let's hope for the best. There it is."

Standing outside the door to the bar, Shane, who was a bit nervous after seeing the doorman, then told Shaun to act big.

"Okay, Shane'o, but we can also bribe him, can't we? How much have you got?"

"I'll put in R10 and you put in R10. That's R20, Shaun, if he stops us."

They walked to the door but were stopped by the doorman, "Sorry, boys, you are underage!"

"Olah mpinchi *(Hello friend)*. We just need to drown our sorrows. We lost at soccer. We won't cause trouble," said Shane.

"Here's 20 bucks for you if you let us in," added Shaun.

Taking the money, the doorman said, "Now you are talking my language. Just behave, okay!"

"Sure, sure, mpinchi!" they both chorused as they were let in.

"See, I told you it would work, Shane."

"Yes, money talks, I must say," agreed Shane, and they entered the bar, laughing, making it straight for the bar counter, where they ordered two beers.

There was an empty pool table, so they took their beers over to play some pool.

"I need this to relax, Shaun. Maybe at least I'll beat you with this white ball. I hope you are ready?"

"Bring it, bru, I will rock you! Leave the white, always bet on black! Watch out."

Shane laughed and spoke like a girl, "Okay, be gentle. It will be my first time."

Shaun burst out laughing and set up for the game. It was typical de-stressing. They had a few more beers, and with both of them being good at pool, they each won a few games. With the basic goal being to de-stressing, the score did not even matter. As they had each had their few beers and felt relaxed, they started to make their way back upstairs to their room.

When Shane reached the exit, he felt a tap on his shoulders and heard a girl's voice saying, "Are you guys leaving so soon?"

He turned around and saw a gorgeous, young Coloured woman. Putting on his charm, he flirtatiously replied, "Aha, do you want to come with us to have some fun? Or do you want to have a jol here?"

She giggled. "I like to party. Are you a joller?"

"We both are," replied Shaun. "Do you want to find out?"

"Maybe. My name is Debbie, and you guys are?" she asked.

"My name may be too hard for you to pronounce, so I will spell it out for you," responded Shane. "The first letter is a Y, the second an O, then a U, then R and then S."

Debbie realised what the word was, giggled and added her own touch, "Okay, how do you say it? MINE?"

"Well, if you want to, or you can just call me Shane. This is my best friend, Shaun."

"Nice to meet you guys," responded Debbie. "Where are you from, and what are you doing here?"

"We are from Newcastle. We came to play in the soccer tournament," replied Shaun.

"And I made us lose. That is why we are drowning our sorrows."

"Now you are talking my language. Don't leave so soon. Let's have a bigger jol. You and Shaun buy us some drinks," said Debbie, pointing to her two girlfriends.

Shaun jokingly proposed to Shane, "One for you and two for me?"

"A strong single and a double? Well, good deal," responded Shane, pulling Debbie closer to him.

"You are too cute, Shane. I like you. We are going to have a paarrtee tonight!" smiled Debbie as they went over to her friends.

"Hey, guys, these are the cutest, funniest and friendliest guys I have ever met!"

"That may be coz you live in East London, Debbie?" joked Shane, and the girls giggled.

"Very funny, Shane," retorted Debbie. "And this is his friend, Shaun. They are here to play soccer."

"We have played," said Shane. "Our tournament is over. We lost, so we wanted to party to forget."

"Well, you are with the right group. My name's Sandra. This is Veron."

"Nice to meet you, ladies," winked Shane and then putting his arm around Debbie, naughtily stated, "Well, I'm taken, but Shaun is all yours."

"Okay, what are we drinking?" asked Shaun.

"Let's get a bottle of brandy and some Coke, Shaun. What do you girls think?" Shane asked.

"Whatever, we don't mind," chorused the girls.

"Nice, not even fussy like typical women," smiled Shane and continued, "You guys are living life just to have fun. I like it."

He grabbed Shaun by the shoulder and turning around, said in a *Terminator* accent, just as Arnold Schwarzenegger does in the movie, "We'll be back."

"Hurry up. I'll miss you," laughed Debbie.

"Hi ho, hi ho, off to the bar we go," sang Shaun.

"Tara rum pum pum, tara rum pum pum, rum pum pum pumm!" added Shane, and they burst out laughing as they walked off.

Sandra smiled, "You were right, Debs. They are so full of fun, aren't they, Veron?"

"Yeah, we are going to have a jol," replied Veron. "You'd better behave, Sandra. Lesley won't like it if you spend too much time with us. At least he let you come. Think of him, hmm?"

"You got that right, Vee," said Debbie. "Don't play with my brother, Sandra!"

"I won't. That does not stop you single girls though, so enjoy!" winked Sandra.

"It looks like that is where this is headed," agreed Veron. "You can have Shane, Debs. Shaun does it for me; he has such a great body!"

The girls were giggling when the boys returned. Shane placed the bottle of brandy and three glasses on the table. "Here you go, girls. I hope you like?"

The girls replied together in their sexy voices, "Mmm! Thank you."

"And you'll need a mixer to dilute it. Plus, here's the ice!" announced Shaun, putting down the Coke and a bucket of ice.

"Why only three glasses? Are you guys not drinking?" asked Debbie.

"Well, we are, but we thought about it and decided to stick to our beers," answered Shane. "We don't want to mix hot stuff. Right, Shaun?"

"Yep, you'll get sick very easily. We're sticking to our beers, thank you."

Shane added, "Just a few more, as we've had some already. This bottle is just for you three. Enjoy."

"Okay, then, bottoms up, my girls!" toasted Veron.

The girls opened the bottle and poured the drinks and then Debbie announced, "Okay, everyone, raise your glasses. Here's to a great night out!"

Everyone raised their glasses, "Cheers!"

The night wore on, with everyone having a good time, chatting away and sharing jokes as they drank. Debbie's flirtatious behaviour with Shane was very obvious. If an outsider looked at them, they would think that the couple had been together for a long time.

Veron sat with Shaun. Noticing that Sandra looked out of place, Debbie said to her, "Hey, Sandy, why don't we phone Lesley to come and join us?"

"You know how he is, Debs. My baby is too much of a goody-two-shoes. He won't come."

"Who is Lesley?" asked Shane.

"He is Sandra's boyfriend, Debbie's brother!" Veron replied.

"Okay, call him. The more the merrier," stated Shane. "But this place is going to close now."

"We can always go to the nightclub," suggested Debbie.

Enthusiastically, Shane said, "Now you're talking my language. I love dancing!"

"But how are we going to get there? Do you guys have a car?" asked Shaun.

"No, my mum dropped us off," said Veron. "We'll easily get a lift. That club pumps. There is also a chill-out place called The Joint, for the car boot parties down the road from here."

"Ja, and there's even a samosa lady there. I'm hungry. This bottle is almost empty. Let's move. What do you think, boys?" asked Debbie.

They both nodded. "We're game."

"I think my brother David will be at The Joint as well. I can chill with him," said Sandra. Winking at Veron, "You guys carry on, Vee with Shaun and Debs with Shane. Lesley will never come out. I'll back out."

Veron responded, "Sounds like a working plan, Sandy. Last one for the road? We'll carry the last part of this bottle for the guys there. Bottoms up!"

"So far this has been a night to remember, but it keeps getting better. All thanks to you guys. You guys rock," smiled Shane.

Holding his beer up, Shaun added, "Aha. Thanks. Cheers!"

When they had downed their drinks, they walked down the hill to The Joint. The parking lot was full of cars and young people. The cars had powerful sound systems, and the music was pounding as the drinks flowed.

"There's David, my bro's car next to Neil's. Come, guys," said Sandra.

They all followed her to find David with his friends. When they reached him, he asked her in a Cape Coloured accent, "En nou, wat gaan hier so aan, Sandy?" *(And now, what's going on here, Sandy?)*

"Hoesit bru. Jy mos weet Debs en Vee. Hier is ons nuwe vriende, Shane en Shaun." *(How's it, brother. You know Debs and Vee. These are our new friends, Shane and Shaun.)*

"Orright, van waaraf kom julle, my maaitjies?" asked David. *(All right, where are you from, my friends?)*

"Ons is van Newcastle. Ons speel sokker. Kan ons Engels praat asseblief?" asked Shane. *(We are from Newcastle. We play soccer. Can we speak English, please?)*

David burst out laughing, "Sorry, my friends. We talk our own Afrikaans like that to each other."

Shane winked and thanked David, who then asked his boys to give him and Shaun beer.

"And here's some brandy from earlier. While you boys drink, the girls and I will get some samosas. Let's go, girls," said Debbie.

Meanwhile, the boys poured their beers and talked sport. It was a typically clichéd men's discussion.

The girls returned with the samosas. "You guys have to try them. They've got tinned fish inside," stated Debbie, passing the samosas over to the men.

"Yoh man, they are big!" exclaimed Shaun.

"The real Indian ones are smaller," added Shane.

Veron sniggered, "We are only talking about samosas here, Shane, right?"

The girls laughed out loudly as the boot party continued. Everyone was very friendly, and there was a lively buzz in the air, but one by one, they all started leaving. David was not going to the nightclub because his wife and child were at home, and Sandra was going home with him.

David arranged for his friend Neil, who was not drinking, to take them with him to the nightclub, then drop the boys off at the hotel and take the girls home. He wished them well and told them to "party hearty".

Neil opened the doors of his modified Japanese car, the Datsun.

"So, are you all ready to party?" he asked.

As she was jumping in, Debbie shouted, "Dancefloor, watch out. Here we come!"

Shane laughed, "You have to keep up with me also, my dear," as he squeezed in next to her.

Shaun and Veron jumped in on the other side because Neil's girlfriend, Shanice, was with them as well, and the front seat was hers.

"Okay, let's pump some tunes to get us into the groove!" she said as she turned on the radio.

Starting with the latest hit at the time, which would become Shane's favourite song of all time – 'Show me love' by Robin S, Neil drove off. They reached the club and stood in the queue to enter.

It was the boys' first night out in a real nightclub, and holding their dates helped them escape with their age. Amazement filled them as they entered the packed space. They wasted no time in getting onto the dance floor. Shane did not stop moving to the music. He really could 'shake his ass', their slang for dancing.

Debbie complimented him, and when a circle opened for the dancers to do their thing by showing off and challenging each other, she pushed him in. His skilful break-dancing soon brought approving shouts and clapping from the enthralled crowd.

At about 05h00, it started to get bright, and even though they were still partying, Shane asked Neil if they could leave as he wanted to see the sun rise over the ocean for the first time. Neil agreed and they left.

At the beach, the sun was just about to rise. This was also the first time that Shaun would see this beautiful spectacle. Their faces were alight with joy. This was also because of having beautiful girls with them. Their thrill at seeing the sunrise spectacle was tangible.

Removing their shoes and rolling up the legs of their jeans, they all walked hand in hand on the beach with their partners. Each couple was separate, sitting down and cuddling on the sand as they watched the sun slowly rising.

Shane's amazement was apparent on his face, and Debbie told him, "The expression on your face makes you look even more handsome. Your green eyes are also lighting up with the sky."

"A new day, with the sun to warm everyone. Maybe even warm their hearts?" responded Shane, blushing.

"Well, mine is close to yours, warm enough!" she said and kissed him.

It was a long and passionate kiss that literally took Shane's breath away. It was the perfect ending to a perfect night.

After a while, Neil shouted out, "Okay, guys, we have to leave now."

Shaun and Veron had also been making out, but they stopped, and everyone went back to the car. As they reached the hotel, the girls climbed out to give them a goodnight kiss.

"Thanks, Shane, you are an amazing guy. Pity you are going back to Newcastle. This could have gone further," said Debbie.

"Yes, it is a pity. But it was a night to remember at least. One last kiss?"

"For you, my dear, any day, anytime, anywhere."

They kissed again, for the last time. Glancing over at them, Shane saw Shaun and Veron also sharing a passionate kiss.

The girls left, and Shaun and Shane now had to backtrack to the balcony and across the ledge to their rooms, but floating on the high of the perfect night, it was now much easier!

They crept into their rooms, and when woken up by Vusi for breakfast, they were still so tired that they skipped it and stayed in bed. Their night out had been 'perfectly' tiring for them because that morning, the team could lie in to rest, so it was lucky for Shane and Shaun.

At lunchtime, Vusi woke them up. The plan was to watch the final match of the tournament and drive back home through the night. They woke up with a typical 'morning after' feeling.

To cure the hangover, Shane and Shaun had a cold shower and then had lunch for breakfast. Feeling slightly different, they then packed their luggage to put in the Kombi for the drive home.

South Western Cape, their enemies, beat Northern Transvaal and their team was crowned the kings. Shane's heart sank again when he was not selected to join the national team. He accepted the sorrow though and joked with Shaun, "Well, they don't know what they lost out on, neh?"

"That's right. You can score on any field."

"Especially on the beach!" exclaimed Shane.

"Exactly!"

They laughed as they marched back to the Kombi for the long overnight trip back to Newcastle. For Shaun and Shane, their minds still reliving the events of the previous night, a night to remember.

Chapter 11

Durban, Twice as Naughty?

(Okay, I know that I said that each year would have one focus, but for this particular year, I want to describe two incidents that took place in Durban, the port city about 300 kilometres from Newcastle.)

Round One

Back in Newcastle, after the tournament, it was soccer season in the park with friends. Moonshine *(my lightie bro, remember?)* often used to join them in his normal goalie position.

After a thrilling match, all the boys sat on the pavement together to have the losing team's provided prize. As it was then winter, it was not a bottle of Coke but lemons with salt and masala.

Feeling the cold in the air, Moonshine exclaimed, "Aye, man, Ladysmith is warmer than this joint in winter!"

Anesh stated, "Must be because it is closer to Durban,"

"Yes, Durban is nice and warm. I am going there for July classes," put in Atish.

"Huh? What are those?" asked Moonshine.

"Well, Moonikes, it's like extra classes for learners in Standard 9 and matric," explained Shane.

"Like tuition," said Atish.

"But you have to be rich vye? It is expensive going to Durbs, Atish."

"In a way, yep, Moona, but you can make a plan if you want. Like three of my class friends are going with my mum and dad."

"How, but why you have never asked me to go with you, Atish?" asked Shane.

"That is because you do a different course, Shane'o. I am going with my classmates."

"Oh, okay. I would really like to go though. I've heard that Durbs is more of a jol," said Shane.

To which, Atish smiled, "So have I. That's why we're going."

"Why don't you make a plan, Shane'o?" asked Moonshine. "It sounds great. I wanna vye. I have only been to Durbs once, when I was small. Can't you organise something for us?"

"Yes, sounds really good. Let's go, Moonikes, I'll con my dad."

As they are walking home, Moonikes asked Shane what they were going to say. Shane responded that his father always spoilt him and would agree now, especially as it was for school. When they entered the kitchen, Shane told Moonshine to wait there for him while he went to sweet-talk his father.

"Sharp, Shane'o. All the best," encouraged Moonshine.

Going into the lounge, Shane greeted his father, "Good afternoon, Mr. Praag."

"How's it, Shalendra," replied Manny. "Did you have a nice game?"

"Yep, Dad, it rocked, thanks. Dad, there are tuition classes in Durban, and Atish and some of his mates are going. I want to go. It will improve my marks."

"Where, my boy?"

"At the University of Natal, Dad."

"How are you going to go?"

"Moona and I will get a lift with my friends."

"Hold on, Shane. Moona too? Where will you stay?"

"Haven't my loving sister Shanaaz and her husband, Abdul, got a flat there?"

(Oh, forgot to mention that my older sister Sunitha is now known as Shanaaz because my loving father agreed for her to marry a Muslim man, Adbul, that she met while studying hairdressing.)

"Yes, she won't mind, but you must ask her first," Manny replied.

"So, we can go then, Dad?" asked Shane with a big smile.

"I'll give you money for the classes and the trip. How much will it cost?"

"I'll find out, Dad. You're the best!"

As Shane went back into the kitchen to tell Moonshine the good news, Manny smiled when he heard them roar with excitement. Many always provided the best for his son; whatever his son asked for, he got.

"How are we going to go though, Shane'o? How?" demanded Moonshine.

"We'll make a plan. It won't be too hard, my lightie!"

Two weeks later, they were fully packed and ready to leave. Just before 09h00 on Saturday morning, one of Shane's friends, Ajay, who had just got his licence, picked them up at home. They bid adieu to Manny and Sheila, who assumed that they would be fine, as they jumped into the car.

Ajay just drove them to the outskirts of Newcastle and dropped them off there. Shane's plan was for them to hitch-hike to Durban.

Holding their 'Durbs' marked signs out, they stuck out their thumbs with smiles on their faces. After about half-an-hour passed, just before 10h00, a huge truck stopped, and Moona and Shane quickly ran up to the driver, who asked them where they wanted to go to in Durban. They answered that anywhere in town would be fine, as Shane's sister lived in town.

The driver told them that he was heading to the industrial area, which was a bit out of town. They assured him that they would make a plan and excitedly jumped in.

Being a huge truck, the normal three-and-a-half-hour trip by car, now took five hours. When the truck drew to a halt in the industrial area, they asked the driver which direction they should take to get to town.

"Are you walking?" he asked.

"Yep, we'll be fine," Shane told him.

"It will take a long time, but go to the main road over there and just follow it," explained the driver.

Thanking the driver, they put their backpacks on their shoulders. They did not realise that it would be a two-hour walk to where they wanted to go—the shopping centre known as The Wheel, where Shanaaz worked as a hairdresser.

It was now 18h00. Thinking that they were not coming, she had already left for home as her lift club left at 17h30.

The sun had already set, and Shane had no idea where Shanaaz stayed. They were stuck. Using the payphone, Shane called Manny to tell him that they had reached Durban but then lied. He said that they were late as they had had a flat tyre, with no spare wheel, and had to mission to get going again and that when they had finally reached The Wheel, Shanaaz was not there.

Manny said that they must wait there and that he would phone their uncle, Dan 'Mama', Sheila's brother. After about 45 minutes, Dan Mama picked them up, and they went to Reservoir Hills, the area he lived in, to spend the night.

Annie 'Mami', their aunt, gave them a lovely supper. They filled their tummies. Then it was off to bed.

The following morning, Dan Mama took them to The Wheel to leave them with Shanaaz. They thanked him when he dropped them off and then entered the big shopping mall. It was the first time Moonshine had been to The Wheel.

There was a huge escalator at the entrance, and Moonshine just stood there, shocked, as he had never seen one before.

"Hey, Shane! What is this? Do we have to jump on it? I'm scared!"

"Don't worry, my lightie. Just step on it and it will take you up," Shane reassured him.

"Okay, if you are sure. You know I trust you, my bro."

He stepped on with Shane behind him, and they went up to the first level. Having been there once before, Shane knew exactly where to find the salon. Moonshine's shock and amazement were evident on his face. This was the biggest building he had ever been in.

When they reached the salon, Shanaaz came out.

"Shaynoo, you made it. What happened yesterday?"

"We were late. My friend had a flat tyre, so we were delayed. We came here at about six-ish, but you had already left," Shane told her.

"It's okay. You are here now. But I'll be busy till five. Why don't you go and hang out with Tom? He works at Hip Hop, the clothes shop at the bottom of the level. I'll meet you there. Oh, here's some money for lunch for both of you. Get a pizza or something."

She handed them a R20 note and kissed them goodbye. As they went down the escalator, Moonshine commended Shane, "Aye, you can really tell lies, my bro, even face to face. No one would say though."

"It just comes naturally. White lies are easy for me. Look how I conned Dad too! That was dangerous though. Thankfully we got lucky."

Moonshine giggled, "You are Shane'o, the main'ou, the lucky one!"

Shane mimicked the warnings he'd heard on TV, "Ladies and gentlemen, please don't try this at home," at which Moonshine burst out laughing.

"No, seriously, my lightie. I am lucky, please don't follow my example. Promise?"

Moona nodded his head as they walked off. They reached Tom, who was Shanaaz's brother-in-law from Newcastle, but he now lived with her and Abdul. He had just finished matric the year before and had come to the big city to find work.

Sitting at the entrance of the Hip Hop shop, he bawled out, "Wazup, Shane? You made it, my boyee? Well done, son!"

"What a mission, but we are here now! You know Moona, neh?" Shane asked Tom.

"Yep, yep. How you doing, Moonikes, my boyee?"

"I'm super excited. This place is big, Tom. There must be lots of cherries here?" replied Moonshine.

"What do you think, boyee. Chill with me for a bit, then we'll walk around just now in my lunch break. I don't want you boys to get lost, you know what I'm saying?"

"Sounds good. We'll cool off here. I love Hip Hop," replied Moonshine, gazing in the window. "This gear is slamming! Yoh, Patrick Ewing shoes! Baseball and basketball shirts! This shop has some kit!"

(My attempt at trying to cater to everyone includes hip hop heads as well. I love old school rap, especially as an art form. Nowadays it seems as if it's all about the money though. The essence, the roots of rap, seems to have been forgotten.

Originally, there was no background. There was beat-boxing—making vocal sounds, group rapping or free-styling verse on street corners, addressing oppression. True RAP seems to be lost due to popular chart-topping HIP HOP! ☹

The roots of break-dancing—pushing the limits as to how the body can move—and turntable scratching—pulling back the vinyl record on the turntable to make a sound—are all from original hip hop.

Some commercial artists are trying though. There's also underground rapping that's a huge cult, unknown to the commercialised masses, as it is just that—underground free-styling.

Here's a personalised rap on the relevant issue – ☺ *I love to verse! My sporadic rhyming is not just limited to poetry. Let me free 'type' a rap, coz I can.* ☺

Maybe it will be performed by a rap artist? Or even rhyme some of it myself? Why not? You gotta dream big! That is the essence of life.)

Essence

Verse 1:

Nowadays, hip hop's commercialisation,
has kinda tarnished the roots in fabrication!
Rap was formed to reflect,
by the niggaz that were rough necked!

BUT now it's ruled,
by a money-hungry dude.
The essence is lost,
And rap's paying the cost!

Chorus: (sing italics, rap straight font.)

What have we done? Who has won?
Have we lost the essence, as true rap lessens,
It seems to pay the cost, frozen in a frost?
We'll melt it though, so rhyme to what you know,
Shit must stop,
Bring back hip hop, hop hop hop.

Verse 2:

It started on the streets,
rhyming to beat-boxed beats,
versing what you felt,
about the shit hand you were dealt!

Now there's no shit
coz on a throne you sit,
rhyming for the money,
coz your days are so sunny!

Chorus:
Verse 3:
You have forgotten the roots,
with your fancy, expensive boots?
Now money's rolling in,
so you **think** *you win?*

But the rap game has lost,
even tho' it covers your cost!
All about the money?
Capitalised game is quite funny.

Chorus:
Verse 4:
Watcha gonna do?
Wear a trendy shoe?
At least you can,
not like the poor man!

You bask in your glory
And don't even feel sorry!
Fancy cars, a big mansion,
money is your passion!

Chorus:

Verse 5:

Mint it in,

it ain't no sin!

Who is to judge

your sweet fudge?

You have made it,

so why the fit?

Your own commercial rap brings the bucks,

but you forgetting the root sucks!

True rap is dead,

coz of your commercialised head!!!

(My hip hop is done. It actually felt good to spill what was on my mind. Rap attack? Any artists interested? Holler back, yeah! But back to the Hip Hop shop.)

The speakers in Hip Hop were blasting some Snoop Dogg, the now-famous rapper, who was just starting out then. Tom, who also loved 'S', 'N', double 'O', 'P', rapped with it, *"La dee da dee, we likes to party, we don't cause trouble, we don't bodder nobody, we're just some niggers who on deh mike, and when we rock up on deh mike we rock deh mike wild, wow oh my dawgs keeping your'll in hell, just to see you smile and enjoy yourself!"*

"Yoh, Tom. Word for word. I bow down to you, my boyee," applauded Shane.

"You're good, Tommy T. Slamming," agreed Moonshine. "Let's go check what this shop has got, Shane'o."

Tom laughed, "Okay, browse around the shop. My lunch break is just now. I'll take you around the mall. Check the stuff and dream."

During his lunch break, Tom took them around on a tour. Moona, especially, was astounded at the things on offer. They had a pizza at the takeaway, and when Tom was returning to work, their tour continued.

They found the games arcade, and he left them there. Shane and Moona played the fancy games and tried to also socialise with the other youngsters in the arcade. The afternoon passed.

At 17h00, they went back to the salon, and Shanaaz jokingly said to them, "Welcome back, explorers!"

"Aye, what a place to work, Shanaaz? You are so lucky!" said Moonshine.

"Well, we hardly get to really enjoy it. It is normal work for us," replied Shanaaz.

"Pity, hey, but working life has not changed you, my darling. You're still so sweet," said Shane.

"Thanks, Shaynoo. I organised a lift in a bigger car for us. Let's go. It's leaving now," said Shanaaz.

They got into the car and went to her flat, which was in a nice building in the city. Abdul was home and showed them a few things on his computer, while Shanaaz prepared a lovely kebab and roti supper. A wonderful day just got better.

The following morning, they took a bus to the university campus for Shane's extra classes in the cafeteria. It was full of schoolchildren, Standard 9s and matriculants. It was a perfect environment for the thrill-seeking adventurers.

They spent their whole week on campus in this way, making friends, but that Friday, the 'schooling' ended. Being their last weekend there, it had to end with a bang.

At The Wheel with Tom on Saturday, Shane asked, "Hey, Tom, what's up for the weekend? It's our last night here. We must return home tomorrow."

"Yep, real school starts on Monday," said Moonshine.

"Well, I can organise for us to get into the nightclub," suggested Tom.

"But we are underage," moaned Shane.

"I know the bouncers there. They come to the shop. I'll see what I can organise."

That evening, after a tasty chicken curry supper Shanaaz lovingly made for them, they dressed up in their best clothes and set off to Bassline. It was about half-an-hour away, so they walked through the city. Even at night, the atmosphere and buzz in Durban was astounding.

When they got to the nightclub, they stood in a long queue, with Moonshine and Shane behind Tom. As they reached the door, Tom winked at the bouncer, who winked back. They calmly started walking towards the pay counter, but then the worst thing happened.

Shane was allowed in, but Moonshine was stopped, "Hold on, Tom. This guy is too young. He looks it!" Pointing at Shane, he added, "This one is sharp, he can get away with it, but this other one, nah."

"I tried, now what?" Tom said regretfully to Shane.

"Too bad. If not both of us, you go in and enjoy yourself. We know the way back. We'll be sharp," said Shane sadly.

"Don't worry, Shane'o. I'll park off here outside. You both go in and have a jol!" Moonshine assured them.

"I don't wanna just leave you here alone, Moonikes, but..."

"It's safe here, don't worry, Shane'o. You wanted to hang out in a Durbs club for the first time, so go and have a good time for me too. I'll be sharp!"

Tom and Shane went in and had a smashing time. Shane came out a few times to check if Moonshine was okay. He seemed to be fine outside; he had made a few friends with the other underage children, and they were socialising together.

When the action in the club started winding down, Tom and Shane went outside. But there was no sign of Moonshine. In a state of shock, they searched for him, calling his name. Their stress levels built as they asked everyone outside, without any luck.

"Do you think he pushed off somewhere, Shane?"

"Nah, Tom, he'll never leave me. Keep checking. Let's just look for him. I'll try that alley, you go that way," he said, pointing to the bottom of the street.

Tom rushed off down the road, while Shane entered the dark alley. Lighting a match, he could see something in the back. He walked towards it cautiously, calling Moona, Moonshine, Moonikes! There was no response.

The match went out, and Shane fiddled in the dark to light it again. In the dark, he could hear the lump move, and his heart beat rapidly as he lit the match to see what it was. In the dim light, it moved again.

Too scared to speak, Shane tried to take off the dirty blanket that was lying over the lump, but then a cold draught blew through the alley, killing the match. In a panic to light it one again, he dropped the matchbox. Quickly, he stooped down and scrabbled on the ground, searching for it.

In the dark, he could not see much, but then his hand touched it. It moved; it was alive. Shane's heart was in his mouth, and a cold shiver went up his spine. Turning, he tried to run out, but his ankle was grabbed from behind. He screamed as he fell over, his heart pounding as he cowered in fear.

The shadow blurted out, "Shane?"

It knew his name? In his shock, Shane had no idea of what was going on.

Then it asked, "Shane'o, what's wrong?"

Taking a deep breath, Shane softly asked, "Moonikes? Is that you, my lightie?"

"Yep! Why? Who else do you want?" asked Moonshine.

Now standing, Shane gave him a small kick, "Bloody hell. You gave me a fright."

Getting up and walking to the alley entrance, Moona said, "Hell, sorry, Shane'o. I was tired and things were dying out. I was scared to crash outside there, so I came in here."

Holding him, Shane said, "Never do that again, my lightie. You could have told me that you were tired when I last saw you."

"I didn't want to end your fun. You were having a jol. I didn't want to spoil it and be bad."

"You bad? Never! I am. The jol meant nothing. You mean more. Don't do that again. Say it straight, neh? You let me go inside in the first place. Why did you not say you were tired, Moona?"

"Okay, sorry, Shane'o. I'll tell you the next time. Let's go now; it's getting cold. Where's Tom? Has he found a chick?"

"No, he's looking for something more important than a chick. Let's go, he must still be there."

"Yoh, more important than a chick? What?"

Shane smiled and hid his tears of relief as he silently walked down the main street. Tom saw them and came running towards them.

"You found him! You're lucky!" he exclaimed.

Still sniffling, Shane replied, "Yep, I was lucky, wasn't I?"

"What were you looking for Shane, Tom? What?" demanded Moonshine for the second time.

Looking at Shane and shaking his head, Tom said, "That's Shane's secret."

"Oh, okay. Let's go home now. I'm tired," Moonshine replied.

Shane winked at Tom, and they started back home together on their half-an-hour walk.

They slept and woke up to a lovely Sunday brunch. After they had eaten, Moonshine and Shane thanked Shanaaz for the lovely time they had had. She assured them that they should not even think about it; it had been no trouble.

At about midday, their lift back to Newcastle arrived. Shane had managed to organise a lift for himself and Moona back to Newcastle with his friend Skeech, who had space for them now, as not everyone he had come with was going back to Newcastle.

They thanked Shanaaz again for her hospitality as they got into the car for a safer trip straight back to the Mandir. This journey was enjoyable, and it was also quicker. They reached home just after 15h00.

As they entered, Sheila said, "Welcome back, my boys. I hope you learnt a lot?"

"More than ever, Ma," smiled Shane.

"Durban is big but not bigger than us," added Moonshine laughingly.

"Ja, Moonikes, nothing can come close to our bond, neh?" He giggled and under his breath added, "Even when you gave me the biggest fright ever!"

"Oh, really now?" laughs Moona. "Nothing will come between us, Shane'o. Nothing!"

"As long as you enjoyed it," said Sheila. "Now go and shower and come and eat supper. You've got school to prepare for tomorrow. Now move it or I'll belt you! Don't think you are big now just because I allowed you to go to Durban. Move it!"

They scrambled off together, laughed and said, "Yes, Mam!"

Round Two

It was the normal morning assembly at Lincoln Heights Secondary School, and after the generalised prayer, the principal announced, "Okay, students, India is coming to Durban for a one-day cricket match in two weeks from now. The school is making a trip to the match. First preference for the 75 tickets at R50 each will be given to the five cricket teams from every division."

"If anyone does not want to go, the remaining tickets will be given to other interested supporters. The cost of the bus will be covered by the school, but you will have to carry your own food and refreshments."

"Please see the sports teacher in the grounds outside the changing rooms during the second break if you would like to go. Okay, you are dismissed. Head back to your classes quietly in an orderly fashion."

The lines of classes filed off. Shane's 9A line was next to Skeech's 9C line. Full of excitement, he broke the rule by softly asking Skeech, who was actually the fastest bowler in the school, "Yoh, boy, I'm excited. You gonna vye?"

Skeech, seeing the prefect coming towards them, just nodded his head.

During the second break, a crowd of boys gathered outside the changing rooms.

"Okay, boys, as the principal said, the teams are given first preference. Is everyone here?" asked the sports teacher.

"Looks like it, sir," replied Vinay.

"Thank you, Vinay. Now, who from the teams cannot make it?"

A few of the boys, including Skeech, raised their hands.

"Skeech? You? Why?"

"Well, sir, I want to come, but I can't afford a whole R50. My father's not too well and is not working now," Skeech informed him.

"I'm sorry, Skeech. Is this also why you other boys cannot make it?"

A solemn sound of agreement was heard. Shane then said, "Well, sir, I'm just thinking that because we all love the game, we are all one team, even though we are from different age divisions. Why can't we make a plan? There are 75 of us here."

"You are right, Shane. Okay, how many can't afford the trip again?" he asked and counted the raised hands. "Okay, that is 15."

There was some discussion amongst the boys. Vinay, the maths boffin, took out his scrap notebook and pencil and started scribbling. Finally, he announced, "Sir, I'm just doing some calculations. If they can at least pay R30, and if the 60 of us others can afford only another R5 each, it should cover their outstanding R20."

"Are you sure, Vinay? How did you get that?" asked the sports teacher.

"Well, sir, if you multiply the 15 of them by R30, you get R450. At R50 each, they will need a total of R750, so the deficit is R300. When that R300 is divided by 60 boys, it gives you exactly R5 each," Vinay explained.

"Well done, Vinay, my boy. Okay, so can the 60 boys who can go afford R5 more and agree?"

They all said, "Of course, sir, we are one team."

"Nice, but can the 15 of you afford at least the R30?" enquired the sports teacher.

Everyone said that they would make a plan. For a start, they could collect more bottles from the girls.

"Okay, so that is sorted. Just take your own food and refreshments. The other teachers and I will actually check your things. Besides, no alcohol is allowed to be taken into the grounds anyway, but there will not be any on the bus as well. Watch out, naughty boys, we'll just leave you behind," warned the sports teacher. "The match starts at 10h30, which means that because it will take four hours, we must leave here at 05h30 at the latest. The meeting time on Saturday morning is 05h00. We should reach there by 09h30 and stand in line. Then return after the match. I will collect money from tomorrow; R55 from the 60 and R30 from the 15. Enjoy the rest of the break."

The Saturday of the departure to Durban, they all gathered in the assembly area. They were told to make a line so that their lunches and carry bags could be checked as they boarded the bus.

Mr. D, the nickname of the biggest of the teachers, saw Shane carrying a blue polystyrene cooler bottle and polystyrene cups. He asked, "And what have we here, Shane?"

Shane responded, "It's just orange juice, sir."

"I know how mischievous you are, Shane. You think that because we can't see inside, you'll get away with it, don't you? Open and pour me a cup to taste," instructed Mr. D.

Opening the bottle, Shane claimed, "I promise, sir, it is juice."

Mr. D tasted it. "Hmmm... Shane, you are actually following the rules this time. I know you always sneak around, using those brains of yours to outsmart the system." Then jokingly he added, "Better luck next time," and downed the cup of juice in a big gulp.

There was a jolly mood on the bus as they set off on their journey. The bus had a big sound system and CD player, which was rare in those days, and it pumped the tunes.

When they were almost there, Mr. D asked Shane for a sip of the juice because he was thirsty. Shane freely passed it to him. Mr. D jokingly winced on the first sip, "This vodka is strong, man! It's burning me. Give me more orange juice to twist in it!"

Everyone burst out laughing. Mr. D smiled and thanked Shane for the screwdriver, the nickname for vodka and orange juice. At exactly 09h30, they reached the cricket stadium and queued at the entrance with their backpacks and cool drinks. They stuck together as a big group, and after entrance, they were instructed by the sports teacher that the seating in the section they were in was not reserved. Therefore, there were different areas to go to. He also added that to make sure that they stayed on the grounds, the teachers would keep their pass-out stubs.

Shane and his crew headed to the back of the grandstand, so they could sit high up away from the teachers. Shaun was with him again, as well as Trevor, Vinay, Tatty, Bruce and Ashin. They took their seats and focused on the field as the match was about to begin.

Because he was on the extreme right-hand side, the seat next to Shane was empty, and he laid out six cups stating, "Okay, guys, I'll pour the juice here, then will send to you to pass along."

Nobody was really interested because their attention was on the field, as the match had just commenced. Shane put the cooler bottle on his lap. He removed the lid and put his hand into the bottle. Then he removed a plastic bag from it. He closed the bottle and shook it. Then he poured into the cups and passed them one by one to Bruce, who was on his left, to pass along. Quite subconsciously, the boys all just held their cups in their hands because no one was really thirsty, and they were there to watch the cricket.

Shane stood up and said, "How, guys, must I propose a toast to get your attention off the match and to the drink in your hand?"

Laughing, Ashin responded, "Toast with orange juice. Nice one, Shane, nice one!"

All the other boys burst out with laughter. Shane also laughed and said, "Just smell your orange juices, you fools!"

Vinay was the only one who did and exclaimed, "Yoh! This is more like vodka than orange juice."

"Oh, you also part of the scam, Vin?" joked Ashin. "Let me play along and down it."

He quickly put the cup to his mouth and gulped it. But then he choked and gagged, "How, Shane, how?"

"Well, I thought about it long and hard. Then the light bulb flashed on in my head. I took a bottle of vodka, poured it into a plastic bag and sealed it with tape. Then I filled the cooler bottle with the juice and put the bag in. Undercover in a cover?" he giggled. "Only thing is now, after Mr. D drank most of the juice, it is a real screwdriver. A strong shot. It's a burner!"

Ashin added, "Real burner! Let's down them quickly to get more dizzy here in the sun."

After about half-an-hour, the bottle was down. Being so far behind, the match now looked too far away.

Trevor said, "Hey, guys, I'm buzzing, but we are in Durban, and we are stuck in the stadium."

"And we can't even see the match properly. It's too packed down there to go and sit there now," Vinay added.

"There's nothing else we can do. Let's take a walk around and check out one-two chicks?" suggested Bruce.

"Sounds good, let's go," agreed Shane.

Walking around on the grass, Shane noticed a group of children playing under the stands near the cool drink stall, and he had a bright idea.

He told his crew, "Hey, guys, why don't we con these lighties into giving us their pass-outs?"

"Now you're talking my language. We are wasting Durbs," said Trevor.

"How true, bru, how true. We are in Durban," agreed Shaun.

They approached the boys, and Bruce, putting his drama training to use, told them, "Okay, boys, gather round. We have a special promotion. If you have your ticket stub on you, you win a cool drink. First six in my hand."

Quite quickly, six were in his hand, and the crew escorted the six lucky youngsters to the vendor, where they each got them a cool drink.

After the formalities, they took their pass-outs to the security at the gate and were checked out.

"And now what?" asked Tatty. "Where are we going, Shane? This is my first time in Durbs."

"Well, I was on this same road three months back with Moonshine. The beach is that way," pointed Shane.

"That will be nice. I have never seen one before either," added Bruce.

"I'll be the tour guide then," said Shane.

On the approximately hour-long trip, he pointed out the landmarks he knew as they went. At the beachfront, the boys were stunned. There was a mini town and fountains, as well as a funfair with rides and games. They chose not to go on any rides but walked along the pier to let Bruce appreciate the sea. At the end of the pier, the waves came crashing in, sending spray right up in the air.

Sticking out his tongue, Bruce said, "Salt water. Nice. Now I want to feel the sand between my toes."

"Okay, then, down we go. It's a race," Shane shouted as everyone laughed and scrambled down.

Trevor won and joked, "Come on slow-pokes. Catch up."

Huffing and puffing, they all reached the beach. Breathing heavily, Ashin remarked, "Yoh, I need a skyfe now."

He lit up a cigarette, which they shared as they removed their shoes and rolled up their jeans. When they finished the cigarette, they all waded into the water to wet their feet. It was fun to feel the sand between their toes. They kicked water at each other, then decided to make a sandcastle, but it collapsed.

When they had had enough, they made their way back to the main area to wash their feet. They saw a cable car high above their heads and found out that it was R5 a ride.

"You guys want to go?" asked Shane. "I've been. You can see the whole beachfront from up there."

Bruce, Trevor and Vinay went. The rest waited for them while they walked around and saw the arts and crafts stalls. They also stopped and gazed at the rickshaws. After about an hour, the cable car ride ended.

The boys then made their way to the park to have their lunch, even though it was about 15h00. At the park, they opened the small backpacks that they had been carrying all along and took out their lunches.

Shane took out his wallet and asked the guys what two-litre cool drink they wanted.

"Now don't be so silly, Shane'o. We owe you so much for this," protested Bruce.

"Ja, bru. It's all coz of you. Pity you never figured out a way to sneak our fishing rods here though," said Tatty. "We could have caught some big ones."

Everyone laughed. Trevor said to Shane, "Ja, we'll get the Coke and a box of skyfes just for you. Come, guys, let's club in."

They all reached into their pockets and went off to get the Coke and a box of cigarettes.

After they ate and drank, Shane opened the box of cigarettes and announced to everyone, "Okay, I did not complain about you all getting the Coke and cigarettes for me, did I? So now I am giving everyone one cigarette to smoke on their own; no more passing around one cigarette. Enjoy."

He passed the box around. Everyone had a cigarette and relaxed silently in the park. Shane listened to the sounds of nature—the clichéd 'birds and bees', seagulls, pigeons and even other birds further out over the ocean.

A day truly well spent. The smiles on everyone's faces were portraying just that. They got up and started making their way back to Kingsmead, the cricket grounds.

As they reached the gate, they had to take out their pass-outs to get back into the grounds. They took them out of their pockets, and as they were about to enter, Shane noticed Tatty fiddling around in his pockets.

He couldn't find his pass-out. Now what? They had to be inside and gather for the roll call first before they could go out the gate; that was the instruction. This was going to be a problem. While wondering what to do, Shane noticed the toilets just before the exit.

Handing his ticket to Tatty, he said, "Tats, take my ticket. When Mr. D asks where I am at the roll call, just say I went to the toilet at the gate. Go in now. The match is almost over. It's 17h00."

About 30 minutes after that, India won the match as they bowled South Africa all out with an over to go, winning by 28 runs. It was time to leave, and all the boys gathered at the gate.

Mr. D counted. "Only 74. Who's missing?"

"Sir, Shane had to use the toilet. He's at the gate. He said he'd wait there," explained Tatty.

"Okay, let's go out. Stick together," instructed Mr. D, and everyone marched out.

As they reached the gate, seeing Shane outside, Mr. D asked him, "And now, Shane? Why did you go out without us?"

"Sir, I had to go to the toilet urgently. After I finished, they instructed me to wait outside because the queue was building up inside."

"Oh, okay. You wait there."

Everyone came outside. As Mr. D stood at the bus door, checking them all in, he got a sniff of Shane's breath. "What is that I smell, huh? That's dop, isn't it? How, Shane, how?"

"How what, sir?" asked Shane.

"You know what I'm talking about. How?"

"Well, that's my secret, sir. I like beating systems."

"I can smell these boys too. You are all too naughty. But luckily you monkeys never perform in the grounds. Couldn't even see you. Well done, Shane. You beat me. How?"

Laughing, Shane responded, "Why, thank you, kindly sire. I will teach you someday."

"I know that. It must have had something to do with that cooler bottle thing of yours. I forbid it for any other school trip. Now get in and go and sit. We have a long trip back home. You all are lucky I am tired, so I'm not gonna do anything coz you all never really make problems here. You all just had a drink, illegally, and watched the match."

The boys smiled because no one knew that they had also left the grounds. As the bus started the homeward journey, with the music quieter now as most of them were having a nap after their tiring day.

The trip back seemed longer for those who were still awake, but those who had passed out did not even take note.

Chapter 12

Life Presentation

(Okay, I'm staying here in Standard 9 still. Here's a very short chapter description: later in that year, there was a practical task for English with Miss Chetty. We had to do a speech or a dialogue, reflecting our lives and what we envisaged for the future, our destiny.

All learners at school have these, but here was where my mind opened more. From my side, with Bruce, a speech and drama student, we did an act, a duologue. It was based on a movie at that stage, 'Wayne's World'. Mike Meyers, the talented star, these days known as Austin Powers, played Wayne, a clichéd surfer and sixties hippy.

Synopsis of Wayne's World for those who don't know – Wayne and his friend, also a surfer dude, hacked into the cable network and were broadcasting a daily show on MTV to homes across America. It is a fun movie. Get it if you've never watched it yet.

Now Bruce and I had fun with our dialogue that we acted out, sticking to the accents and language of the hippies. How we want to be 'cool' for the rest of our lives. The class gave us a huge round of applause as we bowed.

Let me give you the conversation we had afterwards with our teacher.)

"I somehow knew you guys would be different. But all the slang was not proper English."

"But, Mam, that is how we and those guys talk. We are role-playing," explained Shane.

"Also, we are just being us, as we are, it is our life presentation," added Bruce.

"Nice cover-up," remarked Miss Chetty. She smiled and asked the class, "Do you guys think it worked?"

The whole class agreed. They were actually happy to see the difference.

"Okay, then. Thank you, boys. Keep up your individual personality traits all the time. Just be you. Everybody, there is no correct answer to life. Just be yourself, in a good way, of course." Miss Chetty told them.

(*Thank you for that advice, Mam. I am me, but all I can be is not what I am supposed to be.*

Do you remember my 'Life riddled' poem from earlier? Be true to what's in you if you can.

Yes, there are rules for us to follow, especially for young readers who are still under their parents.

Parents, your children are your creations, but they are individuals as well. Let them find themselves. Yes, obey the rules that our societal structures have created, but inside, find peace and happiness.

Yes, children are under the parents' control, BUT they are not robots. I love my parents more and more, as they give me the freedom to be me. I am in charge of my life.

Young readers, there are rules, yes. Stick within them. Show your parents how responsible you are. Do well at school. You're carving and shaping your life, to find your true you when you grow up.

What led me to these thoughts is the same task that I did with Bruce; Suraya's speech broke my heart. No, not only from the main romantic side but more from a logical mindset.

It got me thinking, especially about how as children, we are trained, indoctrinated into believing in an honestly ridiculous normality that, when analysed, escapes logical thought.

No offence again, readers, straight talk, this is my book and my thoughts. I am just analysing our world and our indoctrinations.

Suraya was one of the top students, whose vision in life was to be covered in 'pardas', the name of the outfit that Muslim women wear, that blocks everything from the world, especially other men.

This was because, in her mind, she was destined to be her husband's possession, providing for his every need.

Her envisaging this as her destiny, to me, was not only unjust but is a total wasted talent. A remarkably top student, friendly, caring and beautiful, seeing herself as destined to be an item, encased for her husband. To hide herself, follow and abide by his needs. Come on!

What about gender equality that feminism has brought about? Why must women be under their husbands? Also, why must they hide under that pardas thingy and kept as prize possessions that no one else can see?

Suraya was indoctrinated into formulating those beliefs. Factually, that pardas outfit was for the very first Muslim women in the deserts, where the blowing sand particles stung them. Makes sense, BUT what is the logical necessity for wearing it where there is no harmful sand?

From a young age, she was sculpted and moulded into a formulated belief system. Her mum and grandmother wore pardas. This system was formulated a long time back.

Yes, men ruled; it was a patriarchal society, only they could view their woman. But now? We are equal. That is my qualm with all religions in a way. Honestly, we are indoctrinated, in every belief made on a flat Earth, that 1 + 1 is 3???

In my Hinduism, which I was born into, a woman having her menstrual period—for me, the most important ingredient to life, as the old egg cell is passed on for a new one—is seen as dirty and she cannot pray!

Yes, in the uneducated times, they would see a woman's period as dirty, but come on, my thinking people!

Life is not the same as it was then. We know so much more now, especially about equality. We know the answer of 1 + 1 is 2, but we are still, respectfully, indoctrinated and trained to say 3 on our spherical Earth, which is basically a mere grain of sand in the universal desert!)

Okay, Shane, too much now, that comes later. Proceed to that painful reality that got you into this thought pattern.

(Sorry? To readers as well, humble apologies, but my last thought is: THINK if you are 'allowed' to!)

Chapter 13

Matric, Finally!

January 1994, after a thrilling December holiday, Shane was given the normal lecture by Manny, "Okay, my son, your time for playing is now over. This will be the most important year of your life. Study hard, get a good pass, and I will take out a loan and send you to university."

Shane understood and agreed, but in a way, he still wanted to have fun. Being almost 17, he took his learner driver's test at the start of that year (driver's licence only for 18-year-olds).

The learner's licence test meant that you could drive a car but only with a licensed driver next to you, teaching you.

As spoilt as he was, Shane could now use the family car to drive to school. This was all the more reason for Shane to take more advantage. It was all sorted, plus a free car.

He would pick up his older friend Arnu, who had a driver's licence, from around the corner and drive to school. Driving the Camry, aka 'cruise mobile', to school, Shane's ego rose to a different realm. He was the first and only schoolboy to drive himself to school in a car. His friends had actually cleared a spot inside the schoolyard for parking his car.

Life was now full of fun for Mr. Popular. His new best friend, Walter Msinga, was the first native African child at his school, and he welcomed Walter with open arms.

Shane had the clichéd everything: the wheels, the chicks, the friends, the drinks and the cigarettes.

(Again, fun, yes, but I wasted my potential. Please, readers, please, this may sound like fun but is detrimental, especially the smoking. If you do, quit for your own good; if you don't, good for you, don't start!

Regarding smoking, let me just entertain you with another humorous incident regarding smoking at school.

In the school breaks, the boys that smoked would go to the toilets to sneak a cigarette. But because our minimal schoolboy budget did not cover smoking, we all used to 'run the loose' or passing it along: while smoking, you would receive a tap on your shoulder to pass the cigarette on.

One day, towards the end of the year, I went to have a smoke in the toilets. I had just lit up and started to urinate at the long silver urinal when someone tapped me on my shoulder. I did not turn but responded, "Wait, man, I'll give you just now."

Tap tap again. Still facing the urinal, I lost it, "What the hell is your problem, man? I just lit it!"

Tap tap again. As I turned to take off with him, my mouth dropped as I turned to stare at my 'friend', who smiled and said, "Well, my boy, now we are settled!"

It was the headmaster, the school principal. I thanked him for keeping to the agreement he had made with me on the school excursion. Shocked? Okay, let me feature this excursion as the main story in this matric chapter.)

At Lincoln Heights Secondary, it was customary for the matriculants to go on a school excursion early in the year. 1994 included an excursion to the Eastern Transvaal to visit God's Window, the Sudwala Caves and Kruger Park.

The three-day trip was planned to depart on a Friday morning and return on Sunday. With their bags packed, the excited lot jumped on the bus for the long trip to the Eastern Transvaal, which would take about six hours. Shane had a chance to sit with Suraya on the bus, and they spoke a lot.

This was their first time together out of school. They were just two teenagers, dressed in casuals, having fun. Their love was evident but pitifully unattainable. Shane had accepted that they would just be friends but still tried. At least he was fulfilling his flutter just by talking to her like this.

During their together time on the long trip, he asked her to switch off the real world for five minutes, while he played her a song on his portable CD player.

Knowing that Suraya was not actually allowed to listen to western music, because she came from a strict Islamic background, he had to let her hear his favourite Tevin Campbell song, which was as if it was sung by him to her.

Putting the earphones on her ears, he told her to listen to the words as they were so relevant to what he felt for her.

(With today's digital world, you can get the lyrics. I can't put them here due to copyright rules. I can, however, type out a few words. I hope you don't mind, Tevin? I am not infringing, I hope? Besides, I'm not singing it! You, my dear reader, are lucky or rather your eardrums are.

Basically, the guy asks the girl, "What do I say, what do I do, tell me, baby, just to get next to you?' He says that they have been the best of friends and he does not want that feeling to end, but he wants to possibly take it further. It was truly kind of written for me to play for her.)

Tears were in Suraya's lovely eyes as she listened. After the song, she took off the headphones and said to Shane that he had said it all along, and she had heard him and was even asking herself what to say.

This was because her heart had said 'yes' a long time ago as well, but her mind, due to her religion, had said 'no'. She told him that her love was true, like she never knew even existed, but that she had to conform to keep her parents happy!

Shane opened his palm and took her hand to one side. He hid the contact from everybody. Even though everyone knew all along how they felt for each other, he did not want to make things worse. Even just squeezing her hand took his emotions to a place he never knew existed.

The power of love was immense because her hand returned the love as well. In his mind, he knew that nothing would ever compare to this.

He let go of her hand because the five minutes he had asked for had passed. They had to get back to normality. He then had the typical 'bus trip fun' with everyone, playing games, laughing, joking and blasting music while they danced in the space between the seats.

The route was very scenic, and in the afternoon, they reached God's Window. Shane was taken aback by the beauty of the scene, aptly named as it was almost a full vision of Mother Earth.

(Curious? Well, lucky we can now Google, hey? – 'God's Window – images')

Being in Suraya's group, Shane felt the utmost bliss as they just stood together, next to each other, and gazed out, sending their visions and dreams on a journey out the window, even without holding hands.

Back in the bus, they proceeded to the chalet campsite, which was not that far off. They were allocated their chalets by being split into groups of four, with each group sharing one. This was because the chalet was a 'rondawel', a round hut, with four single beds.

Shane, Trevor, Vinay and Walter formed a group. They were given time to freshen up and walk around and were told to re-join at the main site in 20 minutes, as they would have a braai.

Everyone checked in to their chalets, chose beds, freshened up and proceeded to the main area. Just as he got there, Shane noticed that the elderly school principal was sitting by himself at a concrete table and was just looking out emptily.

Knowing that this poor man had just lost his wife after her long battle with cancer, Shane had to go and speak to him. He told his friends that he would catch up with them later as he walked to the principal.

Shane sat down next to him, but he did not even notice; he just gazed out into the distance. Shane could feel the pain in the air, and from the side, he saw the tears trickle down his cheeks.

To make conversation, Shane tapped him on his shoulder and asked him if he was okay. The principal took a deep breath, wiped his eyes and said to Shane that he felt really lost and totally empty. He added that he thought that this trip away would help him to forget about his loss, but even though a month had passed, he was still torn apart.

Shane said that he understood and that he wished there was something he could do.

The principal smiled and said, "Well, I know about how you sneaked some drink for the cricket trip last year. Pity you never did it this year now because I would love to have a drink to settle my soul. Even we teachers made a pact not to carry liquor on this trip."

Now a few tears came into Shane's eyes. He breathed deeply, got up and told the principal to wait for him as he would be back. The principal just nodded his head and stared into the distance again.

Shane stopped at the food table and told the teacher in charge of the catering, Mrs. Aboo, that the principal had sent him to get a two-litre bottle of Coke and a glass.

When he got the Coke, he said he had forgotten something in his chalet and had to fetch it. He walked quickly to his chalet, took out his bag and smiled that his plan had worked.

This time, he had concealed some brandy in his bag. When the bag was checked, he was asked why he had carried so many socks, and he merely replied that he had a thing for smelly feet and often changed his socks.

He now had 'smelly feet', so he took his socks out. They were single socks containing small plastic bags of brandy folded inside, making them look like a ball of socks.

Opening the bottle of Coke, he poured out some of it to make space in the bottle for the brandy. Emptying the packets into the bottle, he picked up the glass, drank the Coke, rinsed it and took it along with the bottle back to the principal.

Sitting next to him, he placed the glass on the table, passed him the bottle and presented it, "Here, sir. This is for you. Please do as they say and drown your sorrows. You still have the whole weekend here, out in the beauty of nature. It is no time to feel down."

"True, Shane. But I'm not a good actor like you, I can't pretend that this Coke is alcohol."

Pulling the glass closer and opening the bottle, Shane said as he poured, "Well, sir, that is what everybody here thinks."

The smell was already evident to the principal. Shaking his head, "Yoh! How? Anyway, thank you, Shane, but I can't."

"Why, sir?"

"I cannot drink alone. I am not an alcoholic, my boy."

"But I am giving you the bottle, sir. It is all yours."

The principal quietly shook his head and called to Renell, who was walking past, to bring him another glass.

Shane smiled when she brought the glass. He did not even tell her what was in the Coke, but knowing him, Renell already knew what Shane had in the bottle. She winked at him when she calmly left the glass on the table.

The principal thanked her as she walked away. He then took the bottle, winked at Shane and poured him a drink. Shane used his verbosity and proposed a toast, "Sir, the brightest future will always be based on a forgotten past. You cannot go on well in life until you let go of past failures and heartaches."

"So true, my boy, so true!" said the principal as they clinked their glasses together. Sipping his drink, he added, "Well, you have a super way with mixing alcohol, as well as your words for a toast, as well as using your big brains to break rules in the first place."

"Thank you, sir. This is a very memorable moment for me. Imagine having a drink that you sneaked in with your principal!"

"Well, my boy, it's not imagination. It is real. I always saw something different in you. I could not put my finger on it, but now I know."

As he sipped again, Shane asked, "What is it, sir?"

"You, my boy, are so different to everyone I know. I am your 53-year-old principal, and I am sitting and having a drink with you as if you were my best friend. That is what you do, my boy. You make everyone your friend. You respect them for being themselves and make adjustments yourself to fit in with them."

"Is there anything wrong with that, sir?"

"Nothing at all, my boy, nothing at all. Just remember that you must ask yourself the same question. If you are happy, then who's to judge?"

"Thank you, sir. I'll remember that."

The drinks were finished, and Shane took the principal's glass to pour another one. But the principal said, "No, thank you, my boy. I just wanted one. You take it and have a nice night, but behave, hey? I saw Renell wink at you."

"Just good clean fun, sir. Thank you for trusting me. We behave but socialise. Not get drunk-drunk. This is all of the very little alcohol I sneaked in, and it's only for a few of us, just for camaraderie."

"Good, my boy. I wish I was in your shoes, but my young days passed me a long time ago. So, you live for me. I know I would love to join to have fun, but I am still in charge here. You just behave though. I know I can trust you."

"Sir, that means more to me than you will ever know."

"I am not saying it as payback, I owe you one. Take the bottle back to your chalet and let's go and eat, it looks like the braai is ready."

"Okay, sir, I am so happy I cheered you up."

"This simple 20 minutes with you, my boy, has changed my whole trip around. We switched roles. You taught me that I need to forget and move on. I'm still living."

"That's right, and that's when your brightest future will come."

As he got up, the principal said, "Good stuff my boy, good stuff," and winked at Shane as he made his way to the braai stand.

Shane smiled as he headed to the chalet to leave the bottle there. Returning to the braai site, the learners seemed to have formed into two different camps. Shane had always noticed this but had no problem with it. Each to their own, but he had a problem with the reasoning as to why. To him, it was almost as if he was not worthy of joining the Muslim learners and eating with them.

Why is that? It was the same food. Feeling the same way, Suraya didn't mind if he joined her and her friends to eat when he did. The looks from the others though were very belittling. They were friendly on the outside, but he knew that they had been indoctrinated to feel exclusively superior.

After he had eaten, Suraya told him that she could see the way he felt and that he should go and have fun as she knew that he was naughty and would love to join him but couldn't. Then she winked and added that he must do whatever he wanted to escape their reality because at least he could.

Shane understood. He smiled, thanked her and wished the exclusive group well for the night. He knew that they would be off to bed in a short while, but for him, the party had just begun.

Going over to the 'dark side', he got into a huddle with the 'inferior group' to plan the party. Their companions would be Chivani, Sharon, Denise and Renell, who were sharing a chalet on the outskirts of the main site. Perfect, as they wouldn't be heard. His crew would head to the girl's chalet about an hour after they were checked in by the teachers.

Assumed to be asleep, the boys snuck out. It was quite a party, but they were still extremely well behaved. Shane had already made the rules. They played music, had only two drinks each as it ran out, danced a bit and retired early, as they did not want to get caught.

The next day, they journeyed to the Sudwala Caves. Here Shane's mind started ticking for the first time scientifically. He had only read about these things before, but the books had not gone into much detail. He could actually see how human intelligence had grown through creations of art, like these cave paintings.

On the tour, he was also shown pieces of fossilised dinosaur bones and the striations in the rock surface, which showed how the different layers of rock were formed over the centuries. To him, this was like seeing Her, Mother Earth, with Her estimated approximately 4.5 billion years of past history exposed.

His knowledge of his reality, in his closed, confined culture, was now being questioned. His main issue was that his religion had no idea about this. Man's evolution was so factual. Dinosaurs existed. But it was not in the books, the prayers or anything from any religion?

At lunchtime, they went on to Kruger National Park. Shane had been there before with his family when he was about eight, solidifying his love for Mother Nature.

The bus was allowed to drive around on the main tarred roads, and the children clamoured around the windows to see nature. They proceeded to the campsite. There was a similar setup at this site, which also had four-sleeper chalets, and the groups from the previous night were told to stick together.

A special final night was planned by Mrs. Aboo and the female teachers who were not part of the tour. They had gotten there earlier to begin preparing a huge pot of chicken biryani for supper. There was also a hi-fi for playing music and an area for dancing.

After filling their tummies, the party started. The music was blaring, and the 'inferior' children were having fun, dancing to fast music, like on the bus, while the 'superior' children clustered together and just stood around.

This was strange territory for Suraya and the other Muslim girls. In hers, and in most of their households, western music was banned.

Shane loved music and dancing. He had asked Suraya to dance earlier, but she had said that she did not know how to as she had never danced before and that he must enjoy himself.

While having a break and changing CDs, Shane was approached by Suraya. She told him to put their Tevin Campbell song on to play so that she could at least have one dance with him.

That CD was taken out faster than lightning speed. Renell, who was also close to the hi-fi, told Shane that he must lead Suraya to the dance floor and that she would play the song. She shouted to everyone to clear the dance floor.

Shane's hand went out. Suraya clasped it, as it did not have to be hidden this time. Even just holding her hand was enough for him. He did not even lead her; they just walked together.

Still walking side by side and hand in hand, he stopped close to the centre of the floor without telling her. Suraya still stepped forward. He squeezed the grip on her hand, turned her around and pulled her into his arms, holding her for the first time. His arms encircled his world.

She folded her hands and held them together on her chest as he pulled her close. Her head turned to the side. She was inside of him, in his heart and now in his encircling arms. They closed their eyes and embraced.

Some time passed before Shane realised that Renell had already started the music, and they began to move slowly. Even though this was her very first time dancing, Suraya just moved to the beat.

Their own beat, their hearts' flutter beat with their bodies intertwined. Not led by or following, it was in total unison; a true match. It was just a pity that different 'Gods' sent them to live on their same Mother to be apart.

Within their blissful state, time passed quickly. They did not even realise that the song had finished until Renell tapped Shane on his shoulders saying that the song was over. He stopped moving and opened his eyes, looking at Suraya and all around.

In the silence, a few sobs could be heard. All the girls, and some of the boys, were crying. Suraya opened her eyes as if waking up from a dream. Still encircling her, Shane just moved slightly back. She looked into his eyes as he stared at her. And then it happened for the first time.

Their tears came together. She held her tears back even though her eyes were brimming with them. Her gorgeous green eyes were bloodshot, just like his. Even their eyes matched. With her hands still together on her chest, she raised her left hand, as she was left-handed, and dried his tears.

(Yoh, as if I am not quivering enough, having to type this book with my left hand! Did I 'come back' left-handed to dry my tears or to match her in every single aspect? ☹ Tears are falling, and even my left hand cannot stop them from coming. ☹☹☹)

There was still silence all around them. Still looking Shane in his eyes, Suraya said, "Our tears will not help; we can't change it. Just let go."

Shane breathed in and sobbed, "I know, Suraya, but I will never feel this way again."

"Neither will I, but we can do nothing about it."

"I do know that, but how do we get over it?"

"Why do we have to get over it, Shane? I will keep this memory forever!"

"So will I, my love. I felt this way for six years, and even though I tried and am still trying, I will never get over you."

"We have to stop this at least. Everybody is watching us now. Let's give them their music. I will never dance again because dancing with you has caused me more pain, Shane, because I know that I will never feel this way again. So, what's the point?"

"Our love is denied, my dear. You are right. We cannot have the life we want, but that does not stop the others from having a good time."

Suraya took a step back, but before she turned and walked off, she whispered, "Or you from having a good time too. I also fell for you when I saw you that first day at the debate, but I told you in the note that I cannot enjoy my life but that you must enjoy your life and party, and all the best with the other girls. I know you will find someone else."

"And in the note, you ended by saying that this is not meant to happen. I must keep my spirits up!" added Shane.

"Yes. For me, for us, like you were doing, enjoy your life at least."

Shane put two fingers of his left hand together and kissed them. He asked her to do the same and to place her fingers on his as a goodbye kiss because their lips would actually meet when they put their fingers together.

She did as he asked, and at least their fingers felt their first kiss. Shane saw a tear trickle down her cheek as she turned and stepped away. Still facing away, she raised her hand to wipe it away. He so much wanted to go to her. At the same moment, he stopped himself from taking a step, she shaped her hand into the stop sign. Their spirits knew that it was pointless. Shane stood and watched her walk to her friends, who were still in tears.

When she reached them, Thasneem passed her a tissue. Drying her eyes, she turned to look at him and pointed to indicate that he must go back to his friends.

He walked away. Also sobbing, Chivani gave him a tissue. Hugging him, she said that now she knew why he hadn't asked her for her bottle all those years back.

Shane just nodded. Taking a deep breath, he smiled. He would, he thought, follow Suraya's wishes and asked for the music to be turned up so Suraya could watch him having a good time as he danced the night away.

The following morning, the trip back home on the bus was more subdued as everyone was tired. There was idle banter but nothing memorable. The trip seemed longer than it actually was, but they finally reached Newcastle.

As they got off the bus, Shane gave Suraya the CD and asked her to be naughty and listen to it at home with headphones. She thanked him but took just the cover. She told him that he must have music in his life, and she would keep only the empty cover, which would maybe be filled by another CD but would not play the same song—their song.

Suraya further said that he just needed to hear the music, not encase it, because his spirit was free. Walking away, she told him that he should fill his spirit with more music and search for another 'song' to dance with.

The following morning, before school could start, he went to her, asking her for a private moment. They stood under a tree. He presented a rose and a bookmark that he had made at home the night before. He knew that she loved to read. It had half a broken heart with S.M., his initials, on one side, and S.N., her initials, on the other half of the heart, on the other side.

The words on the one side read, *"One heart torn apart, to different sides, pierced by a dart, time abides, but..."*

On the other side, *"...like this, they will never meet, to live together, and beat, forever and ever!"*

It was made from cardboard. There was no laminating procedure and he wanted to make it last forever, so Shane spent time sealing it himself with plain cellophane tape.

Suraya shed tears and thanked him, saying that she would keep the bookmark but find a place in the garden to make a tomb for the rose. She would bury it and at least let the earth, Mother Earth, feel his love, which she felt as well.

She then stated that because she could not absorb any nutrients from it, at least Mother Earth could, and even dead roses would live in Her forever.

"Suraya, this love in our hearts will live forever and ever, my dear!"

"Aha, just promise me that you will always enjoy your life, Shane?"

Their fingers went to their lips and came together as they parted with their unique goodbye kiss.

For the rest of that year, Shane's party spirit thrived. It was almost like he wanted to prove to himself that he could have fun. The rush that he got from thrills filled him, so much so that he never studied much for his matric trial exams, which are used for university acceptance, relying on his brainpower instead. But, for the first time in his life, he failed something. ☹

(This had to happen, didn't it? Too much playing around. How I wish it had been different. To all my younger readers who are still in school, I know you are probably tired of my preaching by now, and all adults, but scholars, please focus on your studies, for your own good.)

Shane then steadied himself. Knowing that trial examinations were important for his university applications, he approached the principal to write a note for him.

The note said that he did not write the trial exams because he was ill but that he was an exemption candidate with all higher-grade subjects and to therefore put his name down on their preliminary list for acceptance.

Once again, the principal cautioned him about wasting his life away and did it to show Shane that there was hope of changing. It was his second payback, and all he asked was for Shane to pass matric and get his life into order.

Shane sent the letter out to various universities. Only the University of KwaZulu-Natal in Durban, where he wanted to go to join those friends of his who were studying civil engineering, and Rhodes University in Grahamstown, to study for a B.A., gave him primary acceptance.

His ship was back on course. He was fortunate that Skeech, Trevor and Vinay formed a study group to teach him the matric syllabus. This was because his notebooks were incomplete.

The study crew would stay over at the Mandir, and Sheila, who was now home, would cater for them. Shane finally knew how important matric was and put his head down.

When the exams finally ended, Shane's crew had a big party planned for immediately after the final exam that would finish at 11h00. Shane had other things on his mind though.

That morning, before the final exam, because this was the last time that he would ever see her, he asked Suraya to take a walk with him after the paper. He had something planned and told her that she had given him five minutes when he had asked her on the excursion, but now he wanted 20 minutes. She agreed.

After the paper, the matriculants displayed their normal party spirit, signing each other's uniforms with felt-tip pens, and some even messing up each other with polish and breaking eggs on their friends.

Shane and Suraya, after signing a few friends' uniforms, left together. Looking back, they saw everyone was still at school, so Shane just held out his hand. Her hand automatically grasped his. Emotions surged again.

(Let's make it happen now. Bring the past to the present, at least in tense.)

Holding hands, they walk to the river bank that is not that far off. On the way there, he tells her what the plan is and what he wants her to do. She agrees.

Reaching the shallow bank, they take off their shoes and wade in. From his pocket, Shane takes out a needle. He kisses the fingers of his left hand and pricks them to make them bleed. Suraya does the same.

Their fingers kiss, and they let their blood mingle as they both recite, "Mother Earth, we give you our blood together because we know that, at least in your world, our flutter will flow forever on its journey to its true destiny."

The blood mingles and drips into the water. A teardrop or two are added. They embrace as spontaneously as they had for the dance, holding each other tight, his arms around her, encircling her, and she has her hands together on her chest.

Their heads are turned to the side, with their eyes closed. The sound of nature fills the air, birds chirping, and they hear the river trickling away, carrying their love with it.

(I need some poetic relief now for the tears which are trickling from my eyes.

Sorry if yours are as well.)

Trickle

Where does it start?
It is like a true work of art.
A true bond
from way beyond.

Joined from forces in a different realm,
to even overwhelm,
its tricky nature,
if the merged blood's stature
is to look beyond
and trickle into a future pond?

To flow its own way,
and away from indoctrination, stray.
Inside it, it is so true
a trickle stuck like glue.

But it doesn't flow,
as it seems to know
that it has to abide
and stay away from a destined glide.

Suraya and Shane's hearts beat together, pulsating the blood through their veins, on its destined glide. Time stands still during their close embrace. Only the water trickles.

For Shane, even though he is holding his true love, it was the same as when he was sitting on the rock in solitude at the Rotary trip when he wrote his *Flow of Life* poem.

He truly feels alone. Their dream life is only flowing in the water. They are together, holding each other, but only their merged blood is truly together. He takes a deep breath and pulls his chest back, sliding his arms to grasp her waist.

Suraya places her hands around his waist as well. With her head back, their eyes open almost at the same time, and they stare into each other's eyes. Shane is amazed that eyes can hold such a look of desperation, of longing and yearning.

That look into his eyes pierces his soul. He needs to feel her kiss, her lips, so badly but knows it will hurt even more. It would just be a physical representation of true, restricted love.

Suraya takes his left hand to her lips. His fingers join as she touches his hand to her lips, kissing it and then transferring the kiss to his lips.

Shane is in heaven as he kisses his fingers with her kiss on them. He does the same for her, and his heart skips a beat when his lips make contact with her fingers.

They embrace again, this time with both hands holding the other. Holding her tightly for a while, Shane glances at his watch. He can see that his time is almost up. Even though they feel that they could remain there forever, they know that they have to leave. It is a solemn walk back into reality.

(And the tense changes. Back to the past. ☹)

Shane walked Suraya to her home, which was close to the school. Leaving him at the gate of the yard, she told him that she wanted to watch him walk out of her life, so he must just turn and never look back.

Their fingers kissed goodbye, then Shane breathed in deeply and turned and walked back to the school.

He never even thought of looking back. He did not want to see her and remember her waiting for him to return home because she had once told him that she would be a housewife who would wait for her husband to come home every day after work.

In his mind, he knew that he must search for the same feeling again, to feel as though he had met someone made just for him—a true match, another flutter for his heart. In his mind, he planned the formula—the holding of hands, the finger kiss and the embrace. These actions would be his own formula for success, his testing ground.

His legs involuntarily broke into a run as his tears poured. Running into the school, he went straight to the tap, opened it and stuck his head underneath. The cool water helped his mind to cool down but did not reach his heart.

(*Yoh! Hold on, Shane. That's not how I remember it.* What do you mean? *Yes, my boy, you did walk her home and make your solemn parting with the special kiss, but that trip to the river with the blood?*

Oh, sorry, Shane, that is what I remember. *But it never happened like that!* I know that, but that is how it should have ended. *But it didn't.* Come on, give me credit for my imagination at least? *Okay, credit where credit is due. I was even in tears.* And I'm sure the readers were as well.

Maybe, but some people who are not as sensitive read as well. Well, the rest of my story with Suraya is 100% true. It was just my final parting that I made up. I only walked her home, and the rest is true.

I know that, I felt the flutter. And how! *Anyway, let me not linger here anymore, go and toast your completion of school at the party with your friends.*)

Everyone was in an extremely cheerful mood, so Shane put on a happy face for the party, clouding his true heartbreak. He had officially lost his soulmate and would never see her again.

Shane distracted himself further because he and his crew actually planned their own matric ball, which Suraya was not even allowed to attend. The Standard 9 learner's, traditionally responsible for it, plans had gone sour. This would be the first year without a matric ball, so they took action.

On 28 December, Shane and his friends stayed up for the newspaper so that they could get their results as soon as it came out. He got an exemption. He could have done so much better than a mere 'C' aggregate, but he still obtained the points required for the University of KwaZulu-Natal as well as Rhodes University.

In January, he submitted his full applications and waited. The school also had its awards function, and for the first time in high school, Shane actually received an award. This had been normal for him in primary school, but with lost focus, he never received any book awards for academic excellence in his five years at high school.

At the awards ceremony, he climbed onto the stage to receive his first award as the best...

(Unfortunately, soccer is where I had chosen to put my energy and focus on, hey? I know that I chose to let my academics decline, and I will forever hold that regret.

But it was my past, and I cannot go back and change it. More importantly, maybe IF I had done better at school, or something changed, it would not have led me here to write this book?

Do your best at whatever you do, but do not waste time regretting not doing anything in the past because your present makes your future.

Second Half, here we come!)

BOOK 2
SECOND HALF

Chapter 1

R U Ready?

(The reason I ask is that I'm going to use clever pun and speed through my five years there in one chapter! Lucky, no speed traps!)

In the middle of January 1995, Shane received his letters of acceptance from the University of Natal and Rhodes University (R. U.). At the University of Natal, he was accepted to study the subject his school course had been designed for, civil engineering.

Paging through the information booklet, he finally realised that a civil engineer could be out in the bush for about six months building a bridge. Nah! Not for him. Plus, studying technical subjects on a city campus was not sociable.

He chose the Rhodes option for the university lifestyle, that of living on campus and socialising. Also, he assumed that he would achieve his father's dream in some way at least. He would be studying to become a mind doctor in a sense.

He was accepted for a B.A. degree and planned to major in psychology. He confirmed his acceptance, and in the first week of February, he excitedly started preparing to set off on his adventure.

On a Saturday, the drive from Newcastle to Grahamstown took the family 10 hours, with Ricky and Manny sharing the driving. When they entered Grahamstown in the afternoon, Shane felt as if his calling for a university was met.

He wanted more of a clichéd 'old school' feel, a bit like Oxford University in a sense. He wanted to socialise in a student community.

Rhodes was perfect. He made the registration in time and was directed to his residence, Cullen Bowles (C.B.), which was on a hill. At Rhodes, the prerequisite is that all first-years must live in residence.

After leaving his luggage in his room, he went with his family to spend the night at their hotel. They had a quiet night of rest, and the next morning, Shane was taken to his residence. His family then left to return to Newcastle.

There were about 20 first-year students, along with roughly 60 seniors, in C.B. that year, and the spirit of camaraderie at Rhodes, regardless of race, made it easy for them to make friends.

With Nelson Mandela as president, everybody could feel the need for unity in the new South Africa. Shane was realising his dream of a mutual community with fellow students.

Up on the hill, around his dining hall, there were eight other residences in addition to his; five for men and three for women. That made about 500 students from the different houses. There were 12 other dining halls around the campus with a similar setup. The socialising was amazing, and getting to experience different people, different races and different interests fitted Shane's bill perfectly.

That first week, the orientation week was held to register students for different subjects. On the first day, Shane registered for psychology, but for his Bachelor of Arts degree, he needed at least three other subjects as well. He, therefore, took sociology, as it seemed to go with psychology.

The next subject Shane chose was to enhance his verbosity and dexterity, to heighten his intellectual prowess and portray a verbal splendour.

(I have a way with words, even when speaking to myself. ☺ Layman's terms – I chose to study English.)

It was a long day. Shane still did not know what to choose for his fourth subject. He walked around and searched without any success, and by late afternoon, he decided he would go back up the hill and rest his feet in his room.

Being lost, he did not know where he was on the campus, and as he walked past a huge building, he heard a woman screaming inside. Thinking it was a desperate plea for help, he stormed in. Expecting the worst, he clenched his fists as he pushed the door to the building open.

It was dark inside, but he could see a boy lying on the floor. A beautiful white girl, with blood on her clothes, was down on the floor with him. Shocked, he rushed forward to assist. Then a voice boomed, "And now? Get him out of there. Bloody first-years!"

The lights came on, and the audience burst out laughing. Shane realised he was standing on a stage.

Waking up, 'Juliet' smiled and asked him, "Are you signing up to do drama? We are doing a presentation to interest the first-years."

To cover up his embarrassment, Shane winked at her and then said loudly to the audience, "We caught you guys with some real acting. This was to show you that nothing in this course is set. This was planned."

She added, "There is freedom of thought, so join! We love breaking the norm."

The applause built as they joined hands and took their bows. Juliet didn't let go of Shane's hand as the curtain closed.

"I'm Laura, and who might you be, fine squire?"

"Well, pardon the interruption, me dear. I be Shane, I be sent from yonder to save thee as thy love for Romeo was in disdain."

Laura smiled and enticingly said, "Shane'o, Shane, wherefore art thou, Shane'o?"

On one knee, he rhymed, "Can thy not see, this bending on one knee, as a plea, for your company?"

"Well, thou art more temperate and lovely than anyone here for me, so I will oblige."

It was a dream world for Shane. There were so many women to charm, from all the different races, and with so much fun to be had, he hardly attended his lectures.

He would wake up late and miss breakfast at his dining hall, so he would walk down to the cafeteria for a doughnut and some milk. Then he would spend the morning there, playing pool and using his charm to 'con all the cherries'.

At lunchtime, he would walk back up the hill and then attend whichever tutorials he had in the afternoons. He had to attend his 'tuts', as they were compulsory, with a marked register.

Some evenings he had soccer practice. He had joined the H.S.S. team (Hindu Students Society), who had training sessions and matches in the week. It was a sociable soccer league.

There was also a nightclub, The Vic, for him to 'shake his ass' on the weekends, while in the week there were various pubs to visit.

It was an eventful first year, but he hardly focused on work. He was also now a campus radio DJ, Insane Shane, the 'I.S' of the airwaves. The station was RMR – Rhodes Music Radio. Insane Shane also started to DJ at The Vic. His popularity stakes, especially with females, rose even higher.

That first year sped past. Even with his playing around and hardly attending lectures, after borrowing notes and cramming for the exams, Shane had passed all four credits.

It was the end of year break, and he was at home in Newcastle when he received the good news. Also, he had just got his driver's licence, so Manny bought him his first car as a reward. It was an old, metallic green 1983 Volkswagen Golf GLS. *(I think what we call a 'Golf' is called the 'Rabbit' in America?)*

'Golfy' was a typical student's car. He made a sticker for her back glass: 'INSANE SHANE'. In February 1996, the starting of the academic year, Shane drove Golfy down to start his second year.

He registered for Psychology 2 and also chose Drama 2 to major in because he had always loved it but did not realise before this that he was cut out for it. He dropped English because the workload to cover was interfering with his social calendar.

Sociology was dropped as well because there were too many assignments to hand in. A Bachelor of Arts degree needs 10 credits in total. He had four already.

With second and third-year psychology and drama adding four more, as they were his majors, he needed two more credits to complete his course. As he saw these as play around subjects, he would, therefore, do one a year.

Having a love for art, in that second year, he chose the easy visual communication course. All set, the second year 'studying', aka partying, to him began.

He was still at C.B., his old residence, and was now the 'Entertainment Representative', which fitted his attitude of enjoying life.

The year began in much the same way as the previous year, but he now had some seniority. Relationship-wise, he decided to grow up. He had noticed Trish, a beautiful first-year student, around as they had their meals in the same dining hall. She lived at the Walker Residence for girls.

What caught his eye was that she had the latest Toni Braxton short hair look. She also wore Levi jeans, which were very rare at that time because the sanctions against South Africa had just been lifted. They shaped her so nicely.

Coming from Durban, she was the typical clichéd 'city chick'. She was full of fun and loved to go clubbing. On the dancefloor at The Vic nightclub, Shane noticed her smooth moves from the deejay box.

After his DJ set, he pursued her, just to explore his possibilities like he normally did. His charm worked, and that night they made it back to his room. She was very decent though, and they did not go too far.

Shane had never felt this way before, so he decided to retire from player mode and get into serious mode. They started going steady, and it got serious. It was a 90% flutter replacement.

The relationship grew stronger, and as a result, Shane's spirit became more settled. His love for performing also grew, and he auditioned for a part in the university's big annual production.

In 1996, it was Junction Avenue Theatre's play, 'Sophiatown', about the apartheid government 'relocating' (force moving) the African people from Johannesburg.

Sophiatown, in the sixties, was an inner-city, functioning like the much-known American Chicago with its Mafia. Shane got the role of Mingus, the lead gangster.

Manny and Sheila even drove down to Grahamstown to watch their son perform. They were thrilled at his immense, unrealised talent and were also very happy that he now had a serious girlfriend.

At peace and pride, they returned to QwaQwa, in the Free State Province, where they now lived. This was because Manny and Ricky now ran their own RIX Motor Engineering company there.

The 10-day annual National Arts Festival that Grahamstown hosts, mainly for performance artists, was taking place in July that year, and Shane had heard from friends, who had attended it the previous year, that it was a party. Because it was in the middle of the July vacation, he went to QwaQwa for the first week, then planned the trip for the festival.

Trish could not make it because it was her first year, and she missed home terribly, so she preferred to remain behind in Durban.

Shane invited Ricky to join him with his girlfriend, Ashrin, and friends Vini and Rishi. They booked a lodge in Port Alfred, the coastal town, and had a great time. Shane was deejaying at The Pyramid, a special nightclub that only opened for the festivals.

It was during this festival that Shane first got a shock regarding his destined future. It happened when he went into a gypsy's caravan to have his palms read.

Putting his hands on the table as instructed, the palm reader actually screamed and backed away. It was almost as if she was in shock. She said that in her many years of reading, she had never seen anybody with two such distinctly different palms and therefore refused to read his hands.

(If only I knew then what I know now. You'll see what I mean later. Never mind though, as I told you I believe in timing. Everything falls into place. My other life will start when I finish this book with my special hand.)

Ricky and the others left after enjoying the festival as well, and university resumed for Shane. He had no problems with results at all, and neither did Trish.

She was doing a Bachelor of Social Science because she wanted to become a human resources manager one day. In fact, everyone on campus had set career goals except for Insane Shane.

He was basically there to have a party, and if it led to an additional bonus, all the better. His love for soccer was also met as he played in the league.

(Just remembered an incident. I was kicked in the face once at one match when I went to 'head' a ball that the opposition player was trying to kick, jumping up and using the 'bicycle' trick.

His foot and my head met, breaking my nose. Trish lovingly took care of me and did not mind if I looked like Zorro for a bit, with the patch over my nose.)

Shane and Trish were having a great time in their residences, but they felt restricted in a way. In addition, the residence fees were going up. For the following year, they wanted their own 'digs', the name given to private accommodation away from the campus. Searching around, Shane found a two-bedroom flat at Loyola Heights, which was not far from the main campus. They took it.

That December, in Newcastle, Resh, who was now working after studying in Durban, bought Shane a lounge suite and fridge. He took the TV set from their Mandir, as it was not really being used, and was gifted a television cabinet from his uncle Chunky.

Sheila provided Shane with dishes and bedding, while Manny bought him a small two-plate stove. Being the loving father he was, Manny also spoilt Shane by buying him a mobile disco set because he had realised that he could make money in Grahamstown as a deejay.

For the rest of that holiday, he was like a typical 'charou', kitting Golfy. He was busy pimping his ride.

Shane's mother's brother, Anil Mama, was running a panel-beating shop in Newcastle. They firstly changed her old rear-facing front fenders and replaced them with the forward-facing ones of the new model, Citi Golf. Her old bumpers were also replaced with modern ones. They did a total colour change and re-sprayed her from her original metallic green to blue and colour-coded her mirrors.

Then her boot/trunk lid was sprayed black around the back windscreen, and a three-piece, black, back roof spoiler was also added.

Ricky gave Shane a set of 15-inch mag wheels, and they rolled her fenders and changed her springs to shorter ones to lower her. Then, finally, Shane put a black spoiler under the front bumper. So, her exterior was complete. Her brown cloth interior was changed to a matching black and blue one. Completing her interior, a blue and black steering wheel was added as well as a gear knob and blue and silver pedals.

Her engine was left standard as she was an 1800, fast enough for Shane, but her exhaust was changed to a free flow with branches and an upturned double tail piece. For a campus car, Golfy was a real looker, ready for Grahamstown.

(How, like one typically clichéd char lightie and all and all, smiling so big coz he smaaks his wheels too much ekse.)

Shane transported his furniture and other goods for his digs by train from Newcastle to Grahamstown in a shipping container. In two weeks' time, Manny and Sheila followed him and Trish to Grahamstown to help set them up in their flat.

It was a big mission but was successful. They blessed Shane as they left to go back to QwaQwa. They were happy that their son was now finally sticking to one girl and moving in with her.

It was Shane's third and final year for his B.A. in Psychology and Drama. As he needed one more credit for his B.A., he and Trish decided to do a subject together to make the workload lighter.

They chose anthropology. The study material intrigued Shane, so he actually attended lectures for a change. He was finally realising a whole lot more about his world, Mother Earth and his own existence from evolution. This fire was sparked by his trip to Sudwala Caves.

That year, he did mobile discos as well. Golfy would be fully packed with her back seats folded down, as Shane used her to transport his huge disco speakers and deejay equipment.

He loved to deejay, and having the equipment meant that he could hire venues and use his deejay set for his own pay parties, charging a cover charge at the door to get in.

(At that stage, it was a whole R5 per person to enter.)

Luckily, petrol was just over a rand a litre, and Golfy's tank cost only about R70 to fill. *(Good old days, hey?* ☺*)* She was even broken into a few times as she had no alarm. The back quarter glass window was smashed in so that the doors could be opened. Her radio was pinched the first time. Nothing was taken the next time, even though her window had been smashed in. The only difference was her front seat being put back to recline.

(I am assuming that to fight the winter cold in Grahamstown, a homeless person sheltered himself inside Golfy. If I had known this and they had knocked on my door before breaking into Golfy, I would have given them shelter.)

Living as a student is tough financially, so to cover their extra costs, Trish also got a part-time job at a coffee shop, waitressing. It was a very strenuous life, but they managed. They clicked and were very happy.

Shane had also started performing with the First Physical Theatre Company. Physical theatre is the way in which the body expresses emotion through movement. Shane's love for performance continued to grow, and his body's movements now extended into a different realm. He did ballet classes as well.

In that year, he performed for the company during the 'Grahamstown National Arts Festival.' He also deejayed on RMR as well as at The Vic. Trish stayed back with him for this festival because she had work at the coffee shop. It was a smashing time.

The year passed and Shane completed his B.A. Degree in Psychology and Drama.

Even though Shane had graduated, Trish still had her third year to complete the following year. They were a team, and he was not prepared to leave her alone in Grahamstown.

They were inseparable. Seeing furthering psychology as a bit too much work, even though he did well and was accepted for honours, he stayed back to do his honours degree in drama instead. He loved it because it was studying while having fun.

With his study schedule now lighter, it was the start of him spoiling Trish. He had more time, so he was doing more of the cooking, cleaning and driving her around. Trish used to pull her weight the best way she could though. The relationship balanced itself.

In that year, even though he was a student, Shane's performance realm expanded. This was because he was now a professional dancer as well. He toured the country, taking First Physical Theatre Company's most famous work, 'Lilith', to theatres.

(Picture proof with my VERY proud parents.)

It was a touching story about the myth of Adam and Eve being the first man and woman to be created. In the finer print of the very first Old Testament Roman Catholic Bible, the original story, Lilith was the first woman but was not worthy as she had sinned.

Because of this, she was cast out of the Garden of Eden to live on the outskirts. Obviously, Adam was a lonely man then, so God obligingly made him Eve to keep him company.

(Honestly, this also got me really thinking about how even written mythology changes, and we follow the version that we know. I am assuming that devout Christians do not even know of Lilith at all.

Hmmm... We seem to divert from the original and make changes, then those who do not know about the older version assume the diverted version to be original?

Which storyline do you want to follow? Go with the masses?)

This storyline made for a very moving performance, which toured to Cape Town, Durban and Johannesburg.

(Featured story in The Star Newspaper during 'The Dance Umbrella' festival of dance with nationwide entry, held at the Wits University in Johannesburg.)

'Lilith' also featured at the National Arts Festival as First Physical Theatre Company's attraction.

That year, Shane and Trish rented out their flat and bunked with friends. It was 10 days of a great time again, with Shane performing as well as deejaying on radio and at The Vic.

In the latter part of that honours year, Shane and a few other students toured to schools and held drama workshops to get scholars in touch with their creative sides. Shane's love for dispersing education flourished.

At the end of that year, with Trish now having completed her degree and him his B.A. Honours, Shane was ready to leave university and venture into the working world. Trish, however, wanted to study further and do her honours degree in industrial psychology. Given this, they stayed for the following year. Shane felt obliged to do so, and also because he loved performance, he wanted to further it with a two-year master's degree in drama, as he had passed his honours year exceptionally well.

(The most moving moment of my life with my Ma'jee was my final practical exam performance for my physical theatre course in that honours degree year.

It was called 'Membrane'. Let me give a description of this incident, in which my mum and her spirit never leave my side. For now, put on theatre attire – including a symbolic brain? Now picture this...

The curtain rises. Through the dim reddish light, in the smoky background, we see a transparent square 'eggshell'. It is of different colours of cellophane paper and is in the centre of the stage. It also looks like a metal cage, roughly about one and a half square metres in dimension.

As it is transparent, the interior is visible. As the spotlight focuses on the egg, it starts 'warming' it up. There is a white membrane inside the egg that starts to writhe, in readiness for its birth.

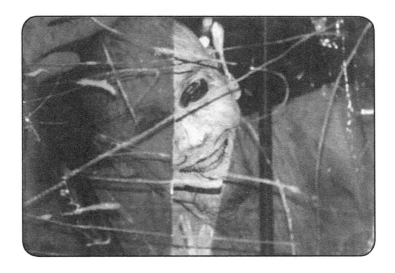

Above and around the egg are fragments from other 'cracked' shells. Behind the egg, a three-metre-high ballet bar runs across the whole stage.

The stage looks extraterrestrial. Solo piano music starts playing, and the membrane starts moving rhythmically around inside its shell. It is searching for a way out. It is trapped inside. After searching for a while in swirling, surging movements, it collapses.

The light dims. The music changes as the eggshell rises on its own and opens, leaving the membrane on the floor.

Blackout.

As the light warms, we notice what has been born. It is a collapsed white human body. But there are two heads—two masks, one at the back and one in the front.

The body is prone to the floor, and its first movements keep it spread-eagled on the floor, like a crawling toddler.

Blackout.

When the light warms up, the toddler is now grown and is standing up but with swapped masks, and the other face is on. It is now free to move. The music changes.

The body moves around rapidly, energetically, searching for something, in a haste. Then as the music builds, the movements become more frenzied.

The music suddenly stops. The membrane crashes to the floor.

Blackout.

On the left of the stage, we see a burning lamp (diya) being brought on stage by an Indian lady in a sari, carrying a 'thari', the holy tray.

The light slowly gets brighter as the lady reaches the membrane being on the floor. Soft tinkling music is heard in the background. When the woman turns the thari with the burning light around the being, it slowly rises and stands facing her.

We see them from the side. The lady removes the mask and circles the thari around the being's head, like the ritual practised by Hindus.

After that, she takes its hands, turns them over and puts a red powder in both hands. She then symbolically puts her hands together in prayer to say 'namaste' (bless you) and walks off.

The being is now free, no masks; it does not have to follow any indoctrinated pre-requisites, just enjoy the joy of existence—freedom of the inner soul.

He goes to the front centre stage, where he freezes for the audience. The music starts again, loudly this time.

Both his hands are rubbed down his face, staining it red from the powder. He 'finds' this as his true mask—his life, his choice.

Now he dances with reckless abandon, swooping down to the floor, up and down and all around. He goes to the high ballet bar at the back.

He jumps up and grabs it with both his hands, pulls himself up on it and actually does a tightrope walk across it to go offstage as the lights dim and fade to black!

The end.

That was my honours exam piece symbolising my life. The birth depicts my sporadic attempts at trying to find myself. On the floor with first 'identity', never work. Then trying the other identity standing up. Still never work. Then finally finding it by being blessed by my mother. No masks, just my face with my mum's added blessings.

Then I'm free to be me, a physical theatre performer on a ballet bar. Read again if you did not see these things the first time, as it is hard to describe.

Now the moving part is that Mum was always too scared to look at me walking on the bar, which was three metres in height. A tightrope act. She always kept her head down.

What follows is proof to me of her power. On the final night of my performance, while walking on the bar, I could feel myself losing my balance and falling.

I pushed against the hanging eggshell that should have swung back easily. I should have fallen. But it was as steady as a rock. I pushed against it and regained my balance, and I could then complete the walk across.

The light faded. As it brightened, I usually made my way to the stage, and holding hands with my mum, we would take a bow.

This time, Mum had tears in her eyes and squeezed my hand. We bowed. When the applause died down, I held her tight.

I knew in my heart that she looked up for the first time ever when she intuitively sensed that I was in trouble. Her mere glance to me made a free-hanging structure as solid as a rock? Coincidence? Only time! You decide.

Her spiritual power is immeasurable. Thank you again to my mother, my Goddess, who is everything.

Now, where were we? Oh, starting of 1999.)

With him being accepted for a master's degree, as Shane was not yet ready to leave Grahamstown, that year seemed shaped. He still kept up his tutoring, performing, deejaying and workshopping though. And he played soccer too. Everything was linked. Life was good for him.

How Much is Too Much?

In the middle of 1999, for the festival, Shane and the DJ prodigy friend he had trained, Jessy, opened a nightclub they called X.S. *(My witty way to say 'excess'. Come party to the max! ☺)*

They built it up themselves from two abandoned shops with a mutual garden in-between. It was basically two separate dance floors, one playing faster house music and the other playing mainly slower R&B.

This was Shane's busiest festival ever. He performed for the First Physical Theatre Company in the mornings and at lunch. Then he deejayed the afternoon away on the radio and then made sure that the nightclub's bar was stocked, opened up and deejayed the night away.

This was every day. It was extremely physically demanding, especially staying open on quiet nights. He was pushing himself and only got about two or three hours sleep a night over the 10 days.

In the early hours of the final Sunday morning of the festival, Shane smiled as he closed up X.S. Everyone involved in the club worked very hard, especially Trish, who managed the bar. Aside from Jessy deejaying with him, Shane's other close friends also gave him a hand. Dillon, Ravi, Edwin, Sharlinie and Reshma helped. It was a productive effort.

It was finally over. Feeling drained, Shane and Trish went straight back home. Grahamstown in winter is very cold, and because Trish came from Durban, she did not like the cold. She, therefore, jumped straight into bed, where fatigue made her pass out. Shane did the final packing away of his things. He felt relieved. He had done so much and managed.

Sitting in silence in the lounge, he reflected. After a few minutes, he smiled to himself as he stood up and went to their room to finally rest his weary head. In the passage, he took a deep breath. Then the fatigue finally caught up with him as his legs gave in, and he passed out on the floor.

After about an hour, Trish could not feel him next to her, so she woke up. She called his name. No response. She got up to look for him. As she entered the passage, she saw him lying on the floor.

Shocked out of her mind, she ran to him and screamed, "Shane, Shane! Wake up, my love! Wake up, please!"

There was no response. She checked for his pulse. He was alive but out cold. She grabbed the phone and called Edwin.

He and Ravi came in about 10 minutes. They carried Shane to the car and took him to the hospital. He was still out cold. Edwin and Ravi put him on a stretcher and then sat down in the waiting room, while Trish accompanied him to the ward.

Trish, with her mind spinning, held her breath while the doctor examined him. After about 10 minutes, the doctor said to her, "Well, my dear, he is lucky you brought him in when you did. It would have become a whole lot worse if it had been any longer."

Breathing in deeply to calm herself, she asked, "But is he going to be okay, Doctor? What's wrong?"

"Well, according to my diagnosis, it is extreme fatigue. His body needs a break. He should be back to normal in two or three days. I have checked him in."

"Okay, Doctor, but he is supposed to go to Cape Town to perform with the Physical Theatre Company on Tuesday, plus he has a practical exam."

"He will never make that at all, let alone normal physical activity for at least a week. He should come out of the blackout in a few hours, and then it is purely bed rest for his body to recover."

"Okay, Doctor. Can I sit here with him till he wakes up?"

"Of course, my dear. I am around. I will check up on him later."

"Can you please send those friends of mine in now, Doctor?"

"Will do, my dear. Be strong."

The boys came in, and Edwin, looking at Shane in the bed, asked Trish if he was going to be okay.

"He will be. The doctor told me he is suffering from fatigue."

Ravi, who was studying pharmacy along with Edwin, blurted, "See, Eddie, I told you it was fatigue."

"Well, obviously, neh? This monkey worked like a dog. For what? Heavy rent, plus club was dead in the week!" exclaimed Edwin.

Ravi added, "This hospital visit will take most of the few bucks he has made."

"Well, that's just the way my baby is. You boys know. He pushes himself to do his very best. Never mind the end result."

"Tell me about it. You know how he goes on at me on the soccer field when I miss the ball?" asked Ravi.

Edwin laughed as he answered, "That's mainly because you play like a sissy. You must be a man."

"Whatever, Eddie. He's checked in. Must we go now, Trish?"

"We can if we want, but I will stay with my baby. You guys go."

"He's our bru too," retorted Edwin. "We'll stay with you. We'll come back now. Come, Ravi, let's go get some coffee and have a skyfe."

Trish pulled a chair up to Shane's bed, grabbed his hand and held it tight. In the silence, tears filled her eyes, and she felt she could now let them out.

Sobbing, she saw him sleeping peacefully, almost with a smile on his face. He always smiled, never frowned, even now. She smiled. The boys returned with the coffee and sat with her.

When Shane finally awoke, it was mid-morning, but they were all asleep on their chairs. He looked around, uncertain at first as to his whereabouts, but then put it all into place.

He could feel his hand being held tightly by Trish, even though she was fast asleep on her chair. She never left his side. He just squeezed her hand tighter as he went off to sleep again.

Aside from Shane having to miss performing in Cape Town with the First Physical Theatre Company, it was also a practical exam performance for his master's degree for an external examiner in Cape Town. He obviously could not make it.

Because he missed the exam, he had to give up studying. For the latter part of that year, he tutored students and deejayed on the radio and in his night club, X.S., which he kept open for students. He also carried on doing educational theatre programmes, touring around to different universities to show and teach physical theatre.

(Hey, Shane, how could you forget the Whacko incident? You're right, but I'm shortening. *Then why don't you just do a 'short' short story?)*

Whacko...?

Towards the end of that year, Shane surprised Trish by giving her one of her ultimate dreams. He just asked her to keep the weekend free, finish off all her work and be ready for the surprise. He said he would collect her after her lectures at 3 pm and then take her somewhere important.

That Friday, Shane packed her a weekend bag and then picked up his close friend Dillon and his girlfriend, Sharlinie. Next, he drove to campus to collect Trish, who was waiting anxiously.

Always being a gentleman who opens doors for ladies, he jumped out of the car and opened the back door, next to Sharlinie, for Trish. In his 'British pilot' accent, he stated, "Your chariot awaits, madam. Upon entry, please strap on your seatbelt and enjoy the view. Idle banter will be accepted, but please refrain from any enquiries about the destination."

Adapting to his role play, Trish said, "Why, thank you, sir. I will brace myself for a memorable journey with you, regardless of the destination," and climbed in.

Sharlinie and Dillon knew the plan but just said their normal hellos as Golfy headed off. It was quite a scenic route to Alicedale, the small town just outside Grahamstown. They laughed and joked on the way, especially the girls in the back. Then Shane pulled into the driveway of the railway station there.

Trish noticed and asked, "And now, my love?"

"We have a gravel road ahead. I think Golfy's shocks are gone," said Shane. He opened his door, climbed out and announced, "We should all go and push on a corner. Please get out."

As he always did, he walked around the back and opened Trish's door for her.

"Thank you, my love. Where do you want me?"

Shane had already stuck an envelope on the lid of the boot/trunk before opening her door and said, "You stay here, baby, I'll go to the front."

Dillon and Sharlinie knew this plan already and went to their respective places.

When Trish climbed out, she was still looking at Shane.

"Don't look at me, babes. Go to the corner to push down when I say."

Approaching the boot lid, she asked him, "Here?"

"Yes, babes, push that boot lid down."

Trish finally looked at the boot lid and saw the envelope with her name on it stuck to the lid. She pulled it off and asked, "And now, my love?"

"What? It's not for me. It's got your name on it. Open it."

Holding it, she walked towards him and excitedly said, "I think I know... You never... How could you... We have no money!"

"Never mind how. Just open it, my love."

She ripped the envelope open and took out two train tickets to Johannesburg, plus tickets to Michael Jackson's concert. Whacko Jacko was her God, and she was left speechless.

They had a memorable 16-hour trip by train to Johannesburg through the Friday night and then got picked up by Avinaash, Shane's close cousin.

Then, at night, the amazing Michael Jackson concert that left them both in awe. *(It was amazing because he was my God as well.)*

After the concert, a stayover at Avinaash's home with Shane's other loving mother Shanti Mosi, then back on the train on Sunday.

(Upon reflection, I suppose that this trip was typically me—sporadic, eccentric and erratic? At the end of that year, with Trish's honours degree completed, we left Grahamstown to go to Johannesburg in order to find work. Trish sought the fortune of a good life. And me? Just the FAME!

Manny's friend lent him a truck, and the 8-hour trip took him 12 hours from QwaQwa. Need I say more? The man I love more than life itself packed my furniture and things, spent the night on the floor and then we left Grahamstown for good.)

Chapter 2

City of Gold

(So, that was not too bad, hey? You were ready and handled 5 years in one chapter. Now I'll throw you some gold as a reward. Catch!)

In Johannesburg, also known as the City of Gold, Shane and Trish first lived together in a flat in the suburb of Parktown. They were sharing with her cousins, a young couple and a friend, who owned the three-bedroomed flat.

Trish had a steady job at a recruitment agency in another suburb, Rosebank. Shane used to drive her to work in Golfy, then return home and sit on his lonesome as he paged through the classifieds for part-time work.

Mr. Shalendra Manilal's 'Domestic Science Degree' continued. He remained free because there was no relevant work for a qualified Indian actor. It meant nothing in the market. He may have had an honours degree in drama, but he had no specific field to enter into.

He would prepare supper and go to collect Trish from work, who used to be deservedly tired. They would eat, sit for a bit and then she would go to bed early to prepare for a long working day the next day.

The romance in their relationship was dying because, in the real world, there did not seem to be enough time for it. Justifiably, with her busy work life, there was very little time him.

The free-spirited existence they once shared was no more. Campus student life had to change. The life that Shane loved, and was still living in, needed to be tied down. He needed to enter the commercial world!

Trish was in this world because she was, after all, supporting them. Shane, therefore, put up with this change in their roles. Life in Johannesburg was expensive, so he needed to stop dreaming and help to pay the bills and pull his own weight too.

The stereotypical pattern, to be like a typical working man, needed to be followed, so much so that he even tried a steady salesman's job. This was a challenging task, as it was selling a car fuel-saving device from door to door.

He tried his best, but he found going into different neighbourhoods, knocking on doors and delivering a sales pitch, if he got in, very stressful. His performance training paid off in a way, but only if he was let in.

Then, one day, while they were on their way to visit Shanti Mosi in Benoni, they passed a billboard on the M2 highway with a huge punchline saying, "Live your life to the full, BE DRAMATIC!"

The eccentric dreamer in Shane took over. Seeing this as a literal sign, he quit the sales job and dreamt again. He basically asked Manny to sponsor him to do a TV presenting course because he wanted to enter the industry that he had, in a sense, already studied for.

It was a six-weekend course with Media Concepts, a company that trains TV presenters. He enjoyed it, did well and was told that he was a natural talent. Gloating, he was directly referred to an established talent

agency, Contractors, who readily signed him on because he had potential.

(A talent agency finds artists work. They are the key, as the performance industry first contacts all the agencies to let them know what they are looking for.

Then if the agency has an actor that fits the profile, the actor is sent for the audition, where other hopefuls are in line as well. This is the performance industry's challenging formula.)

It was more challenging for Shane as a newbie. Also, what worsened it was his different look. Here was an Indian, with green eyes, leaning towards Coloured. It was not in demand.

Contractors got him some day jobs as an extra, but not a main role. It wasn't much, but his trivial contribution helped him and Trish financially.

Also, close to where he lived was the Johannesburg Youth Theatre. Shane did the set design and stage-managed their production of 'Beauty and the Beast'. Through them, he was also in charge of organising and conducting educational programmes at schools. With another actor and two actresses, they performed at schools for an Aids awareness show. It was a performance that made Shane feel good.

He was helping to carve and shape youngsters and teach them to be cautious, as the rest of their lives hung in the balance. They needed to be aware that even though sex was seen as something 'cool' to do, a simple error could mess up their whole lives.

Golfy, who now had an engine modification to match her good looks, took them around. *(Car fundis, put on your overalls again. I never forget you. I gave you the 'pimping', now you get the 'tuning'.)*

Ricky suggested to Shane that he modify the motor. Ricky had a passion for modifying engines. His hobby company is called RIX Racing. He put his 'tweaking' *(slang for modifying)* gloves on and rebuilt Golfy's motor. He modified her 1800 engine into 2000. Then he gas flowed her cylinder head, put in bigger valves and fitted Weber performance carburettors with bigger throttle bodies and a 300-degree camshaft. She also had a racing filter with a drilled air-box to produce a louder tone.

Basically, he made Golfy go very fast. The problem with her in Johannesburg though was that she grumbled with her rough idle, especially with being stuck in heavy traffic. Idling was too much work for her, and her clutch burnt out. Shane realised that she was not suitable for Johannesburg as a commuting car, so Manny towed her back to QwaQwa to repair and sell.

Mr. Mover and Shaker was now stuck in Johannesburg without wheels, so he borrowed his flatmate's car to drive around and look for a car.

After a long day's search, without any luck, Shane gave up and was on his way home. On the other side of the road, at a car dealership he never went to, he saw an extra neat red Golf 2. He could not make a U-turn though, as he was on a one-way road, so he just parked on the side and walked to the auto lot.

It was a Mark 2 Golf GTI 1800 8 valve. A 1991 model with big bumpers. It was totally stock standard with a factory leather interior and air-conditioning as extras. There was no power steering or electric windows, but in his books, it was executive class. Even though it was 9 years old, it was exceptionally well maintained. Totally authentic and untouched, as if from the factory.

This Golf was therefore priced a bit above his planned price bracket. Loving his son and knowing how much he wanted it, Manny financed it through the bank for him.

Shane named the car 'Jumanji' because 'his' standard number plate was JMJ 343 GP. This was to balance the gender, as Golfy was a 'her'! He was perfect for Johannesburg—stock standard, reliable and remarkably clean.

Being a 'Golfaholic', Shane was ecstatic. The RIX Racing modification was not allowed though. ☹ Trish had warned him to just leave it the way it was and not fiddle.

Jumanji still turned a few heads though because he was in mint condition. Shane's happiness increased because he was building up a bit financially now as well.

He had the work from the Youth Theatre and was getting some exposure through his agency. Trish and he were now a little more comfortable financially, which enabled them to move into their own two-bedroom flat in Randburg, another suburb of Johannesburg.

Shane hired a trailer because Jumanji had a tow bar. He went to QwaQwa to load up and returned with his furniture load from Grahamstown. This suited the new flat perfectly. They also bought a new TV, washing machine and tumble drier on instalments from a big retail store.

Trish's career progressed, so they were able to build up some funds. However, Shane's work with the Youth Theatre ended. He had time on his hands for domestic chores again, after driving Trish to work in the mornings.

Shane was leading a subservient life. He did, however, have time on his hands to change it, so he decided to follow his other passion.

To balance his creativity plus commercial satisfaction, he decided to study advertising. He joined the Vega Advertising School and completed their six-month course in copywriting.

Sadly, even after doing well, it is hard to get into advertising. He tried but had no luck as a newcomer. The clichéd 'full' potential he did have but was not given the opportunity to enter this dog eat dog world.

Freelancing in the performance industry was basically his only work. So far, he had only landed featured extra roles in TV advertisements. He modelled in a group shot of the stereotypical interracial friendship for beer, showing a group of buddies in a 1960s Volkswagen Kombi, The Love Bus, at a dam. It was a print media advert as well as a poster for bottle stores. In addition, he did industrial theatre work too. He was the big green dinosaur for the health, yoghurt and fruit juice combination. Also, he played a celebrity model for an exclusive deodorant. They would pitch up at all the nightspots in a limousine and promote the cologne.

Another highlight was that he was used as a body-painted model for a make-up launch at an upmarket clothing shop in an exclusive shopping mall. He had to shave his whole body and was then body painted while wearing only a blue G-string. Using the mime skills he had learnt at university, Shane stood still on a platform. Then he would suddenly move to surprise anyone viewing what they thought was a statue of him.

At the beginning of 2001, he got a steady acting job with a company called Hooked on Books. It was an educational show aimed at primary school children to inspire a love for reading.

The performance was done by a trio—another actor and an actress, who became very close friends of his. Very early in the mornings, Shane would first leave Trish at her workplace and then drive to the storeroom for the performance props and his friends.

Jumanji would be filled to the brim, and Hooked on Books toured to schools in Gauteng and acted out a workshopped version of the first part of the story. They would then stop in the middle. In order to complete the story, the children had to get the book and read the rest themselves.

It was fun. It paid the bills as well as ignited his passion for performing, coupled with education, through inspiring young minds to read. It worked so well that he felt sorry to leave in the middle of 2001. This was because when visiting Contractors, his agency, he saw a casting brief wanting a Black music show presenter for a new programme called BassIQ *(A clever way to say 'Basic', implying bass as in music and IQ – Intelligence Quotient).*

When he called, he found that it was Zubz, his friend from Rhodes. *(Yes, the same Zubz who is now a rap artist. We used to DJ together at Rhodes.)*

Zubz was working for Dzino, who was also from Rhodes. Dzino had done journalism and had started South Africa's first internet magazine, Rage. Rage had won the rights to do a new TV music show to replace Studio-Mix, the current show.

Even though Shane's call was late in the afternoon, Zubz set up the audition at the Rage offices in the Norwood suburb. Shane was totally himself for the audition, impressed them and got the role of co-presenter of BassIQ, with two lovely young ladies, Bianca and Azania.

(Hola Magazine) (The Star Newspaper entertainment)

Shane's celebrity status beckoned, and his head grew bigger. He was finally a celebrity. This was the good life.

His as well as Trish's financial situation improved as she was promoted. Their future was settled in their heads. It was more logical for Shane now though.

The 90% flutter he had was wearing thinner though. This was their fifth year together. They were officially engaged at a beautiful engagement ceremony organised by her family in Durban.

Being ready to enter married life, they decided to buy their own place. In his free time, Shane searched and found a beautiful three-bedroom townhouse in Kew, the suburb below Sandton, that was in their price bracket.

It was a small complex of only 10 duplexes. It was ideal. It had a fully fitted kitchen, built-in cupboards, two bathrooms, a garage for Jumanji and a carport. Plus, it had a small outside garden.

They took a joint bond from the bank and bought it. With only Fridays occupied shooting BassIQ, Shane had lots of time. He hired a trailer and gave two men work to help him move.

He also renovated the house as there were a few things that were not to Trish's liking. Because he liked giving destitute people work, Shane got a tiler from outside the tile market, when he bought new tiles for the bathrooms.

Then he went to the paint shop. Once again, he hired a painter from the streets. With his new crew, Shane slaved away, clearing out, adapting and making it appropriate according to specifications.

After a full week, he had created a perfect house. It was not a happy 'home' though, as there was still some subtle stress. Their relationship was on the rocks.

Trish's controlling and bossy stature had been apparent from the onset. They ended up living in their new house for only about three months, but the tension was there throughout.

Her financial yearnings took over. This did not match Shane's needs, as he was a free-spirited freelance actor and eternal dreamer. They had lost their initial compatibility.

Those initial free-spirited university days with Trish were, understandably, a thing of the past. They were now in the real world, and with Shane still dreaming of his fantasy world, he could not cope.

October 2001, he and Trish parted because she was dominating him too much. It was not the balanced relationship that he had dreamt of.

(Now that the cat was away, the mouse wanted to play. Pimp Jumanji?

Dropped suspension and bigger mag rims. Also tuned Jumanji with another RIX Racing modification to make him faster. Ricky made him from 1800 into a modified two-litre with a gas flowed cylinder head, running a 288-degree cam, lifting balanced pistons and a full free flow exhaust. ☺)

Chapter 3

City of Insanity?

After leaving Trish in the latter part of 2001, Shane was now single and living his minor celebrity life in Johannesburg. It was very lonely though. City life is very antisocial. Everyone is suspicious of others.

People seemed to stick to contact with their work friends and not open to meeting others. Shane's work friends, the BassIQ crew, were only seen on Thursday night to plan Friday's show. He was very lonely and needed someone.

Ricky was now married to his teenage sweetheart, Ashrin, who had a single cousin, the beautiful Arthi.

(This pseudonym, pronounced 'Ahrr-thee', appropriately means 'light' in Hindi, which Arthi actually is.)

Shane had met Arthi once before, a long while back. He admired her beauty but was in a relationship with Trish. Now he was free and so was Arthi, who lived in Pietermaritzburg, Kwa-Zulu Natal Province. PMB is, quite funnily, renowned for its mental asylum.

(Please do not think I am mental because sometimes I am too lazy to type it in full, and just like Johannesburg was also 'JHB', Pietermaritzburg is 'PMB')

Having just completed her studies with a degree in law, Arthi was serving her articles at a legal practice, before she could be accepted as a fully qualified lawyer.

In the ending of October that year (2001), Ashrin gave Shane Arthi's telephone number and told him that she had informed her that he would call.

With Shane being in JHB and her in PMB, around 500 kilometres apart, it was a telephonic relationship at first, for two weeks. They seemed to click and got on extremely well.

Sadly, within just two weeks on the phone, Arthi lost her dad. Sheila, who was with Shane in JHB, knew him, and like most Indians who respect people that they are connected to in a sense, even if in the smallest way, they had to attend the funeral.

Even in the solemn state she was in, Arthi, the eldest of two sisters and a brother, held her composure and controlled everything. After the funeral ceremony at her home, even though very tired, she acknowledged that Shane had come all the way and sat alone with him in a corner of the outside marquee to thank him.

They truly connected. Arthi was gorgeous, witty, emotional and understanding. Shane's dream. To add to it, her family were very simple, humble and loving.

Officially then, they started a long-distance relationship. Shane did truly feel an attraction to Arthi, who was the kind of woman he always dreamt of.

In January of 2002, unfortunately, BassIQ's broadcasting contract with the TV station expired. Shane was then back to ground zero.

He played the featured extra role on a hit South African soapie, but this was merely for three episodes.

Also, assuming that maybe his experience on the airwaves would count, DJ Insane Shane recorded and submitted DJ demos to radio stations as well. They highlighted his trained voice that does various accents as well. No luck though, as the radio industry is more of a who you know industry. Without a direct lead, those demos were probably never even listened to? This was the experience he received at the nightclubs as well. He was not even allowed to touch their equipment.

(I honestly suppose that even though me not getting anywhere then can be seen as negative, it would not have led me here, would it? Song lyric in my head: 'Everything happens for a reason, they say, everything happens for a reason!')

Trying to balance his need for employment, coupled with his creative side, Shane did motivational work in industrial theatre for a bank. The aim of this workshop was to get serious bankers in touch with their inherent creativity by helping them to dance, sing and enjoy themselves.

The show toured around the country, and with Shane loving to travel and explore, it worked well. Touring around, he felt free to spread his wings, matching his spontaneous nature.

Also, as freelancing gave him lots of time on his hands, he would drive down to Pietermaritzburg to spend every second weekend with Arthi.

Arthi's open and loving family—her mother, two sisters and brother—opened their doors to him. He was already part of the family and spent the weekends at their house, sleeping in the lounge. They all got on so well, even her sisters' boyfriends.

Shane had always longed for closely-knit family and friends like these. For him, even though there was no total flutter in his heart, like he had with Suraya, from a logical point of view, this one was 100%.

He was still based in Johannesburg though, so the long-distance relationship was hard. Arthi even used to take the bus on some weekends to come and spend some time with him.

(A memory flashback now to the most amazing night of my life – the first time she came to see me in Johannesburg!)

A More 'Special' Night to Remember

Shane and Arthi had been together for about three months, and their relationship was now serious, but they had not yet spent a night together in the same bed yet.

To give Shane a special birthday, Arthi took time off work and took the bus alone for the first time to Johannesburg to spend the weekend with him.

Before she arrived, Shane got everything ready at his home for her first night there. He picked her up at the bus stop in the evening, and love was in the air as he drove her home to his place.

She was rightfully tired after the six-hour bus trip, but it was soon overlooked. Reaching his home for the first time, Arthi was amazed by how nice it was and gave Shane a complimentary kiss.

When they went inside, Shane said that he would put the kettle on for some tea for her. Arthi said that she first needed to shower so that she could refresh and renew herself. She then took her bag and went straight into the bathroom. While she was showering, Shane went into the bedroom to add the finishing touches and then changed into the satin gown Arthi had given him for Valentine's Day.

(Oops, I totally forgot to put that day in. Okay, I'll give it to you in italics very quickly, and we'll come back to the story.

I was in Johannesburg for our first Valentine's Day. I had no work and was still freelancing as an actor. Valentine's Day was on a Thursday. I had an audition for the lead role on another hit TV soapie that Friday. Arthi understood and said she'd see me if I could afford to go to Pietermaritzburg that weekend. If not, then whenever I could.

On Valentine's Day, she called me at one minute past midnight to wish me. My heart sank after a very long call. I could not sleep. At about 04h00, I showered and jumped into Jumanji.

Luckily, I never got caught speeding, because I really pushed him. Just stopping at the toll booths slowed my trip down to three-and-a-half hours of the usual four-and-a-half. I'm not lying. After the RIX Racing modification that I could do after my break up with Trish, and Jumanji's new 17-inch mag rims with a 40mm dropped suspension, he stuck to the road like glue, burning up an empty freeway.

Reaching Pietermaritzburg at about 08h00, I went to an appliance shop to check for a washing machine. They had one, and I made a special deal with them. I then ran next door to a supermarket and bought gift paper and ribbon.

The shop delivered the wrapped box with a note stuck on the outside. When Arthi opened the door, the delivery men said that they were in a hurry and couldn't take the box inside. Arthi said that it was fine and signed the slip. When the men left, she took off the note. It was from me, saying that I was sorry that I could not be there and that her ultimate wish for a Valentine's gift was in the box. She just sat and cried.

The 'washing machine' could hear her sobbing from inside the box. I could not take it anymore and popped open the box. That look of tears turning into joy on her face cannot even be described.

Even though I was standing in the box, she was in my arms, holding me tight. When that hug eventually ended, I climbed out. That was when Arthi gave me my satin gown and a special lunch, after which I drove back to Johannesburg for my important audition the next day.)

Wearing his gown for Arthi's first night there with him, Shane put on some romantic music. He lit the candles in the room, closed the door and went to wait outside the bathroom door.

When Arthi opened it, draped in her towel, she was surprised to see Shane, in his gown, holding a red rose out to her. Putting on a French accent, Shane said, "Mademoiselle, yorr room eez reddee forr you."

He opened the door. In the soft light cast by the scented red candles, Arthi looked down to see red and white rose petals everywhere. As she lifted her head to look at the rest of the room, her hands flew to her face in amazement. This actually made her let go of her towel in the process.

From the ceiling to the bed, there were alternating crêpe paper strips of red and white, outlining the edges of the bed. This made it look a bit like a cage, just like in those Bollywood movies she loved.

On the wall above the bed was a big heart poster that read: 'Shane for Arthi forever' in red. Red rose petals were also scattered all over the white sheets of the bed.

Shane pressed play on the hi-fi remote. The song was Marvin Gaye's romantic song, 'Let's get it on'. Arthi looked at him and smiled. He swept her off her feet to carry her to the bed. They made sweet, passionate love, and she dozed off afterwards.

Keeping up to his word, Shane snuck out to make her the tea that he had promised her and brought it up to her. She smiled when he awoke her for it, and her eyes lit up.

They had a great time that weekend, but Arthi's visit had to end. She had to head back to PMB. ☹

Shane missed Arthi terribly, but his old dream took preference. He struggled for the next few months. When he received a call from his agent to say that he was short-listed for the major role on the TV soapie that he had auditioned for, his spirits rose.

He excelled at his callback audition and impressed them, and in his mind, he thought that he had got the role. A few weeks afterwards though, he was called and told that his performance was astounding but that he did not fit the bill exactly. Quite politely he was told that his look did not match the stereotypical picture of an Indian man they had in mind. They had found a better option.

With that dream being crushed, Shane went to Arthi for consolation. Also, his industrial theatre had ended as well. That very weekend in Pietermaritzburg, a drama teaching position became available at a government high school her brother attended. It was a School Governing Body post for the third term, starting in July. *(The Governing Body of a school is the actual teachers and parents of the school, who have to run it beyond the government's calling.)*

As Shane was tired and frustrated by his lonely lifestyle in Johannesburg, chasing an almost improbable stardom and challenging radio industry, he applied.

He had no set teaching qualification, only his Honours Degree in Drama, but was interviewed and got the post. The position was not paid for by the government though. The SGB only offered a paltry amount from its own pocket, but Shane took it.

The following weekend, Shane, miraculously, with Ricky's, Ashrin's and Arthi's help, moved his things from Johannesburg to QwaQwa, using just a pickup van and a trailer.

Moving to Pietermaritzburg was all working for him. He was given his own room in Arthi's family home, plus this was no cost board and lodging. Money was not asked for at all. With his meagre salary, Shane covered as much as he could though. Everything was going well for him. He loved teaching and had a great group of friends in the staff as well as the learners.

(There are too many people and stories to mention, but this is a general thank you from me to everyone at the school during my time there. To everyone from the staff, pupils, cleaners and tuck shop ladies, my sincere and heartfelt thanks.

The school also had an 'old boys' soccer team—men who had gone to the school as well as teachers there. I was head striker and had many memorable incidents there as well. Thanks, guys. I thrived on the companionship everyone returned to me.)

The best part of him being in Pietermaritzburg was that Shane became really close to people, especially Arthi's family. Their close family bonds appealed to Shane, who felt wanted and needed because he was the clichéd man of the household.

Her mother was not working, so they all clubbed their small earnings together to support their family. This worked well, and even though it was financially trying for all of them, their love for each other grew stronger.

Shane overlooked the factor of no full flutter with Arthi. This was the closest he had ever come. Everything else was in place. He thought he had found his destiny as it was taking too long to find him.

The spotlight of being a TV star was missed, but he loved teaching, carving and shaping young minds more. With his job set and, more importantly, Arthi and her family, this seemed like his final destiny. That year passed, and he was so comfortable.

(Another romantic quickie follows. 30 November 2002 would be Arthi's first birthday with me. At a minute past midnight, she was awoken and taken to the dining room. A table was covered with gifts and cards.

Because I had missed 25 of her birthdays, I had put 25 relevant gifts around the table, with cards made to form pages to be made into a book. She was to collect the cards as she walked around the table.

The gifts began with a rattle for her first birthday, then different toys and cards until a bracelet for her thirteenth birthday.

Then more gifts and cards until a necklace for her sixteenth birthday, some more gifts and cards, then a key for her twenty-first, plus a watch, card and four more gifts and cards.

Then came a gold dress ring for her twenty-fifth birthday, with a card as the cover page that read: '25 years of a jolly life that is bound to improve'.

A ribbon was there to join the different cards, forming a story, which, when read in sequence, made like a long poem. Turned out amazing. ☺)

Also amazing was that sometime earlier that year, Ricky had bought his first modern superbike, a 1996 Yamaha YZF 600. He had entered the biker world. Shane used to ride it around when visiting him in QwaQwa. The love of the freedom of being on two wheels ran through his veins.

That December, Shane's first cousin Naresh, nicknamed 'Boyks', who also lived in Newcastle, was selling his 1984 Suzuki GSX 1100 because he hardly rode it anymore. Manny and Ricky clubbed together and bought the bike for Shane.

Shane drove up with Arthi to Newcastle to pick 'Suzy' up. He bought himself and Arthi 'cheap' helmets that were R500 each.

(Non-bikers are saying, "In 2002, is that cheap for a helmet?" Yes, my friends, it was the cheapest, and that means poor quality. Please, bikers, invest in a proper expensive helmet that is graded and tested.)

Resh bought Shane one of those Michael Jackson leather jackets from a second-hand shop, and her new husband, Sanj, gave him unused, cut-off gym gloves. Arthi gave Shane a pair of boots for Christmas, while her family clubbed together and bought him bootleg jeans. He was now fully kitted and ready for the open road.

(Again, bikers, kit like this is not protection; it just looks like it is. I was new and just entering the biker world. For us, at that stage, it was the norm, a stereotype from the days of James Dean, but it does not kit you out at all. Invest in proper protection.)

Arthi also had a leather jacket, and on weekends, they would cruise around on Suzy. Living in Pietermaritzburg, cruising around on the freeways was amazing. Also, along the Natal Midlands Meander, the popular inner route along the interior of the province, mostly in natural surroundings, was also divine.

The thrill of being on a bike, within Mother Nature, really enthralled Shane. Plus, Suzy had a flat seat, so Arthi was on the same level as him and was holding him tight as they felt the wind in their hair together.

Being new to the biker scene, which was basically non-existent in the Indian area they lived in, at that time, they were always alone on Suzy.

Shane was now extremely happy. He had a 95% flutter—her loving family to live with, a job, Jumanji and now Suzy as well to cruise around with every Sunday in the outdoors. Life was good.

At the start of that year, 2003, Shane finally decided to use his grey matter. He had messed around too much with his life. He decided that it was time to settle his flight and started a Post-Graduate Certificate in Education through the University of South Africa – UNISA in order to become registered as a full-time teacher. It was a two-year correspondence course, but Shane chose to do all his ten subjects in one year, even though he was working.

Arthi readily helped him, even though she was also working and serving her articles. She wrote out some of his assignments neatly for him from his drafts, helped him to mark his school tests and lightened his heavy workload.

Shane felt that they were totally in sync as he had found a home away from home. The future looked all set. In April 2003...

Will You?

Being a school teacher, Shane used to finish earlier than Arthi and be home waiting for her. When she arrived home from her work, her mum and family were also normally home. On this day though, they were not there for some reason. Calling out to them, she walked around. She was worried because there was no one in sight.

When she reached the lounge, she saw a huge bunch of her favourite yellow roses on the coffee table. There was also an envelope, and on either side of the bouquet, fancy cloths covered some things underneath.

Knowing that Shane was always doing romantic things and was obviously up to something, her concern lessened. She went to the table and opened the envelope.

She found some sort of official instruction note inside, which read:

1. *In order for you to summon your spirit, put these blessed roses in a suitable vase.*

   Looking around, she saw her mother's big ornamental brass vase where it didn't belong. She put the roses in and knew she had chosen correctly because the vase actually contained water.

2. *You will find the hi-fi remote on the table. Press play.*

   She pressed play. It was her favourite song, Whitney Houston's 'Miracle'.

3. *Remove the cloth from the right-hand side of the table and do what is obvious.*

Going to the right-hand side, she removed the cloth. Underneath it was a bottle of her favourite wine. Next to it were two glasses with a note in-between. The note read that she must pour two glasses and place them on the table, which she did.

She was then instructed to lift the other cloth. Lifting the other cloth, there was a box with a note on it, which read, "Take this box, sit on the sofa and then open it." She duly obliged. Opening the box, she caught a glimpse of some kind of bright metallic object. It was a brass elephant statue and with it a small ornamental high-heeled shoe. She giggled because Shane used to mouth "elephant shoe" to her, which, if lip-reading, looked like "I love you".

There was also a small bell, an eye mask and a note that read, "Put the bell on your lap, put the eye mask on, then ring the bell to summon your spirit and leave it on your lap. You must remain silent. When you hear the bell being rung again, remove your eye mask." She put the eye mask on, rang the bell and put it on her lap.

Shane, who was hiding and watching from behind the big sofa opposite her, came out. He was wearing an Indian 'kurtha' *(a ceremonious Indian male outfit. Umm... think Aladdin)* and a turban on his head.

He went over to the sofa, where Arthi heard him fiddling with something on the side of the sofa. Aladdin put something into his mouth, then took the bell and rang it. Removing the eye mask, she saw him standing in front of her.

There was a small note on the turban on his forehead, but she could not read what it said. He smiled, then put his hand on his lips and blew her a kiss. As his mouth opened, a diamond eternity ring, her dream ring, floated magically with the blown kiss.

The ring was suspended in the air. When she grabbed it, he opened his mouth and leant towards her so she could read the note. It said, "Ring the bell if you agree to marry this spirit of yours, then toast to your future happiness."

The bell was rung louder than it had initially been designed for, and Aladin put the ring on her finger. Then he picked up the glasses and handed her one.

They drank a toast to their future together, and he kissed her.

(What do you think? I'm lying about the floating ring? I used to do street magic as a hobby. There's something called invisible thread that's normally used for throwing cards and then 'miraculously' making them come back to you as you pull the string back.

I stuck the end of the thread on the wall behind the sofa, slightly above the centre. I then put a piece of silver foil on the other end. When I 'fiddled', I threaded the ring through it and then put it and the foil in my mouth.

When I blew the kiss, the ring came out as I opened my mouth. I had jammed the foil with the other side of the thread between my teeth, therefore making an invisible line that descended to her.

Now do you believe that it was magic? You are more than welcome to use it to ask your lady; it does not have copyrights, so make your magic.)

Arthi and Shane were now officially engaged because everything was in place. In the middle of the year, Ricky phoned Shane and told him that there was a biker rally near Port Shepstone, a town on the South Coast of the Natal province, around 300 kilometres from PMB.

All excited, he said that he and Ashrin, with his friend Dion and his wife, Pat, were going. Ricky added that it would be like a weekend at an outdoor park, camping out and having a party.

Because, on route there, they would pass Pietermaritzburg, Ricky told Shane to join and that he would meet him at the big service station garage on the freeway just 20 kays past PMB.

It all sounded exciting for Shane as well, and even though he did not know what a rally was, he agreed. Suzy needed a service to change her oil and oil filter, so he took her to a mechanic in town and left her with him.

Arthi picked him up in Jumanji, and they headed to a big retail store. Ricky had told him to take a tent and sleeping bags, so they bought a two-man tent with two matching sleeping bags. They also bought a big backpack for their clothes.

Suzy was picked up a few days after this. That said Friday, they both had the day off work. In the morning, they strapped the tent and sleeping bags on the bike and left, with Arthi carrying the backpack.

The garage was about 20 kilometres away. When they reached Ricky? and the others were already there. As Shane pulled up next to them, Ricky had a shocked expression on his face.

"Yoh, my lightie, didn't you see the oil?" he asked.

Looking down at his legs, Shane was also shocked. They were covered in oil from the knee down. His jeans were soaked. Arthi was fine though, as she had been sitting behind him.

They climbed off, and Arthi went to meet the girls. Ricky and Dion had already figured out why all the oil was spraying out from the engine. Under the petrol tank, they pointed out to Shane that the oil was coming from the breather pipe. That mechanic, who he thought knew his job, had messed it up.

Ricky advised that he must turn back and go home because the bike would not make it. Shane did not want to miss his first rally.

He suggested filling the oil, then putting a plastic bottle on the breather pipe. This was to ride for a bit, then stop and fill the oil that had collected back into the pipe and then ride again till the next stop.

Ricky said that it might work but that he must not ride faster than 120 kilometres an hour. Dion got the plastic bottle, and Ricky went and got some tape from his bike's tool kit. Bottle mounted, Shane ventured forth.

Arthi was fine behind him. Shane monitored the bottle and stopped every now and again and filled the oil back.

They finally reached their destination, but first stopped in town to get Shane another pair of jeans. Thereafter, off to the outdoor park where the rally was.

A biker rally is like a camp out in the open party, with a main tent and bar, and bikers setting up campsites. Kind of like the famous Woodstock in America from the sixties. They had an amazing time that night.

The following day, Shane met a salesman from the motorbike shop in Pinetown, not far from PMB. He told him that a Honda Fireblade was available at a very cheap price. This was because the seller, Richard, was from the UK and was selling it cheaply because he wanted to start a business here in South Africa, where he was settling. Shane took down his details.

After packing up on Sunday morning, they left. He took it easy, so Suzy made the trip back with regular stops to fill the oil. As soon as they reached home, Shane phoned about the Fireblade.

That Monday after school, he and the owner, Richard, who was actually living in PMB, took a drive to Pinetown in Jumanji to see the Fireblade. It was a second-generation Fireblade, the 918. It was black with yellow detailing. Even though it was seven years old, it looked new. Richard had kept it standard and maintained it exceptionally well. Shane knew already that he wanted it.

Because he was now 'UTE', a state-paid temporary teacher, he had a monthly salary slip, so Shane's banking finance for the bike was approved. A few days later, he and Arthi drove to Pinetown to pick the bike up.

Shane rode it back to Pietermaritzburg. 'Blade' was easy to ride and so light. He was just a bit confused as the speedometer read miles per hour. He never went past 100 miles an hour, which equates to 160 kilometres an hour.

Upon reaching home, he called Ricky from his cell phone. Ricky had not known about the whole deal, but Manny knew as Shane had told him. When Ricky answered, Shane made Arthi hold the phone next to the exhaust while he revved it up. Ricky was excited, mainly because there would be no more oil spillages on their future trips.

That year, Shane also made close friends with some of the other bikers he met on one of the organised biker get-togethers. His best friends were now two married biking couples, Jan and Brenda and Deez and Aruna.

The biker lifestyle was perfect for Shane. They would all meet on Sundays early in the morning and ride somewhere for breakfast and even go to various gatherings together. The sense of communion among bikers was exactly the lifestyle Shane wanted.

Towards the end of that year, he wrote his University of South Africa exams, employing his 'no sleep' policy from Rhodes University. He really exerted himself because he had all the exams of a two-year course in one year. Luckily, the loving home and Arthi helped as well.

Even though she was behind him for everything, Shane was still putting off the wedding. This may have been because he did not feel that this was a 100% flutter. Also, Arthi was kind of controlling and taking advantage of him. He could not really handle her typical female moodiness, which seemed to be happening too frequently. She was bossy, telling him how to cut his hair, shave and also what to wear. Shane just followed because if she was not happy, she would lose it with him and then five minutes later it would be forgotten.

He did not like the initial outburst though. Having a retentive memory meant that he kept things inside, but then these would build and build until he could no longer take it. In his mind, he never initiated the altercations. He was always provoked and then would react. His dream of that total ultimate bliss was always looked down upon by everyone, but he just did not want to settle for less, especially when differences like these existed.

As an excuse, he kept on saying to Arthi that he was not ready yet. She was quite obliging and said he must take his time.

Chapter 4

Crash Landing?

Shane and Arthi compromised; Christmas in QwaQwa and New Year's Eve at Arthi's house. They chose to share equally, as it made sense for them to split the holidays with both their families.

For Christmas that year, Manny gave his sons proper two-piece padded leather biker suits to wear for protection. They matched and called themselves 'BikerBroz4Eva'. It as a jolly festive season.

Just before 2004, Shane and Arthi returned to Pietermaritzburg, had a party on New Year's Eve and then a trip to Durban Beach on New Year's Day. This was the routine. Life was good.

At the start of that year, Shane finally decided to accept things the way they were. Time was ticking, and even without the total flutter he wanted so much, things seemed in place for marriage.

He and Arthi went to the pundit *(priest)* to 'open the book' and set a holy date for their wedding. After doing their reading, the pundit said that he would get back to them with the clichéd 'God sent' date two weeks from then. They thanked him as he blessed them goodbye.

School had opened, and all was running smoothly, even the biking. Shane hardly ever took Arthi on it though. This is because the pillion seat on a superbike is higher up and not very comfortable, and she complained. Also, waking up early on Sunday mornings, her rest day, to jump on the bike was not her thing. She needed a break. With a busy week, rightly so.

On Sunday, 25 January, Shane joined The Motorcycle Centre's crew (a leading motorcycle shop) on an annual run for bikers. It was known as the Mystery Run because no one, except them, knew the destination.

The bikers were just told to stick together and follow them to get to the next location because there were quite a few turns and so on. All the bikers rightly followed. Then with the final destination being at least 60-kays away, the bikers were told to open up if they wanted.

There were lots of bikes, even faster than his, but Shane pushed it and outdid them all. He was the top dog. No one could beat him. His ego soared. Insane Shane was the fastest biker in Pietermaritzburg.

The following weekend, Ricky and Dion were going to the 'Stag Biker Rally' in the town of Villiers on the border of the Free State province. They were only leaving from Harrismith, another small town about 170 kilometres away from Villiers, on a Saturday morning though and not the usual Friday.

Shane had not told them that this worked perfectly for him because his school had its 'Speech and Awards Ceremony' for top learners that Friday night. He wanted to surprise them by not letting them know that he would be joining them. His plan was to go alone though because Arthi could not make it, as she was working on that Saturday.

At the awards function, even though Shane was on duty there, his bubbly spirit remained. He was even interviewed by a reporter about his new teaching life after being in the big lights before.

Smiling, he told her that internal happiness was what counted the most and added that he was happy. In conclusion, he stated that people should find this internal happiness and rest their search. The reporter was impressed.

With his duty completed, Shane left early because he had to go home and pack. At home that night, he strapped his tent and sleeping bag onto Blade and then went inside.

The following morning, he was up at just after 04h00. He went for a shower. Everyone was asleep. It was a lovely, warm morning. He kitted up in his leather suit and had an energy drink that Arthi had bought him. It had a love note as well. He smiled, went and kissed Arthi, who was in slumberland, took his backpack, helmet and gloves outside, put them down and locked up the house. Then he mounted Blade, but he never started him in the yard.

Blade was wheeled out onto the road, basically using his feet to move it like a little kid does, and Shane started him, to warm up his motor, before returning to the yard to fetch his things. He threw his backpack on, strapping it tightly to his shoulders, put his helmet and gloves on and walked to the idling Blade.

It was a beautiful summer's morning. Before he jumped on Blade, he breathed in the fresh air while making sure that his camping stuff was fastened and tied down properly on the back.

Jumping on, he fastened his helmet strap and headed off. He really pushed Blade down the N3 freeway as he was internally proud of his accomplishment of being the fastest the week before.

In the early morning, the freeways were totally empty. There were some lovely bends that you could speed through. Bikers call this 'dipping'. Basically, it is when you lean the bike down low into a bend. Shane was really dipping Blade.

It was like twilight because the sun was about to rise. Blade now had a different sounding tone because Shane had his friend shorten the exhaust pipe and remove the silencer canister. Bikers have a craze for loud bikes; they want to announce their arrival.

The sun rose as he passed the Midmar Dam. It was beautiful and filled him with joy. He was really pushing Blade because he now knew the road inside out. He was on the best patch of road on the N3 for bikers. The bends are what bikers call 'sweepers', the best to 'dip' in.

Blade did the 65 kilometres to the tollgate in just over 15 minutes. Shane paid his toll and prepared for an even better patch of freeway before Escourt Ultra City, the garage about 55 kilometres away. He covered that stretch in under 15 minutes. It was a distance of 110 kilometres in like half an hour. Shane's head was swelling; his ego was immense.

At the petrol pump, he took his helmet off and handed it to the pump attendant to be cleaned after filling up the petrol. Still sitting on Blade, he opened the petrol tank to be filled up. After the attendant filled up the tank, Shane took his wallet out from his moon bag, paid extra and told the attendant to keep the change.

He rode over to the shop and switched Blade off next to the concrete outside tables. Then jumping off, he removed his backpack from his shoulders and placed it on the table. His shoulders needed a break as well.

It was 05h45. He took out his cell phone from his moon bag and sent Arthi the following SMS: 'Hey Babe, I'm at Ultra City. All good. Gonna have a jol. 'May' see you again on 17 Sept.? LOL ☺ Brace yourself, you have just over 7 months to prepare. Love you. ☺♥'

He laughed inside because the pundit got back to them with that as their fated date to get married—17 September 2004. Looking at his watch, it was just before 06h00. His plan was all in place. He wanted to surprise Ricky in Harrismith at about 06h30. With only about 120 kilometres remaining, it should take him about 40 minutes with the way he was riding.

He mounted Blade and waved his attendant friends goodbye as he shot off to the Colenso/Winterton (which are small towns) off-ramp just down the road. Taking the off-ramp and getting off the freeway, he headed on the back road towards Winterton.

Getting off the freeway, it was a much more scenic and tarred road to Harrismith that he was on. He would be passing Winterton, going through the scenic Bergville town, then up the Olifantshoek Mountain Pass, with Sterkfontein Dam on the left-hand side.

After this, a 10-kilometre trip to the intersection turning right towards Harrismith, 12 kilometres away. Also, as an added bonus, there was no toll fee. ☺

Shane got to Winterton, covering 20 kilometres in just over five minutes. The road was not a freeway, but he knew it because that was the route he used to get to QwaQwa from Pietermaritzburg as well.

From Winterton, he sped on to Bergville, covering 30 kilometres in only around 10 minutes. He was on a roll. About 20 kilometres outside Bergville is the beginning of the Olifantshoek Pass.

Entering the pass is a sweeper, to the right. When Shane entered the pass, he shifted his weight and dipped it, but as he straightened up, he saw a small car, a white Opel Corsa, right in front of him. He was going too fast to brake in time to avoid smacking into it. Even though he could not see if there was oncoming traffic, he blindly overtook it.

Luckily, the road was clear, and the tight corners were starting. Adrenaline pulsated through Shane's veins. Blade dipped to the left. After that was a downhill with a sharp right-hander. He geared down and tucked in. With Blade dipping right down on the apex of the bend, there was spilt car oil.

Blade started skidding. Stupidly, Shane did not let him slide and kick him away. He forced him upright again, and his hands held on tight as he went off the road.

It happened in a split second. The mountain face was right in front of him. Clinging onto Blade, he went up. His path led to a flat rock, like an open ramp. Blade took the ramp with Shane holding on for dear life.

The driver of the white car saw the bike in the air, heading into the heavens, but without angelic wings, it dropped about 20 metres back to earth.

Shane was thrown off, banging his head on the rock and fracturing his neck on his landing. The helmet visor cracked and scratched his forehead.

The Opel Corsa he had passed seconds earlier screeched to a stop, and the woman driver jumped out and ran to him. She was alone in her car. When she got to him, he was out cold.

(Yoh, I have just got quivers up my spine. This distraction is because as I just typed out 'cold', I heard on the radio in the background that Michael Jackson died. Looking at the time, it is almost midnight on Friday 26 June 2009.

Michael Jackson is my king of pop and has been even before Trish's days. At least I saw him while he was alive.

Sentiment to him: "May you rest in peace, my brother. Wherever you are, I can guarantee you that your soul needed a rest. Mother Earth has changed and was not for someone as special as you, any longer.

Your spirit was not happy with the indoctrination society accepts and was forcing on you. We will hopefully meet 'up there' if it exists and if I am worthy to be with you.

We saw the same, my brother. The song I love the most is 'Heal the world, make it a better place for you and for me and the entire human race... There are people dying...'

You tried, and we still can try. Let's get back to my 'Man in the Mirror' and his 'Black & White' ticket for realising his current vision.)

The woman did not know what to do. Shane was unconscious. His leather suit was intact; it was not even torn, and except for the small scratch on his forehead, he had no apparent injuries.

She grabbed his hand and took off his glove to check his pulse. Miraculously, he was still alive. In her shock, she could hear a phone ringing somewhere in the bushes and ran to the sound. Shane's moon bag with his phone had flown off his waist and landed in a bush. Quickly opening the bag, she took out the phone and answered.

It was Arthi. She had just woken up, read the SMS and decided to call just to check and possibly leave a message, as Shane could not take a call when he was riding.

The woman told her what had happened. Arthi's heart sank as her pulse raced while getting the details. She immediately called Manny, who did not even know that Shane was going to the rally, to tell him.

Manny immediately phoned Ricky, who was in Harrismith with Ashrin at Pat and Dion's. They had spent the night there planning to leave for Villiers in the morning. They left immediately in Pat's car.

Manny was alone in QwaQwa because Sheila was at the Mandir in Newcastle, where Resh and her husband, Sanj, lived. He quickly phoned and told Sanj, closed the workshop and jumped in his car to rush to his son.

Ricky and the others got there first. The ambulance was already there. Shane looked fine. Apart from being on a drip on the stretcher, he just had bandages on his swollen head.

Knowing the protocol, Ricky asked who had taken off Shane's helmet. The woman driver was still there. She said that as she saw blood on his forehead, she removed the helmet so that she could wipe the blood off.

(Note to everyone: PLEASE DO NOT REMOVE A HELMET FROM AN INJURED BIKER. Wait for it to be professionally removed, even if there is blood.

The helmet is designed to hold everything in place, to prevent swelling. That time before the swelling is bandaged makes a huge difference. Please don't get me wrong. I am not blaming that woman at all. If it had been me, before knowing what I do now, I would have done the same. She probably instinctively thought to wipe me and that fresh air was better. The intention was good.

So, thank you, my dear. Pity I never 'met' you when we met. I don't even know your name. If you had not stopped, it would have taken some time for me to be found because the bike and I were in the bush. Much love.)

Manny arrived not long after. He held his son's hand, as the ambulance made the trip to Emas General Hospital in the small town of Winterton.

Ricky drove Manny's car. The trip seemed to take forever, even though it was only about 50 kilometres away.

Shane was checked in at the hospital. Manny's phone never stopped ringing. It was after 09h00 and Arthi pulled up with her family. They had driven from Pietermaritzburg. Shortly afterwards, Sanj arrived from Newcastle with Sheila and Resh. He had told them not to panic as Shane was in a minor accident. Seeing him there though, they broke down.

(Tears come to me now as well. ☹☹☹ This is the very first time I am crying about it. When told about it, I always joked and said that 'luckily' I was out cold, so I did not know what was happening and felt nothing. Okay, Shane, take a deep breath. Let's move on.)

The doctor called Manny aside and told him that Shane was in a coma and that they did not have the facilities there to handle his trauma.

(My body was fine in my leather suit. Not even a mark or drop of blood on my body, except for the scratch on my forehead. And my right leg was broken in three places after landing on the rocks. But the main damage was on my head—frontal temporal lobe damage—because the cheaper helmet could not really take the impact. I also fractured my neck and was too serious for them to treat me there.)

The hospital had made arrangements to send Shane to Grey's, the bigger public hospital in Pietermaritzburg. Manny told Ricky and the others to return to the accident site to clear up and get the bike to a scrapyard and then go to QwaQwa to pack him a bag.

When they reached the site, which was now clear of vehicles, Ricky decided to investigate and walk the route. He followed the bike tyre marks and then took a photo of the clear oil patch on the road, before the bend, as shown earlier.

They had also heard that there were other oil patches on the same road, and two more bikes had also gone off the road. Those accidents had not been as bad as Blade's though.

Ricky saw that Shane went off on the bend and still slid up the steep bank on the side. 'He must have been pushing it,' he thought.

They had arranged for the scrapyard to pick up the bike. They finished at the site and left to get Manny's things from QwaQwa.

Back at the hospital, Sheila was breaking down. In her mind, her child did not look so bad, except for the swelling on his head. But Shane was unconscious.

When the ambulance attendants put him in the ambulance to take him to Pietermaritzburg, it was already about lunchtime. About 7 hours had passed.

Manny and Sheila sat in the back of the ambulance, holding their son's hand. The cars followed behind them. After another lengthy trip, Shane was checked into Grey's Hospital.

His biker friends, as well as some of the Raisethorpe Secondary School staff and pupils from Pietermaritzburg, were already there. The waiting room was packed full. Everyone was distraught.

Manny and Resh both noticed that the doctors were just interns. After consulting with them, they said that they could not perform the required brain surgery. They said that Shane needed to be taken to another hospital in Durban. This was because there was no neurosurgeon in Pietermaritzburg.

It was already evening. Resh phoned a friend who was a doctor at a Durban hospital, and he made the necessary arrangements. Manny then went down to the ambulances, where a new shift was on duty. He spotted a driver and asked him if they were able to make a trip to Durban. The ambulance driver, a young man, asked him what had happened and why they could not use Grey's, where they were. Manny replied that they needed a neurosurgeon, and as there was no one there, they had to take his son to Durban.

The driver then told Manny that Pietermaritzburg had just got one. He was Dr. Scheltema at St. Anne's, a private hospital. This meant that Shane could stay there.

When Manny got back to the ward to tell Resh not to arrange for Durban anymore, there was a big commotion going on. The family were in tears.

Ricky, who was watching over Shane, ran to Manny and cried, "Dad, Dad, I was holding his hand... That heart monitor just flatlined... His heart stopped beating. I screamed. The doctors are inside there with him."

Everybody was in shock. They were not allowed in the room; the door was closed. At least 20 minutes passed.

Finally, a doctor came out and said that Shane was okay. He had flatlined for about two minutes, but they had performed CPR, and his heart was now beating fine.

Breathing a sigh of relief, Manny told him what the ambulance driver had said. The intern said okay and arrangements would be made, but Shane needed time to fully stabilise before he was moved.

When they eventually reached St. Anne's, the private hospital, late at night, Manny had to make arrangements to book Shane in. Shane was not on medical aid. Manny was told that they would need a deposit of R10,000 just to book him in.

Being a Saturday night, they all went to the bank ATMs to draw money and then clubbed all their cash together for the deposit. With Shane checked in, the surgery could only be scheduled for 06h00 the next morning.

St. Anne's waiting rooms became even more crammed. Aside from the earlier crew, the word had spread. Shane's family had come from

Johannesburg, Durban, Newcastle, Ladysmith and even from as far as Venda (10-hour drive). It was a stressful night.

The following morning, Dr. Scheltema called Manny aside and told him, as he had told him earlier, that Shane's chances were very slim. There was only about a 10% chance of him making it.

Manny kept his faith and said that it was all in God's hands. *(Thank you, my Creator. Now I finally realise why you brought me back and refused me entry into the other side. I have a job to do here to attempt to unite humanity to focus on our planet.)*

Shane underwent extensive brain surgery and was still comatose. On Monday morning, Manny was told that he needed to pay a further R150,000 that same day to confirm Shane as their patient. Manny, as well as Ricky and Resh, then went to the bank to take out personal loans.

In fact, all of Shane's family and friends pooled whatever funds they could spare. They would do anything to get money. Pat was even willing to sell her car. Manny thought the R150,000 would cover it, but St. Anne's said that it would only cover a part for now because they were keeping Shane in intensive care. That was all Manny could provide at that stage, so he went to the bank again the following day to take out a R400,000 loan on the Mandir, an extended bond.

(Just thinking, my total medical bills at St. Anne's came to about R550,000; more than half a million rand. At that stage, one night in St. Anne's cost R2,500.

The general public hospitals, who are free, did not have the staff and facilities, so I had to go private. I do understand the cost implications, but my father had to beg and plead for them to even admit me. It was a Saturday night, and there was no way that he had access to the full money for my admittance.

What if they had not felt obliged and did not trust him? What then? The provincial hospitals are not well equipped, and all that moving me around in an ambulance led to my further deterioration.

I do understand that it is a financial world, but what cost does a human life have?

Enough of the financial gloom we have in our world, Shane! *Okay, in closing, an overused saying to you: 'Make sure you have enough for a rainy day!'*

I survived my storm only because of the loving umbrella of my family and friends.)

Shane lay in a coma. His close family and Arthi were on a holy fast, abstaining from meat and not having any salt in the day. Sheila remained behind, living with Arthi's family, but the others had to get back to their lives and return on weekends.

The hospital stayed packed though, even on weekdays. Some of Shane's pupils used to go to see him in the hospital every day.

At the school, there was an open prayer for him in the assembly every morning. Then, during the lunch breaks, Hindu, Muslim and Christian teachers had prayer services with learners in different classrooms. Shane's family even went to temples, churches and mosques.

(That is why I say again that there is a power, but we are wasting too much time and money on our individual religions, trying to approach it, ironically, by separating ourselves from each other.

We are all in the same pot now, not separate ones like when religion was first braised on the flat Earth. I was served divine help, BUT from which kitchen??? Whose holy food did I eat?)

When Shane miraculously came out of his coma after around 6 weeks and was born again, his brain, rightfully, short-circuited. He did not know anyone except for his immediate family and called everyone else by different names. Also, funnily, he spoke in the different languages he knew. He was very aggressive, and his anger at the world was apparent.

Classified as permanently paralysed on his right-hand side, Shane had to be discharged. This was because there was nothing more that could be done there.

Dr. Scheltema recommended treatment with specialists at Pasteur Hospital in the city of Bloemfontein, where he had practiced before.

St. Anne's Hospital discharged Shane and fitted him with a strong drip. An ambulance all the way there would cost too much, plus they would be following in the car, so Manny just put him on the front seat, which had been lowered down, and then drove him to the city of Bloemfontein. Sheila was in the back seat of the fully packed car.

It was a six-hour journey to Bloemfontein. Shane was booked into Pasteur Hospital, and the financial dilemma continued. There, they needed a R20,000 admittance fee as well as R10,000 every Friday. Also, Manny and Sheila had to book into a bed and breakfast because they did not know anyone there.

(Tears return at the thought of what my God and Goddess went through. Also, Dad had to go back to QwaQwa to work, leaving Mum alone for a while. As a replacement, Resh took a week off work to come to Bloemfontein as well.

What they all had to go through all because of me. Now more tears. ☹ I would easily do the same if any of my loved ones ended up in a similar situation. True love is tested when times are tough. Sob! ☹ Let's go for treatment.)

Shane just lay in bed in a ward of four. He was all confused and messed up and on a permanent and powerful drip. Sheila was at his bedside, sitting all alone, lost in her thoughts and spiritually guiding her son, her creation. She never left his side for a single day.

Here, once a day, Shane was wheeled to attempt some physiotherapy, but he could not really move. He could not walk or even stand up because his leg was limp. Holding the bars was also difficult because his hand was disabled. His entire right-hand side did not function.

The majority of his hospital stay, he was just lying in bed. Like a baby again, he wore a diaper and would be given a sponge bath and changed.

(Now my memory kicks in. The preciseness of earlier incidents understandably fails me. Before this, up till now, I was just told about things and did not exactly remember, but typing up to here has got my mind working. No one knows of one particular incident that I am going to tell you about now.)

After being there for two weeks, the nursing staff had had enough of Shane, who was mainly awake at nights. Needing company, he would summon the nurses.

On one night, the nursing staff actually wheeled Shane out into the corridor to give the other patients a rest. In the corridor, he was calling out at first but was blatantly ignored. The nursing staff had had enough by now.

Shane lay wide awake, just staring up at the ceiling and at the notice board. The sides of his bed were raised, and he lay in his cage.

The following day, a complaint was lodged by the staff. It was that Shane was a disruptive patient whose behaviour was not appropriate for the hospital, as he was disturbing the other patients.

Because no private ward was available for Shane, he had to be discharged. Manny arranged to transfer him to Bloemhof Hospital. Not being a private hospital, it was down the scale compared to Pasteur Hospital but had an individual ward for his son.

Shane was checked in. Just as in Pietermaritzburg, Shane's close family would come to see him on weekends. Arthi took a week off from work to join Sheila and be with Shane. She would never have left his side initially, but she was serving her articles and had exams and assignments to try to focus on becoming a lawyer.

They were all hardly eating. Their saltless fast in the day ended when Shane came out of the coma, but they were now being true Hindu vegetarians for him.

(I am not denying that maybe all the faith directed at me helped, BUT why is it only in desperation that most of us adhere to our faith and fast, pray and make vows?)

That weekend, Trish came all the way from Johannesburg with her new fiancé to see Shane. When he was at St. Anne's in Pietermaritzburg, she could not make the trip down, but her mother and family from Durban went to see Shane in the hospital.

When Trish entered the ward, Arthi was there, holding Shane's hand. She never left his side and was not going to, no matter what. At that stage, when Shane was conscious, he actively just thought that he was dreaming and would wake up.

Unlike with everyone else's name, he remembered Trish's though. He regressed into the past. He honestly did not know who Arthi was and why she was holding his hand.

Trish did not stay long and left to join her fiancé, who was having coffee with Manny and Sheila. Manny had maintained contact with Trish even after they had split. He was the one who initially called her to inform her. She passed on her blessings and left.

Arthi was staying over with Manny and Sheila at the bed and breakfast. When the visiting hours were over, Shane was tucked in, kissed goodnight and they left. He fell asleep.

Lying alone in a normal bed without any sides, Shane awoke in the middle of the night. His bladder was bursting. As he did not know that he had a diaper on, or that he could not even stand up, he decided to try and get up to go to the toilet. He leaned to one side and tried to climb out of bed.

The top half of his body obeyed, but the lower half would not move. He fell down headfirst, onto the floor, breaking his front tooth. He blacked out and lay on the floor for a bit until the nurse's next visit to check up on him.

The following morning, Manny was furious with the nursing staff after they told him what had happened. They explained that they did not have the amenities and capacity for the full time, private nurse he demanded for Shane.

Manny then hired a private night nurse to look after his son. This young lady was a student nurse who needed part-time work. Thereafter, he left to go back to QwaQwa as Arthi and Sheila stayed with Shane, who was asleep most of the time.

It was almost a routine now. Shane was awoken, put in a wheelchair and taken to the physiotherapy department. He was trapped in his wheelchair. He could not even stand.

He hated being stuck in his ward, confined to his room. One day, after being wheeled back into his room after his physiotherapy, he asked the nurse if he could just be taken outside the hospital to breathe in the fresh air and then go back again.

She knew how confined he felt being stuck inside and having to breathe that congested air. But she said that she had another patient to take care of, and as they were short-staffed, the other nurses were also very busy.

Shane was so disappointed as he was wheeled back into his room. Arthi could see on his face how much he wanted to go outside, so she went and arranged with the head nurse to wheel him outside herself.

Returning to the ward, she told an excited Shane, who was in bed, that she had made arrangements. Shane was more than thrilled as he was put in his chair and wheeled outside by Arthi.

That gust of fresh air made Shane's insides smile. Arthi immediately stopped the chair and came to hug him. Tears filled both their eyes. He told her that he wanted to go further, to see nature.

Not even sure if it was okay, Arthi pushed him around the hospital grounds. The wind returned and ruffled his hair. He looked like a child who was seeing the world for the first time. He made her stop at every tree.

When they were next to a rose bush, he plucked her a rose with his left hand. Her smile made him feel alive again. In his mind, he was just in a dream, but he could control it.

Then the nurse who had told Arthi that it was fine to take Shane outside came looking for them. She told them that a whole half an hour had passed. It was time for Shane's medication, and they had to go back inside.

Being wheeled back in, Shane asked if it would be okay for him to be brought outside every day. As she wheeled him up the ramp at the entrance, the nurse said that it would be fine.

That night, as usual, Shane's mind was ticking busily. He hardly slept but finally felt alive. It was just a pity that his night nurse was a racist Afrikaans woman who did not really talk much to Shane at all, even though he tried. She just sat on a chair and read her book, and being a chain smoker, she kept on going outside the whole night for a smoke break.

That weekend, Arthi's whole family drove up to Bloemfontein in two cars. They had come to see him and pick up Arthi. Coincidentally, the weekend seemed to be a big gathering as Manny, Ricky and Ash came from QwaQwa and Resh and Sanj from Newcastle.

With all of them being there, Shane wanted to have something outside the hospital, in a different atmosphere. He could not leave though, for obvious reasons. Change was needed desperately. He wanted a social gathering, with everyone who was there with him, and return to his communal upbringing.

Using his inherent charm, he told Sheila that he felt like eating her food because the hospital food was bad. He was now no longer fed through the tube but had to chew and swallow. He asked her to make a big pot of chicken biryani, a savoury Indian rice dish, for everyone. Then Manny was told to ask the hospital if they could use a big table he had seen under a tree while being pushed by Arthi.

Manny accordingly made the arrangements. On the day of the lunch, Shane was outside with both his families. He was wheeled to a spot at the table, and they all sat around it for a family meal.

Shane actually ate on his own with his left hand. Sheila was ready to feed him, but he told her he would manage. He wanted to be like everyone else. It was just like old times. A beautiful day outside, in the open, with the ones you love. He smiled inside.

After lunch, they all sat on the grass. Shane was still in his wheelchair. Quietly, for the first time, Shane pulled his left leg out of his slipper and pushed it forward, pointing it down to touch Her.

At that moment, as he touched the grass with his toes, Sheila looked at him. A tear came to her eye as she walked to her son, her good luck prince, her life. As her hand reached out to him, he clenched it tightly with his left hand and pulled himself up.

He stood up for the first time on his strong left leg that grounded and touched Her, with his left hand holding his other Her. Both his mothers. Everyone looked up as he screamed out his joy.

(I have been so emotional this whole time. I want some poetic ventilation. Why not now?)

Why Not Now?

Why does it seem,
that the zone supreme,
is just a dream?

If timing is your loss,
you are filled with remorse,
the passing of the bright gloss.

Your destiny,
may be,
unreal and too dreamy.

It is real though,
to show,
that you're not down low!
Just take your bow!
Why not now?

Shane stood for less than a minute and sat back down. Ashrin wheeled his chair further back, and Ricky scooped him up, putting him down onto the ground next to Arthi. He could not sit up, but Arthi was already pulling his head onto her knee. He lay back and breathed deeply.

(This is called the 'Berry position' from drama studies. It is a position of total rest. Lying on your back with your knees raised flattens your spine to the floor. You, therefore, do central diaphragmatic breathing, deeper than normal fast clavicular breathing. Everyone slips into this process of breathing just before sleeping. It is a drama exercise for stage work, but it calms you in a way you never thought possible. Try it?)

Resting on her lap was a beautiful way to end a perfect outing. Shane was then taken back into his ward. It was afternoon. That night, he did not want to be left alone. There was another empty bed in his room. That time of solitude was killing him. He wanted Arthi to stay over, but out of respect, he told Resh that she must stay over as well.

Resh made the necessary arrangements, and it was fine. They joined the beds, and Arthi slept in the centre, holding Shane for his first peaceful night of sleep, while Resh slept on the other side.

They were all fast asleep when there was a sudden downpour of rain. Because there was a leak in the ceiling, the rain started trickling through, onto the bed.

Being a light sleeper, Shane awoke first. He then woke up Arthi and Resh, who went out to inform the staff about the leak in their room. But there was not another ward for them. They gave her a bucket, which she placed under the trickle, and Arthi helped her move the beds against the wall. Even the sound of the raindrops in the bucket did not keep Shane up. He slept like a log while holding Arthi.

The following morning, he told Manny that he wanted to go back home. If this was his life, there was nothing special for him there. His body was paralysed but being there was killing his spirit.

Also wanting his son back, Manny tried to arrange this but was told that further testing must be done and the offices were closed. He would have to wait until Monday.

That Sunday, everyone departed, but Manny stayed behind. Ricky and Ash went with Sanj and Resh as they would pass QwaQwa on their way to Newcastle. Arthi went back with the Pietermaritzburg crew. It was a solemn parting, with everyone in tears.

On Monday, Manny and Sheila took Shane to the city for additional tests. When they were completed, his discharge was approved. Manny put the front seat of the car down, like he did for the very first time from Pietermaritzburg to Bloemfontein. The car was fully packed. Sheila was wedged into the back, with a whole lot of things on the seat next to her.

The boot was fully packed with everything else, like Shane's wheelchair, bags, Sheila's pots and other items, as she had been living in Bloemfontein for over a month.

On the barren empty road to QwaQwa, they got a puncture. It was a huge task for Manny, who had to do everything on his own, with a fully packed car. He started to offload the car boot to remove the spare wheel. The boot was filled with a lot of loose, small things. He got a shock when he was suddenly tapped on the shoulder.

From the clichéd 'out of nowhere', our Creator had sent him a man on a bicycle offering his help. The help was really appreciated, as, without it, they really would have been stuck. The wheel was changed, and their helper was gratefully thanked. They then set off to QwaQwa.

When they reached the workshop, Ricky and Ashrin had already organised a welcome home party for Shane. As the car stopped, Ricky scooped him up and carried him to the table. There, he got to blow out the candles on his 'birthday' cake.

Because the medical facilities in QwaQwa were poor, Shane was taken to the Mandir in Newcastle the following day, as there was more access to treatment for him. Resh and Sanj, who lived there, had already transformed it for his wheelchair. The interior doors had been taken out, and in the main bedroom, the bed had been moved against the wall. A spare mattress was on the floor.

Shane was to sleep on the bed, and Sheila would be on the mattress next to him, just as they had done in QwaQwa the previous night.

As Manny carried Shane into the Mandir, the first thing he said to Sheila was that he wanted her to give him a bath, the way she used to bathe him when he was a baby. Willingly, Manny carried him into the bathroom and put him on the pan.

When the bath was ready, Sheila took off his clothes. He was just skin and bones. She helped him in, gave him a bath and washed his past away.

Shane was reborn in the Mandir. His new life had begun in his holiest of places.

Chapter 5

Born Again

Even though he was in the Mandir, Shane secluded himself in the main room those very first few days. He lay in bed the whole day. He was depressed and did not get out of bed.

Like a baby, he was in diapers, fed, given a bed bath and changed. A television was put in the room as well. There was nothing that he had to do. He sank lower and lower, and he even tried to end it all by putting a pillow over his face to suffocate himself. His spirit resisted though, as he gulped for air. This actually made him sink even lower. He could not do anything, even kill himself.

Loving family and friends visited him in his room, and he would talk to them, but in his mind, he was in a nightmare and just could not wake up. He was delusional.

Within two weeks of no changes, Arthi got a lift from Pietermaritzburg. Shane had been speaking to her on the phone every day. She knew he was down and brought his memorabilia box in which he treasured and kept everything that touched his soul.

On a Saturday morning, Arthi handed him his box and left him on his own to go through everything. He took out his newspaper cuttings from his minor celebrity days, and his mind travelled back.

Then he remembered his going to Rhodes University. Memories flooded in. His childhood, school days, love for soccer and, sadly, the loss of his flutter.

In the box, he came across the most important tool to lift him off the bed—a letter that touched his heart. He was in tears. It was from Nerissa, a girl he had taught the previous year. It had actually been handed to him just over a week before his accident. It was tea-stained and tattered, like an old parchment.

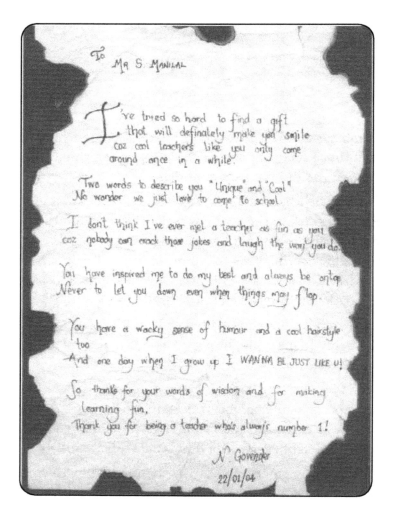

Shane taught her in 2003. His past life came back to him. He now realised that his life was not just his own to take away but that there were others who valued him as well, and he had a message to spread.

Being alone in the room, his hand automatically ripped off his diaper. Then he sat up carefully on the side of the bed. Putting his stronger left leg onto the carpet first, he got down onto his knees so that his hands touched the floor.

Luckily, the room had its own bathroom, and he slowly crawled, on his own, into it, leaving the door open. At the toilet pan, he pulled himself up and sat on the seat. At that very same moment, he looked up to see Arthi and Sheila standing at the room door. They automatically clapped their hands.

In tears, Sheila cried, "My son, my beta, I knew you had it in you. I could so easily have forced you to do just what you did, but I wanted it to come from you first."

"My love, I showed you the life you had, the life you can get back," added Arthi.

"Thank you, my Arthi. I now know I can. To believe I can is everything. Right, Ma'jee?"

With a huge smile, Sheila nodded in agreement. That morning, Shane made a conscious decision to try his best to get his old life back. He arranged for there to be a jug at his bedside to urinate in and a plastic stool in the shower for him to sit on.

Sheila told him that he would remain in the room at first and use the bathroom, but she would still bring his food to him. Also, he would have a nourishing weight-gain shake that he must drink after every meal. It was like the proverbial 'mother's milk'.

(It worked by putting some meat on my bones, but it, along with the Cortezone from my medication, gave me a pot belly as well that I'm stuck with, no matter what I try. Dieting and a varied exercise routine, plus running, have not helped. My tummy is my baby. ☺*)*

Shane was now really born again. His new life had begun. He wanted to do things for himself. For breakfast though, Arthi lovingly asked to feed him. Shane willingly opened his mouth.

His taste receptors were back. At first, he could not taste properly, not even the lovely chicken biryani Sheila had made for the family lunch. But now, probably just from a new mindset, he knew that he was having bacon and eggs, as he could taste it. *(The mind works amazingly!)*

Sadly, on Sunday, Arthi had to return to PMB, but she vowed to come down every second weekend to be with the man she loved and missed.

The following Saturday, she came with another female lawyer for a quick stopover to assess Shane. Before she left, she gave Manny the good news. She had worked on Shane's insurance policy and got them to legally reimburse him for all of Shane's medical expenditure, which the other lawyer confirmed.

(A much-lifted burden off my father's shoulders. He had taken so many loans, extended the bond on the Mandir, plus kept note of all family and friends' donations. My medical bills came close to one million rand. Thank you to everyone who helped, and a very SINCERE thank you to Arthi for shining her light to help my God.)

To surprise Shane even more, that very weekend, on Sunday, his teacher friend from PMB made a trip to Newcastle just to see Shane. He was also a Christian pastor and told Shane about how all the different faiths at the school were praying for him—general prayer at every morning assembly, lunch breaks, three different services in different classes, Muslim, Christian and Hindu.

*(Again, the root of Ethe**real**ism, an almost agnostic belief, is that we are all the same. In fact, all the different religions, cults and faiths have the same goal of peace and unity. Why separate though? Are we not all the same? Why the differences? We all belong to the human race, and our religion should be love!)*

This visit, and being told about the whole school praying for him, added to Shane's realisation that he actually touched other souls. He had to get up and pay them back to pay himself back.

That very Monday, he had an appointment to see a physiotherapist, who was Resh's friend. After analysis, the physiotherapist stated that

Shane's prognosis was not good. His right hand was locked at the shoulder joint. The elbow and wrists could bend, but the palm and fingers did not move. His leg also had no nerve impulses for directing movement or the strength to support him freely. She planned a routine for him and said she would see him on Mondays, Wednesdays and Fridays. She gave Sheila home exercises to do with him and left her with a piece of a flexible elastic cord and routine exercises for him to do with his hand.

After a few weeks of his physiotherapy treatment at home, Shane could stand up but not walk freely. Because he couldn't grip the crutches, he was still wheelchair-bound. He could, if supported, move upright in small areas, but this was mainly by hobbling around.

The physiotherapist then said that she wanted him to see her at her rooms. Sheila couldn't drive, so Aunty Rekha, Resh's mother-in-law, came and picked them up.

Shane was supported by Sheila as he hobbled into the car. She helped him get into the surgery as well. There, he was put between horizontal bars to attempt to walk, but his weaker right hand was letting him down because it had to support his weaker right leg.

Very slow progress was being made. During the mid-year school holidays, another friend who taught with Shane, his best friend, came to Newcastle to visit him with her husband and children. She told him how the school missed him so much and that he must do his best to come back to them. If only he could have, he would have because he loved to teach. Shane sincerely missed his old life where everything fitted in place—a loving home with the woman he loved and a good job where he could make great friendships and help carve and shape the youth to achieve their best.

Also, he missed his social life acutely, especially after Arthi arranged and escorted by car his biker boys from Pietermaritzburg, who rode up to Newcastle to visit him.

Shane was helped to walk outside. He could not keep his hands off the bike. He had to touch it. His biker buddies assisted him to just stand next to a bike to take a photograph because he could not even lift his leg over to sit on it.

After that photo, Shane saw himself on camera for the first time.

Something was missing in his smile. Luckily, his loving brother-in-law Sanj arranged with a dentist friend to cap his tooth and prepare him for his UNISA graduation ceremony, coming up in Durban, for receiving his full teaching qualification.

Smile sorted but body still battered. After his intensive physiotherapy sessions, he could hobble along slowly. He was not using the wheelchair anymore or the crutches due to his hand, but he held onto things or was assisted by whoever was with him.

While studying the previous year, Shane felt like he had finally found his calling of using his intellect. He put in a lot of effort and excelled.

(My regret for wasting my intellect was being alleviated. I was more spiritually ready to collect my degree, even if I had to be escorted onto the stage by my father to lean on, however...)

...at the graduation ceremony, when the announcement was made that Shane had not only completed a two-year correspondence course in one year but got a distinction for it, the applause and praise led to a standing ovation.

As he stood to go on stage, Manny stood up to take him. Shane gently pushed Manny down on his chair and said, "It's okay, Dad, I'll manage."

With his head held high, he walked across the stage. Even though no one from the around 3000 people present knew his history, their mere acclamation of his academic achievement built his spirit enough to surpass even his limp.

(I smiled my now perfect smile, made a solemn bow and walked off the stage and have never defaulted. I walk and jog but cannot run as fast as I used to because I have a pin in my right leg at the shin.

Also, I don't play much soccer, too limited. My only disability is my right hand, but I adapt. I can even type long books with one finger of my left hand.

Learn from your mistakes, correct and move on. Never lose hope. Once bitten twice shy. You can make your mind achieve what seems not possible if you actively try.

*It is your inner mind. Think of a specialist doctor prescribing placebo (mock medication). Patients think it will work, so it works. We are remarkable. We can defy accepted normality. It is all about believing **I can**.)*

This graduation, for Shane, was a defining moment—of the inner spirit defying expectations. He was motivated to walk and did.

Back in Newcastle, he could walk freely, but his hand was still paralysed. Manny brought Jumanji from QwaQwa for him to drive again. It was tricky with one hand and no power steering on Jumanji, but Shane adjusted everything. His actions changed.

He is a person who ponders on something, formulates a plan and pushes on without stopping till he achieves the goal. This is prioritised, regardless of how long it takes him. Using this thought pattern, he adapted accordingly and changed his world. Like tying his shoelaces with just one working hand. After about two hours of trying, he developed a working method.

Everything had a way and not accepting defeat yet, about his hand, Shane was advised by Resh to do craniosacral therapy, head massage, with another of her friends. Joining Resh's friend for Shane's sessions was a new physiotherapist.

They worked very hard, but Shane's hand remained disabled. His bicep, shoulder, elbow and wrist were fine. It was just his palm and fingers that were disabled, and because of muscles not being used, his hand had shrivelled.

In October, the new physiotherapist referred Shane to her friend in Pretoria, who was a hand therapist. Manny took Shane there, and on analysis, she told Shane to give it a full 12 months from damage at least, because his nerves, to take the message from his brain to his hand, were short-circuited from his fractured neck. They needed at least a year for any possible nerve regeneration to take place.

She explained that it was pointless to rush into making a brace too soon. He should use his hand as much as possible and come back to see her in two months' time. They thanked her and left.

During his recovery period, Shane had maintained close contact with his friends from the school he taught at. Another close friend had also driven up to Newcastle with her husband to see him. She expressed how the school missed him so much and that his old teaching post was available for the following year.

He knew in his heart that he missed it as well. His time there was the most fulfilled his restless spirit had ever felt. For once in his life, he was complete, with a steady job, a home and a future wife, just like the clichéd stereotype of 'normality'.

At the end of November, as schools were closing, Shane was told to come to Pietermaritzburg for an interview. At the interview, his keen love for dispersing education outweighed his disabled hand. He had adapted and could now write with his left hand, so he was given the teaching post again. His old life was ready to resume.

His job for a new year in his life was set for Shane. He remained behind in PMB to retune himself. Like in the past, Shane and Arthi went to QwaQwa just before Christmas. Life was back to normal for Shane, just one...

Missing Ingredient

One morning, a few days before Christmas 2004, Shane stood and admired Ricky's new blue Yamaha R1, the top bike of the time. It was outside, shining in the sun. Shane could not help it. He had to at least sit on a bike again. As he sat on it, Ricky came and started it for him.

The Yoshimura, an aftermarket racing exhaust pipe's idle sent a rumbling hunger through Shane's veins. Ricky asked him to try and rev the bike, so he placed his shrivelled hand on the throttle that is on the right-hand side of all bikes, but he could not grip it in order to squeeze it and pull it down.

Then Ricky placed his hand over Shane's hand, squeezed down onto the throttle and pulled it down. The bike roared, rumbling more hunger into his tummy as it vibrated between his legs.

Shane then had an idea in his head. He ran quickly into the office and came back with some tape.

This time, Ricky taped his hand onto the throttle. It worked. Looking at Ricky, before he could even ask, Ricky could see Shane's hunger. He winked his agreement for Shane to go for a spin.

Shane ran into the house and pleaded with Manny and Sheila. They never turn down their children, whose happiness is their existence, so they agreed. But this was only if Shane rode slowly, in full protective gear, and there was a car following him.

Arthi helped Shane to put Ricky's leather suit on. She was trembling more than he was. All kitted up, he went and jumped on the bike. Ricky taped his gloved hand to the throttle.

Vroom, vroom, vroom to the planned town, Kestell, which is about 20 kilometres away! Shane took off excitedly. Just being on the bike again, the clichéd 'wind in your hair', his missing ingredient, made him feel alive again.

Ricky drove behind him in Jumanji, with Arthi, Sanj, Resh and Ashrin as passengers. Shane did not really go fast on the R1. This was because he only had the foot brake for the rear wheel, as a bike's front wheel brake is a lever on the right-hand side, and his hand was stuck to the throttle. He must have been doing about 150 kph *(kilometres per hour)*, half the bike's true speed. Jumanji was right behind him all the way.

When they reached Kestell, Ricky came and laughed at Shane, "How, you made my bike go slowly so that your car, that I modified, can stick up with it, huh?"

Shane smiled and said, "Rix, my bru, this is your baby. I don't want to give her too much of a good time with me because maybe she'll leave you!"

"You can try if you want, my lightie, but your hardest is like my softest. You just do whatever you feel comfy with. She can take it. I trained her well."

Shane winked and took off. He pushed the bike a bit more but did not top 200 kph. It was too beautiful a day to waste the scenery. He reached the workshop with a smile on his face.

There was more of a smile on Manny and Sheila's faces. Their son's joy was all they lived for, and they could see it on his brimming face.

Shane's mind worked as well. The idea of was born.

(Will tell you later. Be patient? LOL ☺)

A New Beginning?

For the new year in 2005, Shane was ready to resume the life that he had pressed pause on. He was still living at Arthi's loving home like before. Nothing had changed. A few weeks in and work resumed at the school. It was really challenging, but he had assistance. Everyone wanted to help him. Even in his form class, duties were allocated to the learners. His only real problem then was that he could not write neatly, but that did not stop him. Learners would mark the register and perform other form room duties. Then his different classes had different learners who would enter marks into the mark book and so on. The main thing was that he was understood to be different.

His personal philosophy was well understood. It is that creativity does not have limits or guidelines, and this was the principle behind the subjects he taught. They were the same as before: Arts and Culture for Grade 8 and 9 and Speech and Drama for Grade 10 till matric.

Yes, his visual art was not the same as before, and nor was the physically demanding role of teaching drama practicals of movement and physical theatre. But his spirit was enough to inspire young minds to greatness.

Shane formulated a new approach to his teaching. It was the principle he made, stuck above his blackboard: 'Work, Work Hard. Play, Play Hard.'

Given the class loads averaging in the forties, with only a 45-minute lesson, of which there was time taken for shuffling around as the school was 'teacher based', where children would go to teachers, Shane's ethos worked.

This principle seduced the learners, like reverse psychology. To them, it basically meant that the sooner the work was done, playtime began. They would enter, almost military style, sit down and begin the work. Shane would deliver his actually shortened lesson to complete the work. The faster the work was done, the more 'play' time they had. Here he became their friend, laughing and joking with them. Even singing and quizzes, etc. It made learning fun.

(I miss it. ☹ My trick really worked. My bonds were very strong with my learners. Affecting and changing their lives, inspiring them to greatness. Anyway, let's not sob anymore; my bigger dream is to reach many more, to inspire them to greatness and also realise that we are actually all the same.)

Teaching again got Shane's life more 'back on track', even without a bike to put on the track. *(sic)* Inside he was still a biker though.

At the end of January, a whole year had passed since his accident, so the coming Stag Rally was like a ceremonious ending, an anniversary of his crash. Even though he drove to Villiers in his car, his groove was back.

Moonshine was there for his first rally. He now had a 400cc bike and was finally fulfilling his childhood passion of joining his biker brothers. With Ricky and the normal crew, it was like the old times—a biker bash. It was not exactly the same though; Arthi did not want to join Shane. Even though he went by car, the memory of seeing a bike understandably sent shivers down her spine.

Shane saw life differently in a way. He now knew that he was not totally happy fooling himself that all was well with Arthi. His dream vision of a perfect relationship hounded him. He knew that the full no flutter factor, which he kept telling himself to overlook, was going to pull Arthi, who loved and cared for him more than he could ever have imagined, down as well. But his full bliss was not there.

Shane was overlooking a lot of things that would be needed if they were to spend eternity together. There was a lot of stress now, especially since he had been complying with her bossy nature for so long, even before his accident.

The obvious stared him and her in the face though. He was not the same. His personality had changed. He now never held back, like he always used to in the past when he always took whatever was thrown at him. Now, due to his temporal lobe damage, he spoke up and retorted.

Things were no more as rosy as they were. Shane needed some space to work out if Arthi was his future. They were still together, but disputes were building, and their personal relationship was deteriorating. It was almost as if it was already established, but that was mainly because Shane was living with her family, with whom he really connected with. Now he needed space for them to grow individually.

The mutual love and care for the family was deflecting Shane from the reality of the personal relationship with Arthi. He asked her if they could move out together to ascertain what they had. Even before his crash, he had always wanted to move out on his own. *(I suppose typical male machismo. Weaker men live with their in-laws?)*

He had seen quite a few outbuildings and flats but always changed his mind at the last minute because she was not so keen. Shane wanted them to live on their own, mainly because they were building a foundation for their lives. He was not so keen for their lives to be built up like that, with her loving family being the main reason for keeping his interest in the relationship. There was no personal freedom for them to establish their compatibility. Arthi was not ready to leave her home just yet though.

But at the end of March, Shane found a place in the newspaper classifieds. Arthi still disagreed but gave him the freedom to move out if he wanted to. With her not really wanting it, Arthi then made it clear there and then that he was doing it on his own and therefore should go and see it on his own. She was not going to join him.

It was an individual outbuilding at the back of a nice main house, with one master bedroom, a lounge and fully fitted kitchen, plus a yard and carport. Shane loved it, and it was in his price bracket. He signed the lease, hoping that it would be shared by him and Arthi in their future married life. That was still the plan. But by signing the lease this time, Shane had thrown the proverbial 'spanner in the works'.

Arthi was not very impressed. She never thought that he would actually sign the lease because, in the past, he had just gone to take a look.

Manny and Sheila lovingly brought his furniture and things from QwaQwa and set him up. Arthi came and saw it for the first time, to help set up, almost out of obligation. Here as well, she was assertive, doing things her way, not discussing and working out a plan. Yes, she might have come to help him move in, but she was clearly not very impressed with Shane going against her wishes.

Arthi was so upset that she did not even attend the housewarming dinner party Shane threw for her family the following weekend. He made a stir-fry in the non-stick frying pan set that they had given to him as a gift.

With Arthi, there was tension. This was why she did not come with her family. But Shane overlooked the headache excuse and even packed her some stir-fry for them to take home for her.

(I am used to the clichéd 'female mood swings',, but this was getting too much. ☹)

Shane managed perfectly living on his own. Arthi did visit and stay over when she wanted to. It was now clear to him that the main discrepancy was the fact that he was undermined and controlled by her.

There was a lot of tension in this classic whirlwind romance. From the time Shane initially moved to Pietermaritzburg, everything seemed to fall into place. He had lived in her home with her loving family, but now he was living on his own, and it was almost as if they were just dating, even though they were still engaged.

They were trying, but it was not working. His deciding factor, regarding her controlling and bossy nature, was when she decided to leave for a holiday in London in May with her sister, who was now working there.

(My problem was not the trip, but the fact that we planned to go together during my school's June holidays. She wanted to go without me, so she went.)

The time when Arthi was away finally gave Shane the realisation that this was not what he wanted. They loved and cared for each other, but the balance had changed. This was mainly from Shane's side though, as he had finally realised that he was under her thumb.

(It was the same with Trish. Why is it that the majority of men, even women, must succumb to their partner's ways and wishes? What about equality, where no one is in charge?

Is my idea of 50% plus 50% giving a whole 100% just a dream? No one in charge, mutual equality.

Quick appropriate joke? During a church service, the pastor asked all the men who were controlled by their women to move to one side of the church. All the men went, except one. The pastor said that it was so nice that at least one man was not controlled by his wife. Under his breath, the man said, "She told me not to go to the other side, Pastor." LOL ☺)

When Arthi returned from London after her month's visit, Shane finally told her that it was over because she was not letting him be himself. He sabotaged the most logical relationship ever, even though it made total sense? Same race and religion and the families knew each other well. Except for the most important thing in a relationship: compatibility and working together as one unit, not really about the family connection.

(Arthi and I are now still close friends. I am also extremely close to her family. I even go and stay over there in the lounge sometimes, without even seeking permission—open door policy.

My bonds cause pain but never break. Yes, I am different, but I would rather stay true to myself, instead of putting on a show. I work with my gut feelings inside of my centre.

Just thinking, as a gift when we were together, Arthi gave me this card with the Piscean explanation of my nature.)

PISCES: February 20 – March 20

In life and in love, impulse propels you like a fin from the start. Because your element is water, you are often dreamy and elusive, your emotions and passions constant. Your desires and needs go with the tide, so to say. Ebb and flow – that's you.

You were born to entertain the world and to indulge others. You are caring and yielding and approach life with no grand design or strategy, yet intuitively know how to touch the right nerve centres at home, at work and in bed.

Your favourite place on Earth is the erotic zone. 'Love' is an expression of faith, and sex is your escape from the harsh side of life. You are lost in a delicious reverie for hours afterwards.

Always the whimsical child of yesterday, dreaming of a better tomorrow.

Chapter 6

One Hand Job?

For some time now, Manny and Sheila had been taking Shane to Pretoria to see his hand therapist. She tightened his band every time, but there was still not much progress for his hand.

His fingers were now bent, shaped like the band, and could not straighten properly. More importantly, his thumb could not bend and move as it was designed to.

(Quick fact: Our opposing thumbs make us unique in our hominid species. We are the only ones with them, which led to our grip on things, sculpting our brains through the different ages—stone age, iron age, etc.

Basically, that is the root of our development as the ruling species on this beautiful planet. That led us to what though? To take over it, be in control of it and, sadly, ruin it? ☹)

Even though he was not fully functional, Shane was still managing well at school, thanks to the assistance he was given by fellow staff members as well as pupils. He really wanted a full resumption of his life though. He was managing fine on his own, free from the relationship tension with Arthi, but he needed some vibration. Time was passing and he had an undying need to be back on two wheels again. To feel that vibration between his legs.

In May, before the school's June holidays, his hand therapist finally said that it was time for surgery. She had done her best. Shane's hand was not going to recover. She referred them to a hand surgeon, who analysed the hand and confirmed that there would probably be no further recovery, as a whole 18 months had passed. His verdict was that the sinews and tendons in Shane's palm needed to be adjusted and shaped. Doing that would leave Shane's hand that way for the rest of his life. There was nothing more that could be done subsequently. He could still wait, but it was unrealistic, too much of a dream to hope for change. It was almost set in stone. He was booked in for surgery two weeks later.

When the time came, Manny and Sheila took Shane to spend the night at a bed and breakfast in Pretoria because Shane's surgery was early the following morning. He was not allowed to eat anything or even drink water 12 hours before his booking time.

Early the next morning, in the surgery, before being anaesthetised, he chatted to the beautiful nurse who was helping the doctor. He then closed his eyes.

When he opened them, he saw her again and asked her when he was going for the four-hour operation. When would it start?

She laughed as she pointed at his hand. Glancing at his hand, Shane saw a cast covering his lower arm, from his elbow down to his fingers, with just his fingertips showing. He had been out cold for the entire procedure, and she was now making him regain consciousness.

Manny and Sheila walked in. The operation was successful, but the waiting period in the cast was six weeks before removal.

Once again, Sheila never left his side. She went back to Pietermaritzburg with Shane to take care of him. He had managed fine living alone with his one hand prior to the operation. Now it became more difficult, especially having to put it into a plastic bag for protection before showering in the morning and getting ready for school.

Sheila would wake early just to put it on for him. She would then make his breakfast and pack him lunch to take to work.

(Aha, time for tears again. ☹ Mum would wait alone at home the whole day until I returned in the afternoon. In her solitude, she would cook and clean for me.

Even my dexterous verbosity cannot describe her love for me and how much she does for me. I told you, I put my parents through hell, but they never ever complained. They just take it, even to the extent of Mum leaving Dad for six weeks, just to take care of me! Amazing, don't you think?

Even more amazing, when I was in hospital, close to three months, sitting next to my bed every day! Mum just does. She never questions, just does.

Thinking about this, she actually even helped me with my extensive school workload. It was the June exams then, and the marking of hundreds of exam scripts was taking its toll on my left hand. Mum slaved away with me, doing the easier questions, like multiple choice and open questions, using my model answer sheet. I did the more in-depth marking.

Then she would help me to add the results that we would double-check. I basically had a personal secretary. ☺ Thanks, Ma. I needed to smile.)

The most stressful part for any teacher is the examination period. Luckily, Shane had support from fellow staff members, along with his mother, and managed.

Shane was still in touch with Sylvia, his friend at Ekerhold Yamaha, the bike shop. A few weeks before the school term ended, she had called him because he had asked to be informed of specials on bikes.

(Told you, my biker dream never left me. I used to frequent bike shops and window-shop. I still had strong friendships with my contacts from before, and even though I did not have one, just seeing bikes gave me a thrill.

I started visualising my dream. I wanted to opt for something smaller, just to be on two wheels again. I, therefore, dreamt of a 600cc.)

Yummy Yummy Yummy I Got...?

The brand-new Yamaha R6 was going on special. It was marked down from R87,000 to R80,000. Shane went to see it. It was beautiful, all black, his favourite colour for a bike, with just a red stripe on the wheels.

Shane fell in love when he sat on it. Even with his hand in a cast, he felt a sense of belonging. He was hungry. Sylvia saw his smile, and knowing that it was just a few weeks more for the cast to come off, she told him that if he left a definite holding deposit, she would drop the price to R78,000 just for him.

He would have to pay a non-refundable holding deposit of R3,000. She was concerned though and asked him what he would do if his hand was not strong enough to pull the throttle down.

Shane responded by saying that he would just have to take the chance and dream, and even if it was not strong enough, he would make a plan because he had an idea after he rode his brother's R1. He paid the deposit.

The school term ended, and on the Wednesday of the last week of the school holidays, Manny and Sheila took Shane on his last voyage to Pretoria to see the hand surgeon and remove his cast. Shane was initially shocked because he finally saw the scars left by the operation.

The palm's minimal movement was the same as before, but the shape was different. After his physiotherapy, his shoulder and elbow functioned, but his palm was going to remain disabled. The change in shape provided some help though.

Shane could move his thumb and index finger slightly and separately, but his three other fingers all moved at the same time. It was not much, as he could not lift anything heavier than a pencil, but this was all that could be done. Plus, the pencil could not even be properly used, but he smiled because even though the hand was shrivelled up because the muscles had deflated, his wrist had movement. This was the clichéd last straw.

With his hand's condition set for the rest of his life, Shane had to now accept it and work with what he had. Yes, there was a lot of adaptation to consider, but as always, he had a zealous approach.

On the return trip to QwaQwa, while Manny drove, like he always did, Shane's mind pondered the achievement of his most important dream: being back on a bike again. He had to ask his God and Goddess for their permission and blessings.

Inside, he knew and understood that they would be against the whole idea, but he had to try. When they reached home, he arranged a meeting for his appeal.

He first told them that he believed in fate determining the outcome of one's life and that he had defied medical certainty. He had been given a 10% chance of survival but had pulled through.

Already sensing where this was going, Manny and Sheila just nodded and agreed that 'thuqdeer' *(fate)* decides where we go. Making it easier, Shane said that because of that, life needs to be lived and the past overlooked and the future not feared. Then he took them where they knew they were going anyway, into his dream to feel the wind in his hair again. He knew from what they had gone through that they were now even more convinced that a bike was dangerous.

Shane told them that he had done some research that may change their belief patterns. The stereotype of a bike being dangerous is understandable, but it is not as dangerous as it is made out to be. According to his equal figures of exact random accidents, the fatality rate in cars was 56%, while bikes only had a fatality rate of 24%. This is mainly because inside a car you are in a potential tomb, unprotected except maybe for seatbelt and now the airbag. In an accident, you are trapped inside.

On a bike, however, you are in a sense free to fly. If one is wearing proper protective gear in the form of a complete riding outfit and a good quality helmet, one is free to fall, and if there is nothing to bang against, one is fine.

Shane was in his normal sales pitch mode. He added that the most serious biking accidents were on bigger bikes at top speeds.

He finally said that he had already put down a holding deposit for a 600cc bike. Because it was smaller, it was not so fast but gave him the thrill he sought.

His desire to be back on a bike again, to feel alive, feed his soul, was seen by Sheila, but from her experience, she rightfully feared for her son's safety and her eyes reddened. Manny just shook his head. Shane understood.

Manny then said that he understood Shane's dreams but that the proper protection he spoke about cost a lot of money and that he could not afford it as well. Shane then said that he had saved up R20,000. He would use R8,000 for the 10% deposit needed for the bank finance and the other R12,000 for proper helmet and riding kit.

Sheila then said that if that is what he wanted and what would make him happy, then she would be happy as well. Manny, insisting that only if proper protection was used, agreed. He told Shane that he must just take it easy.

The joy on Shane's face when they agreed was immeasurable. He could not even sleep much that night as he was brimming with excitement.

The next morning, he excitedly drove to PMB to get his bike and resume his life in totality. Going straight to Sylvia, whom he had called and told the good news, he picked up the forms to take to the bank. The finance of R70,000 over 48 months was approved. All to his liking.

Returning to the shop, he found Sylvia had organised him huge discounts—one-piece leather suit, racing helmet, boots and gloves totalling R12,000. Added to this, Shane told Sylvia to add on a Ventura luggage pack. This is like a removable car boot that cost R3,000, the exact holding deposit. It would serve him perfectly, as he would not be comfortable carrying his stuff in a backpack anymore.

With the bike financed as his, Shane christened it 'Yummy' because it was not only a shortened, cool way to say Yamaha but also brought love to his tummy. *(Yummy yummy yummy, I got love in my tummy, and I feel like loving you!* ☺*)*

(Time for my big idea?) He attached Velcro onto the throttle on the handlebar and his glove. Even though the mechanic there moved the brake lever closer, he could not apply brakes with his right

hand. Another plan had to be made with the front brake lever. The mechanic contacted the biker specialists there to discuss the problem. They arranged a plan to move the brake lever to the left-hand side, under the clutch, for Shane to push with his thumb. Yummy was taken in.

After a week, Yummy was picked up. All the conversions worked. Finally, Shane was back in his groove. Life had resumed—or so he thought.

Loneliness was getting to him immensely. His days started off okay, with him socialising while at school, but in the afternoons and evenings, he was all alone. Yes, he had friends who would visit him to play cards and other games, but they would return to their homes. His heart was lonely.

What was he searching for? Even himself, he cannot answer that. He felt lost in life. He was used to this.

(One of my very first poems ever, written when I was 15.)

Me

I am an artist,
trying to paint my true colours.
How long will they elude me?
Are they hiding behind the canvas of an image?

Or are people just seeing what they want to see?
Maybe it's because
I've already painted a Picasso?
Everybody now just expects them to flow.

How long can I keep it up?
How long will the paint last?
WILL I EVER EMERGE
BEFORE THIS LIFE IS PASSED?

At the school, putting his focus on work, it was now the final term. But with the education system doing radical changes that Shane did not really agree with, he did not want to be the pilot guiding his plane full of learners into the mountain. He resigned, and it was a solemn parting with the school and his friends, but he knew that this was not for him anymore. He was too emotional for this disparity in the system. Also, his loneliness was getting to him. He needed to find something else somewhere.

In his sporadic frivolity, he left Pietermaritzburg and ventured to QwaQwa to try to make a life there with his family and tried to assist in the workshop. It was not his kettle of fish, with his disabled hand, but he helped Ricky as much as he could. Even in the office, he would take calls and do bookings. When the workshop closed, though, there was nothing for Shane to do.

Being in the industrial area, there were only factories. While Shane idled away in QwaQwa, he got the idea of going to teach in England. Even though he had heard that it was bad, he wanted to experience and judge it for himself. Their school year only starts in September, so he had time to organise the trip.

Again, even though they were not happy with him wanting to leave the country, Sheila and Manny allowed it. Manny loaned him the required R20,000 that he would need in his bank account for clearance.

Having got his passport and work visa, everything looked in place for Shane. He joined an international teaching recruitment agency and had obtained a job at a school in London after a lengthy telephone interview.

Arthi's sister and her boyfriend were still living and working there. Shane was still seen as their brother, and they had a place for him to bunk in the lounge of their flat.

It was the middle of August and everything was in place for Shane's departure in two weeks. Deez, his first biker friend from Pietermaritzburg, who was racing his bike on the track that weekend, called Shane and asked him to join him for his first time on the track just to experience...

Track Time

Realising that he would miss Yummy a lot while in London, Shane took Deez up on the offer to race him on the track. This would be his first time ever. It was at the Phakisa Race Track in the town of Welkom in the Free State Province. The main race was on a Sunday, but the plan was to meet on Saturday to get the bikes ready and have a camp out party.

Shane was in QwaQwa, so it was not that far for him, plus he could use Ricky's bike trailer. After a three-hour drive, when he got there on Saturday after lunch, Deez was late, so Shane got another rider to help him take Yummy off the trailer.

He waited for about an hour, but then his adrenaline got the better of him. Just the sound of the roaring exhausts sent shivers up his spine. All excited, Shane could not wait for Deez any longer. He assumed that being on the race track would be the same as the road riding that he had done extensively.

Putting his riding suit on, he eagerly jumped on Yummy. At first, he was just taking it easy. Then, with the second lap, he built up some speed. He was watching his clocks, checking how fast he was going.

Not knowing that the clocks should be blocked out, which is what professional riders do to their road bikes on the track, Shane was getting a rush just from seeing the clocks reach top speeds.

The most important track modification for normal road bikes on the track is that the bike's mirrors are folded down. This is so you cannot see who is behind you. Shane's were obviously still up though. This meant that he had a full view of what was on his sides and behind him. It is good to stay aware as to what is around you. On the road, it is normal courtesy for bikers to give way to another biker who is behind them.

On the third lap, Shane actually got his knee down on the bends while dipping Yummy. He was scraping his knee pads, like professional riders and his idol, Valentino Rossi, the world MotoGP champion. His adrenaline pulsed.

He felt like a real pro, pushing Yummy to her limits. He could not even describe the thrill of drawing that fine line. Still pushing it, Shane looked in his side mirror and saw a faster bike approaching him. His courtesy made him move slightly to the left, to give the bike behind him space to pass him.

He did not realise, however, that he had shifted too much to the left. A tight right-hander was coming up, and he was too wide out to take the bend. Yummy was slightly dipped but could not make the bend. She hit the gravel at a slight angle and landed on Shane, who was trapped underneath and shouted for help.

After some time, the track assistants came and lifted the bike off him. His leg hurt like hell because Yummy had landed on it. He had no other serious injuries though because his kit and helmet had done their job.

Yummy started again and Shane jumped on because there was no way he was going to walk to the pits with his leg throbbing. He rode slowly. When he got there, he parked Yummy and hobbled off. Sitting in the car, he removed his boots. The swelling was huge. Some of the bikers brought Shane some ice to put on the swelling, but the cold did not take the pain away.

Deez finally arrived with his pickup van and bike on a trailer. Seeing Shane sitting on the ground with the ice, Deez already knew that there had been a mishap. He helped Shane out of his leather suit, put his jeans on, helped him into the van and drove him to the hospital. Luckily, it was not that far off.

When they arrived, Shane was put into a wheelchair and taken for an X-ray. He had fractured his right shin bone. This was exactly where he already had a metal pin because he had broken it in three places before with the major crash.

His leg was not put in a cast but was bandaged tightly. He was told not to put any pressure on the leg even by trying to step on it and that it should always be rested. The healing time was approximately six weeks.

The wheelchair had to remain behind, and Deez lifted Shane back into the car. It was late evening and Shane was ignoring the calls on his phone from his family from earlier. On the way back, he finally phoned home. He did not tell them about his crash but explained that he had left his phone in the car when he rode.

As they reached the track, Shane told his mum that he had to join the other bikers as they were having a braai. He wished them a good night and said that he would see them the next day.

The braai was not a lie, as bikers usually have a braai the night before, as they camp out in the pits of the track. They put a chair out for Shane. It was to be an early night though anyway because riding in the morning needed fresh minds. They retired to their inflatable mattresses to get some shut-eye, and Shane was carried to his.

Even though he was on painkillers, the pain, as well as regret, kept Shane wide awake. His leg was throbbing. He was bound again to being restricted. What hurt him the most though was that there was no way that he was going to go to London now.

Everyone was asleep. Shane's mind just ticked. He could not move again as he had no crutches or anything. Memories filled him, especially when he got an empty two-litre bottle to urinate in because he was not wearing his diapers.

In the morning, when Deez awoke, he carried Shane to the toilet and put him down on his one leg and then closed the door and waited outside. When Shane was finished, he carried him to the basin and put him on a chair that he had brought with.

Sitting down, Shane brushed his teeth and washed his face. He told Deez that he was not going to stay and would drive home. Deez put him in the car to test him first. It seemed workable. Yummy was mounted on the trailer, and Shane headed off. The three-hour trip took four hours. It was a long process, but he succeeded.

When he arrived at the workshop, Shane sat in the car because there was no way he was going to get out. Sheila came outside. He opened the window and said he needed to be carried out again like the last time.

Thinking he was joking, as he always did, Sheila burst out laughing. Shane opened the door, and seeing his bandaged leg, she collapsed on the chair sobbing, "Not again, my beta, not again."

Manny heard the commotion and came out. Seeing Shane's leg, he just shook his head and told him not to move, he would come and help him.

Shane was taken to his room, where he lay in bed with his foot elevated, and selflessly served by Manny and Sheila. Ricky organised crutches for him. Both left-hand and right-hand crutches had been fitted with his Velcro glove for easier movement. However, he hardly moved at all as it was so difficult. He was bedridden all over again.

After six weeks, Shane was taken to Newcastle to his biker friend, Dr. Pillay, to have his foot X-rayed again. After the X-ray, Dr. Pillay was astounded. The fracture was still there in full. No bone growth had taken place. It was almost as if the metal pin that had been put in place was a decoy over the broken bone, and the bone did not realise that it was broken. This was because no pressure had been put on it.

Dr. Pillay told Shane that he must give it more time to heal. He added that he must continue keeping it up as he had done before.

When Shane returned to QwaQwa, his mind started working. He thought that if the bones did not know that there was a gap between them after six weeks, because of the metal pin, they would never know.

He then quietly broke the doctor's order and started putting pressure on it. When he stood up now, he put his leg down. Even though the pain was excruciating, he held it down. He still remained on crutches but did not lift his leg; he kept it vertical.

This vertical leg, without putting all his weight on it, made his leg fall in place. He continued doing this and was able to apply a little more force on it every time. After three weeks, when he returned to Dr. Pillay, Shane was told that he had performed a miracle again. First, surviving the major crash, and now the miraculous healing of his leg. It was coincidentally the end of his nine-week fast that he started with his loving parents when he returned from the track. So, maybe that worked? Maybe his healing had a set time limit from above? Maybe his coming back to Earth was also planned?

(Maybe? Maybe? Maybe? We always seem to seek justification, don't we? Even as a thinking humanity, we still question and read meanings into things, especially strange things.

My personal answer to my major return, I know, will all be figured once I'm done here. We just need to work more on our own realities.)

Shane's reality was to get back on two wheels again. Yummy was repaired and given a customised spray job by Shane's close biker friend Ravi, from Newcastle.

From being fully black, she now had silver and maroon. It was a copy of the limited-edition Yamaha R1. 'I.S RACING' was written in the front. *(Guess what I.S stands for? LOL ☺)*

Insane Shane was overjoyed as Ravi also polished Yummy's previously full black rims on the 'lip' and painted them maroon inside. She looked her part.

Yummy even won two concourses, the 'beauty pageant' at biker rallies.

Life was back on track in a sense. With his leg fully healed, the year was ending and Shane's master plan to go to London had failed. He was back in QwaQwa, wasting his life away. It was November.

Then Karishma, his cousin in Newcastle, saw a newspaper advertisement for an 'Arts and Culture' teaching post at a Newcastle high school. She phoned him to tell him.

Shane faxed his CV off and was asked to come in for an interview. He drove down for it and got the job. He went back to QwaQwa for a jolly Christmas with his family again, as he now had work in the following year.

After Christmas, he packed up and moved down to the now-empty Mandir. This was because Resh and Sanj had moved to Johannesburg.

Shane was alone, and his neighbourhood friends helped him unpack. It was his first time in Newcastle after he had left school. Yes, he had been there for his recovery in 2004, but he was now fully mobile. He was excited.

On the morning of New Year's Eve, Moonshine surprised Shane with a visit. Moonshine now lived in Ladysmith and was going to spend the weekend with him there. Together they organised a braai for their cousins and friends. Afterwards, they went out for a nice party at Escape, the local nightclub.

Shane and Moona's strong bond re-established itself. This was their first time actually doing what they used to do in the old days. They were players once more. In his player mode, Shane picked up Marianne, a gorgeous 21-year-old woman. It was a great night.

The year began on a high note. Moonshine returned to Ladysmith. Marianne wanted more from Shane. In the following two weeks, she fell in love with him, but he told her that he did not want a steady girlfriend just yet. He was just doing his usual bungee jumping with no strings attached, his definition of casual affairs. She didn't mind.

Workwise, Shane had just started teaching at the high school but felt the racial tension. It was a former 'Whites Only' school, but apartheid had ended 13 years ago. Why was there still the undercurrent just as when he was in the previously 'Whites Only' hospital in Bloemfontein when recovering from his major accident 3 years back?

(I do dream of a world where we are all seen as the same. Yes, most people state it, but do they act it? It is like Martin Luther King's dream. I know I am dreamy, but that is where everything starts, right?)

Dreams

Changes may come.
Differences may bring some
once remembered happiness,
possibly lost in an abyss?

You have to hope and dream,
save your steam,
as now reality may seem,
unobtainable, 'peaches and cream'.

BUT you may, someday, taste though,
reaping what you now sow,
why then not plant the seed
to make your dream a done deed?'

(Regarding this book, my 'seed', going so well, I can feel it. I never would have thought it, but writing settles my soul so much. Let me start another full chapter because the moon wants to shine.)

Chapter 7

Moonshine

Working at the high school was getting really challenging for Shane. It was Friday, 26th January 2007, the weekend of the Stag Biker Rally, and the third anniversary of his crash. As always, his biker crew were going to attend the rally.

The school was having their sports day. It did not fulfil Shane's expectations of the school sports he was used to. The spirit was there, but this was only among the very few participants. He remembered how, in his time, sports day was the biggest event. Now, it honestly seemed like a formality. The only learners interested were the ones with talent.

Also, Shane was delegated physically challenging work, things like putting up the hurdles, marking the distance of the javelin and shot-put throws as well as organising the high jump site. Maybe this was just because he was the new guy there? Or maybe more? Because he was the only non-white male teacher?

Shane stayed positive and struggled through them, as he was excited about the weekend. He got back to the Mandir around lunchtime and went straight into the shower because he was feeling a bit drained before, but he came out refreshed. Sheila was in the Mandir as well because she had spent the week there with him.

As he was having his lunch with her, he heard a bike roaring up the driveway. He had heard from Ricky on the phone the night before but was not really sure.

Leaving his plate, he rushed out into the driveway. It was true. Moonshine was in the full spare DMD leather suit that Shane had given him, sitting on a second-hand blue Yamaha R1, one of the top superbikes. He smiled with pride as he removed his helmet and chanted, "Talk to me now. Talk to me now, Shane'o!"

Shane was thrilled and responded, "Yoh, my lightie! Awesome, man. Finally, you are with the big boys. When did you get it?"

"I just got it yesterday. I trailered it from Johannesburg to Ladysmith."

He climbed off as Sheila came outside. She kissed and hugged him to congratulate him. She joked and said, "Well, my chota beta *(small son)*, you are finally with your bhurra bhais!" *(bigger brothers)*

Holding her, Moona responded, "Namaste, pooah. Ja, finally, I'm small but I'm with my big brothers. Talk to me now?"

Examining the bike, Shane laughed because 'Talk to me now?' was Moona's usual comment of pride.

Sheila laughed as well and said, "We'll do that inside, come and eat. I made lovely kebabs and roti. Come, Shane, your food is getting cold."

"Coming, Ma. Moonshine, let's go and chow, but I think you must leave your tent and sleeping bag that you got tied on your seat with our bra Naren and them. They are going by car. I'm not happy with them strapped to your seat."

"Ja, I know, bru, but they are on tight. We'll go and check though."

Sheila shouted, "Come, come, boys, the food's getting cold."

"Aye, pooah, I had lunch before I came, but you know I can't resist your food," replied Moona.

They walked in and sat down to eat. Sheila already had his plate dished up. At the table, Moona said that he never told his parents he was going to buy the bike, as they would have objected. He still lived at home and when he came home late the previous night, he just parked it off, almost hiding it in the garage.

Even in the morning, he drove his car to work to finish off early. This was so he could sneak in again, take out his bike and ride it to Newcastle.

Shane then asked him why he did not ride with Uncle Johnny, their friend from the Ladysmith biker bunch to Villiers for the rally, as it was shorter.

"Because I finally wanted to ride with you and stick with you, my bru. No problems. This is the feeling I wanted always. I finally got it. You stuck with me, bru, in my slow days with my small bike. Now you try to keep up. Talk to me now!"

Shane responded jokingly, "Well, you are now in another league, my lightie. You speak a different language."

Moona replied that the language never changes and that a bike is a bike. Shane knew all too well about it. For a biker to be back on a bike again, any bike, is a dream come true.

(You see, Moona was riding before, but never like this. His first was an old 550 Suzuki Katana that gave him more problems than Suzy gave me.

It never even went out of town. Then he bought a cheap 400cc Honda. He pushed it like mad to keep up with the others. He had to because it was the smallest of all the bikes.

On one occasion, he even ran out of petrol because he was pushing his baby to its limit of 200 kilometres an hour to keep up with the bigger boys, who were just cruising.

Condense, Shane. That trip that you guys went on for the Paradise Rally of 2006 is another story! *Sorry?*

Moona sold his 400cc for far less than what he paid for it and invested in something bigger and more reasonable. It was still a scrap though. It was an old Suzuki 750 GSX slingshot that went at a reasonable speed, but it only got him to one rally though and broke down.

Now the joy on his face for finally getting a 'One Rand', this superbike's obvious South African nickname, was so apparent. R1 was a top dog.)

After they ate, Shane kitted up in his full gear. Yummy was packed, and the Ventura bag made carrying everything in it good and secure. It held his normal load of tent, air mattress, sleeping bag and folding chairs, plus his clothes.

Shane kissed Sheila goodbye, put on the headphones for his new MP3 player, then slid his helmet over it and mounted Yummy.

The plan was to meet his close biker friends Ravi and Jody, who also had 'One Rands' at Jody's house.

As they left, Sheila wished them well. The Biker Boyz waved and headed off. They then stopped at Naren's house first to leave Moona's things. On the pavement outside the house, they revved the bikes to signal that they were there.

'Vroom vroom – vroom vroom vroom – vroom vroom vroom vroooom!!!'

Only the neighbours heard it though because no one was at home. Moona signalled a thumb's up and pointed that they should go anyway. Shane nodded, and they rode on to Jody's house.

On the road there, Shane stopped alongside the house of his teenage friend, which was on the way. She loved bikes and had a huge crush on him. He revved his bike.

Nervously, she just peeped through the window, blushed and waved. Shane bubbled inside, feeling like the naughty boys he and Moona were. Looking back, he glanced at Moona, who also cheekily waved back at her as they headed off.

They stopped outside Jody's house to meet him and Ravi, who were waiting for them. Their things were already with Naren in the car. Moona again said that he was fine as his stuff was mounted tightly. He added that it would be fine and that they must get going as he still wanted to buy a bottle of Spiced Gold alcohol to take to the rally.

They hit the road and stopped at the bottle store. Shane bought the Spiced Gold as a congratulatory gift for his lightie and squeezed it into his Ventura bag.

At the garage, they stopped to fill up. Here, Shane told them that because Yummy was only a 600, the baby of the bunch compared to their 1000s, and packed, he would be in front of them, cruising to Villiers, which was just under 200 kilometres away.

They all readily agreed. He was the 'Road Captain', a term that bikers use for the biker in the front who is followed by the other bikers. The small-town route they planned was through Memel towards Vrede, then on to Villiers from there.

It was almost 15h00 and the planned meeting time for the other club members at the rally was around 17h00. Their bikes revved loudly as they left the garage.

Moona screamed out as he revved, "Talk to me now!"

On the open road, they were cruising at a speed of not more than 140 kilometres per hour, which they set. It was a convoy with Shane in front, followed by Ravi, Moona and Jody.

They had only done about 15 kilometres when Shane's MP3 player switched off. It was in his moon bag. Still staying on Yummy, he slowed down a bit to open the bag to put it on again. He signalled to Ravi, who was behind him and was slowing down as well, that it was fine and gestured for him to continue.

Moonshine, who was next in line, also decided to follow Ravi. Shane sorted out his MP3 and got back in line in front of Jody, who liked being the last biker.

Still cruising, they entered the Majuba Mountain Pass. After about three bends into the pass, a sharp left-hander came up. Riding behind Moona, Shane watched in shock as Moona took his eyes off the road to look at his seat. This was just to check the things on the back of the bike, which had obviously shuddered loose.

Looking at the road again, it was a sharp bend that he never planned for, and he couldn't take it. Keeping the bike upright, he went straight off the road. Being immediately behind him, it was too close to stop, so Shane lost sight of him as he took the bend. Jody was behind Shane with more room. He saw the incident and stopped just past the bend.

Shane flashed his lights for Ravi, who was in front, as he slowed down. They stopped and made the tricky U-turn.

In his mind, Shane thought that Moona was just on the side of the road and would only be a bit bruised because he was kitted in full leathers. He was going to get a good scolding for taking his eyes off the road in the first place. That too on a bend, of all places!

Jody was already off his bike, heading down the valley, as Shane and Ravi got there. There was no sign of Moona and his bike; they had expected to see him on the side of the road. There was no barrier on the open patch of rocks.

They quickly parked and jumped off their bikes. As they climbed over the rocks to look down, in the valley, about 20 metres down, they saw Moona. His helmet had flown off, and the bike was lying on the other side of a barbed wire fence.

Jody was already about a quarter of the way down the risky rocks. Shane and Ravi already had their helmets off and were carefully making their way down as well.

It was challenging. Jody reached Moona, whose helmet had flown off. While trying his best to rush down, Shane shouted for Jody to check for Moona's pulse.

Jody screamed up that there was no pulse. He then started slapping Moona lightly to try and wake him up.

Shane was in tears as he reached Moona. He knew how to give mouth-to-mouth and tried for about two minutes. There was no response though. He screamed as he sobbed, "Moona, my lightie, come back, come back. Talk to me now?"

Ravi's hand squeezed Shane's shoulder. It was too late. He told Shane that there was no cell phone signal down there, so they would have to go back up to phone because there was nothing else that they could do.

Shane finally let go, and they started their way up. He did not even look back. He picked up the fallen helmet. It was a cheap one, and the straps had actually broken off. Like with him earlier, the exact same leather suit worked, but not the helmet. ☹

The way up was a bit more difficult than the descent, but they finally reached the top. Shane just sat on the rock. He couldn't talk.

Ravi phoned Ricky, who was about to leave on his bike from QwaQwa. At first, Ricky thought that Ravi was joking because Moona never really told any of them and wanted to surprise them with his new bike.

Ravi handed Shane the phone. "Rix, our lightie is gone, our lightie is gone," he said and burst out crying.

Now Ricky believed the news and then told Manny, and they started making calls. Jody had already called the ambulance and police.

They all sat speechlessly. Naren arrived shortly afterwards in his car. He had been phoned just as he left Newcastle. He sat for a bit and told them that he would go to the rally to tell the other bikers and that he would see them the next day for the funeral.

Within an hour, Reshmika, Moona's 22-year-old sister, was there. Her friend, who drove her car, flew the 120 kilometres from Ladysmith to get there as soon as they heard, in less than an hour. As she got out of the car, Shane held her.

She punched him on his chest shouting, "You are lying, Shaynoo bhaia, you are lying, aren't you? Moona is fine, Moona is fine, neh!"

Shane was speechless. Reshmika broke free. She stood at the edge of the road, looking down into the valley. Seeing her brother's body, she wanted to go down, but as she stepped forward, Shane approached her.

He held her hand and said, "Moonshine is gone, my love. There's nothing we can do. His faith called him."

Now angry, Reshmika said, "You must have all been pushing it, neh?"

Wrapping his arms around her again, Shane said, "I swear an oath on Ma that we were cruising. Trust me. Our bikes can go a whole lot faster, but we were cruising at like half the speed."

Still crying, she said, "Then it's true, Shaynoo Bhaia. You'll never swear on your mother when you lie."

"Exactly. It happened, and now there's nothing we can do. It just needs to be organised properly," said Shane.

"But I can see him sleeping. Are your'll sure? Let's go and check again," Reshmika pleaded.

"We all went down. It is too late," Shane told her softly.

Reshmika's friend, who was speaking to Ravi and Jody, came to them. Shane did not even introduce himself as he would normally do. He just told him to take her back to Ladysmith and to let her tell her parents and then start the preparations.

(This is because, culturally, Hindus do the cremation or burial the following day.)

As Reshmika departed, Shane, Ravi and Jody were alone again. A few cars stopped and their occupants got out to have a look. Also, the boys' phones did not stop ringing.

After about two hours, the police finally arrived. Shane gave them a statement. The ambulance also arrived. It was an extremely difficult task to get Moona's body up the cliff.

A stretcher was sent down and was winched back up the cliff. The tow truck faced the same obstacles to bring the bike up as well. It was getting dark. Shane, Ravi and Jody rode back to the Mandir.

Being alone on Yummy, thoughts filled Shane's mind. He had just lost his soul brother. They were inseparable. The worst was yet to come though.

Sheila was not told about Moona directly. The small-town news had spread to her though. Manny told Shane to tell her personally when he got home.

As he rode his bike up the driveway, tears filled Shane's eyes. Sheila came running outside, just searching for Moona's bike, almost not seeing him.

Then she screamed at him, "Shane, Shane'o, where's Moona beta, where's Moona?"

Shane removed his helmet, sobbed and climbed off. He held Sheila close to him and said, "Moona's thuqdeer called him, Ma. He just never took a bend."

On hearing this, Sheila started convulsing. Moona was her baby as well. He was born prematurely. She used to help Anila Mamie with him and remembered him being put in a shoebox as he was so small. Also, while growing up, he had spent more time at her place than at his own home.

Shane took a deep breath, wiped his tears away, held her hand and led his mother into the house. Ravi and Jody followed.

Taking her into the lounge, Shane saw that some of the neighbourhood elderly ladies were there already, as news had spread. He greeted them and then sat his Goddess down. He then went to make her a glass of sugared water to calm her.

Ravi told her what had happened, but she still did not believe him. When Shane brought the water, Sheila just stared at him.

Just seeing the obvious pain on his face, which he was trying to hide, made her believe it. He left the water on the table and held her close in his arms as she cried her heart out.

Taking a deep breath, she told Shane that she had to go to Ladysmith to help Anila Mamie and Roshan Mama to make the arrangements for the funeral.

Shane said that he had to shower first before taking her. She nodded. In the shower, Shane washed his body clean and also privately cried his mind clean.

He came out refreshed and drove Sheila to Ladysmith. There, more tears were shed by everybody, but Shane held back.

There were questions, and he could sense the undercurrents. He could tell that some thought he and the other bikers were to blame, as they must have been 'going fast and all', but he brushed it off.

The immediate family was there, Manny as well, who also held his tears back and just embraced his son. They cleared the lounge and lit the lamp, which is customary in the Hindu faith. You have to also keep it burning next to the body until the funeral is over the following day.

Moona's body was not there, but the emptiness was felt. His body, his physical vessel for life on Her, spent the night in the mortuary in Newcastle. All of the close family members stayed up the entire night, keeping the lamp burning and chanting.

The following morning, the body was brought to the Ladysmith Funeral Parlour. There, the close family men gathered to give it a bath.

Moona's smile was still there. His body was intact, without even a mark. Looking at him, it looked as if he was just sleeping. He was dressed in his favourite suit and put into the coffin. This was then taken back to his house.

The marquee was already up. Everybody paid their respects by dropping flower petals into the coffin, around his face, and then the pundit performed the last rites.

Shane then gave this speech: "Namaste. We have all gathered here today to sadly wish my dear brother Moonshine, my lightie, well on his journey beyond. I know for a fact that he enjoyed his almost 27 years here. The smile on his face shows that he passed on doing what he loved doing. Let us make his journey yonder blessed by not holding him back. His thuqdeer called him. Yes, we are in tears because he left us, but he was called to venture forth. When it is your time to depart, it is your time. There's nothing we can say or do to change it. Rest in peace, my lightie, rest in peace."

The body was then taken to the crematorium. There, a few bikers who came from the rally had gathered. Shane gave a talk similar to the one he gave at the funeral home. Then the passing rituals were carried out, and a prayer was chanted as Moona's body entered the incinerator.

Returning to the funeral home, the solemn atmosphere was still present. Shane and his family drove down to their Mandir in Newcastle.

As Shane entered, he went straight to his bag and took out Moonshine's bottle of Spiced Gold. He, Ricky and two of their friends each poured a drink and also poured one into another glass. They then walked to the road.

Shane made them gather around. As he raised the glass, he said, "Here, Moonikes, my lightie, this is your drink, enjoy it. I know that you'll want us to as well, that is why we join you for this last one on the road. Talk to me now!"

A tear rolled down as they made the toast and spilt the contents of Moona's drink, the other glass, on the tar. Moona had departed. His chapter was over. They all stood in silent memory as they drank.

(☹ I will always miss you, my lightie, but fate cannot be changed. To all the readers who have lost dear ones, there is nothing we can do. Just remember the good times. I had lots with my lightie. What I know is that he'll probably tell me that I've seen nothing yet. My life is only beginning. So true, my lightie, so true. My life began when I started this book. I am now not the same. Physically, yes, but mentally and spiritually, it is in a whole different realm. I am finally me.)

Chapter 8

A Working Holiday

Shane had a stressful weekend. He was still dealing with losing his lightie. Returning to the high school on Monday, he thought he would work away his personal stress, but he did not succeed. It just increased.

The standard of the work in the textbooks was now way below what he was used to. It was almost as if the authors of the books had not considered the actual conditions or working environment in which the material would be taught. In theory, it looked fine, but in actual practice, it was impossible.

In the prescribed textbook for the Grade 8 'Arts and Culture' learners, the first section was drama practical exercises. Among these were group movements, where learners had to lift each other up and then roll over and onto each other.

This meant that learners needed to change out of school uniform, and into 'working clothes'/tracksuits. Timewise, it would take at least 10 minutes from when the siren sounded to go to the toilets to change and then come back into the class. Then, at the end of the lesson, they would need another 10 minutes to change back into their uniforms.

That left you with 25 minutes to teach 45 learners who were crammed into a classroom, not a dance studio.

As if that was not bad enough, they were also culturally mixed, and the Muslim girls were not allowed to even touch the boys. That aside, these exercises were for mixed-gender teenagers who were just entering puberty. The physical contact, especially for the boys, was thrilling, but the girls were naturally shy.

In the textbook photos, it looked like fun, but realistically, it was impossible. In Shane's university tuition days and even his school tours, the learners were older drama students. Now, ranging from 12 to 15 years old, these pubescent learners had not chosen the subject but were simply assigned it by the Department of Education.

Spotting the discrepancy, Shane had a meeting with his head of department, who was also the acting vice-principal. He explained the disparity in the prescribed learning material.

She considered it and gave Shane the authority to use his old learning material from the PMB school if the 'Kuns en Kulteer' teacher agreed. This was because the school was dual medium and educated in two languages, so the Afrikaans side had to be considered as well.

Shane met with the Kuns en Kulteer educator and showed her the English support material from his old school. They discussed it, and she agreed with him that it was very good. But, as she was busy, he had to translate for her if he wanted to change the curriculum.

He explained that Afrikaans was only his second language. Again, she said that she did not have the time. She told him to ask one of the other teachers to translate his material into Afrikaans and she would gladly hand it out to her Kuns en Kulteer students.

That afternoon, Shane arranged a meeting with his vice-principal again. In the meeting, he was told that if he wanted to use it, it would have to be an exact duplicate. The pictures and length of sentences had to match. With Afrikaans sentences being longer when translated from English, this was impossible! That was the last straw for Shane, as there was racial disparity in the school as well.

Every Friday morning, mass assembly started with Christian prayers only, even though there were also Hindu, Buddhist and Muslim learners in the assembly.

(I can clearly remember how at that stage the school's governing body president, an Indian gentleman, put the whole matter of the morning prayer into perspective. It became a media hype in a small town.

He merely asked why the school could not just change and make it a mutually generalised prayer for all faiths.

Again, that is my ultimate goal with Etherealism. For humanity to unite, value and see each other as one primarily. Why the difference?

I can clearly remember how Muslim boys, who have to attend mosque on Fridays at lunchtime, were out of school. Even though the school was aware of this, lessons still carried on. Teachers taught without them in class.

Yes, I know, Islam makes hard and fast rules in their indoctrination, but even with the numerous discrepancies that I have with all set religions' rules, their ways should be respected and taken into account by all of us. All religions should be acknowledged in our new mass culture.

While I was teaching, if there were missing learners in my class because they had to go to mosque, I would not teach. There were a few who went, but even if there was only one missing, I would not have carried on.)

Shane quit teaching at the school because he did not like the dictatorship and also because there was no way to replicate exactly matched documents. Afrikaans uses more words than English. Why did there have to be an exact duplicate in the first place?

Finding comfort in the *Mandir*, Shane searched for work at other schools and even technical colleges, but there were no openings.

It was now April. He finally decided that he must try his London dream again as he could still be a supply teacher, teachers who stand in for absent ones. Plus, his visa was still valid.

Marianne, the chick from New Year's Eve, was still in his picture though. There was no flutter on Shane's side, but she was head-over-heels in love with him, even though he had pointed out to her that he was just playing, in search of his flutter.

Shane told her that he did not want a steady relationship and that he was wasting his life away in Newcastle. London, his dream from the year before, was calling him. She was disgruntled.

(Quick excerpts from her letter to me before I left. The Indian in me says, "How, coz I am proud and want to show off and all and all." ☺).

- *You not wanting a relationship with me now because you are going to the UK is a load of bull, and we both know it.*
- *I love you and everything you do, but I'm getting tired of not knowing where I stand in your life.*
- *The main reason I fell for you was that you catered for me like I had never before thought possible. You came to me whenever I called you.*
- *You are an extremely lovable guy, and that's just the way you are, perfect. Even though you warned me not to fall in love with you, I did.*
- *You are loving, caring, understanding, a good listener, adorable and divine, and the cherry on top is that you let me be free to be me and did not dictate or control, which gained my respect. Who would not fall in love with you, hey?*
- *You will always be my boo. I will wait for you to come back to me. I will take you back no matter what. To have you is the jackpot of my life that I thought I had but did not. How badly I want it, but I do not want to keep you away from your dream. Enjoy. ☺)*

UK, Here We Come!

Shane's dream to go to London was now different. His direct leads from before, in 2006, Arthi's sister and boyfriend, were now back home. Their work visas had expired, and even though they loved it there, they could not remain. This spirit drove Shane more in a way, just to go and compare and see what all the fuss was about.

Sarisha, one of his school friends, who was just 4 years younger and grew up with him because her home was down the road from his, was now working as a pharmacist there. Sarisha had also studied at Rhodes. They maintained contact. In her stopover visit at home, she told Shane that he must come to London as she would return in a week's time.

Sarisha said that there was no room where she was living but that it was easy to find accommodation there. She said that she would show him around.

Shane updated his visa with the Overseas Visitors Club, who arranged a place for him to stay at a Backpackers Lodge in Piccadilly for the first three days. Arthi's sister told him that her friend Ayesha had taken over where she lived but added that Ayesha could help him to find a place to stay.

All set to go on his adventure, Shane made contact with the teaching agency again to find him supply work as a relief teacher. They agreed, so all was set.

Shane's loving family were cautious but excited for him about his adventure. Resh bought him his economy class plane ticket from the O.R. Tambo Airport in Johannesburg direct to Heathrow Airport in London. Manny got him a digital camera, and Ricky gave him some money to convert to pounds.

His heavy 20 kg suitcase was checked in, and it was a solemn parting at the airport. This is where they all gathered to see him off on his 11-hour overnight trip.

Shane waved goodbye and was all excited as he entered the international departure lounge for the first time. Again, for him, it was a whole new world. He explored a little and was surprised that he was still in South Africa. It was so smart and different.

He got a few magazines to read and a block-word puzzle book, which he loved to do. It was going to be a long flight. He went to the boarding queue.

His final steps on South African soil were made as Shane got off the boarding bus and climbed into the plane. It was a bit crammed, but he was lucky to receive an aisle seat in a row of three.

When the plane finally landed, it was early Saturday morning in London. Because of global security threats that were created by the Muslim terrorists after the 9/11 disaster and his physical appearance, Shane was led behind a cordon and had to go through a security check before he entered London.

All clear with his baggage, he checked the notes and directions Sarisha had given him. She was not there at that time because she loved to travel and was touring Australia. She would be back in a week's time though.

In his notes, it said to board an underground train from the airport to get him straight to Piccadilly. He bought a boarding card at the desk there, and a small instruction booklet regarding the various trains, routes and colours was on it.

It was extremely confusing, but Piccadilly was a straight route on the blue train. He followed the signs, waited and boarded it when it arrived.

Out of the airport tunnel, Shane entered London's clichéd gloomy weather. The people inside with him were even gloomier, not social at all. They had earphones on and were busy on their cell phones or reading the paper.

The train stopped at Piccadilly Square and Shane got off. Pulling his big suitcase, he saw the exit sign and headed towards it. He walked right into more hustle and bustle, but he just pushed on to the exit. Now came the worst.

He had to climb up a huge staircase to get to the outside and struggled to pull his heavy case up the stairs with his one hand.

Not a single person offered him help or even had the time to return his smile.

At the top, he panted in the chilly air that looked more like smog. It was, as he was warned, a typically gloomy morning.

Keeping his spirits up, Shane took a deep breath and braced himself. Getting to the Backpackers Lodge, he just checked his luggage in, as he was too early for his room, and took a walk around Piccadilly with his camera. At a shop, he bought himself some airtime and sent a text message to Manny to say that he was fine and that he would call later when he had access to a landline.

Shane also bought a landline phone card. He walked to one of the first telephone booths in the world, the red call box, and called Ayesha. She told him that she knew he was coming, and she was waiting for him.

Ayesha added that he must come for supper and gave him directions on how to get to Canada Water. This was her suburb, where he would have been living if he had been there the previous year.

Still having time to pass, Shane went exploring before returning to the Backpackers Lodge. It was now clear, and he was put in a room to share with three other men.

The room was empty when he went to it. He had heard of jet lag in the past but did not feel it, even though he had hardly slept on the plane.

Shane went to lie down on the bed but was too excited to sleep. After an hour without any somnambular luck, he went out sightseeing. London was amazing.

In the early afternoon, he made for Canada Water. He sent a text message to Ayesha that he was on his way, and she replied that her boyfriend, Muneer, would meet him at the station.

Muneer was waiting when Shane got there. As they walked to the flat, he told Shane that they were from Cape Town and were also on a working holiday.

In London, accommodation was tight, and Arthi's sister and boyfriend were sharing a small two-bedroom unit with them. Now the room was let out to another lady though, so it was full.

As Shane walked in, the smell of prawn curry filled the air. Ayesha was in the kitchen, preparing supper. It was a lovely night with a tasty supper.

Before Shane left, Muneer told him that he would try to see if there was still a place for him at the first house he had stayed at when he got there a year ago. It was in an area known as Upton Park.

They both walked Shane back to the station. He returned to Piccadilly at about 22h00. Muneer had warned him about the drinking water in London, so Shane went out to get some bottled water. He bought a half litre bottle for a pound.

Then he did what he was warned not to do, which was to compare prices in relation to the rand. At that stage, the exchange rate was almost 15 to 1. That meant he had bought 500 millilitres of water for R15! Back home, the same thing cost about a third.

Well, at least the nightlife seemed to be good because there was a buzz in the air around Piccadilly. There was a pub near the lodge, and even though Shane was tired, he decided to have a beer before going back to his room.

Everyone inside seemed foreign. London is made up of 70% foreigners, so they are all basically strangers to each other, except for their immediate groups.

At the counter, he ordered a beer, which was double the price of the water. He tried making friends, but everyone was on their own boat and extremely antisocial, so Shane had his beer and left.

At the Backpackers Lodge, he took out his sleeping bag and climbed onto the top of one of the bunk beds. As soon as his head hit the pillow, close to 11 pm, he passed out. He needed sleep.

Then, at around 04h00 in the morning, his roommates arrived. They were three noisy, tipsy Germans on a weekend away. Shane woke up and said his hellos, and they did the obligatory introductions and went to bed.

One of the men was sleeping underneath him, and the other two were in the bunk bed alongside his. Shane closed his eyes again, and as soon as he dozed off, their snoring started. He tossed and turned. Too noisy to sleep. It was not a good night.

In the morning, he showered and left for a walking tour. He got himself some raisin bread for breakfast and set off. London is huge. That day, it was actually the start of the Gumball 3000, the cross-country car race for owners of prestige cars.

The cars on display at the starting grid included Ferraris, Porsches, Bugattis, Lamborghinis and Mercedes Benz. Shane had only seen pictures of some of these cars before this. Now they were real, so he took pictures. His camera went wild. Luckily, he had spare batteries. It turned out to be a great day of amazement.

In the afternoon, as Shane got back into his room, Muneer texted him to say that there was space for him in Upton Park for three weeks. The room was going to be permanently occupied after that, but that should give him time to find somewhere else in the interim.

Shane called the landline number that he was given. Sharmilla, the girl in charge, answered. She told him that it was fine if he came right away to have a look. She gave Shane the directions from the station.

She said that it was down one road from the Upton Park station, about a two-kilometre walk. The train trip took about 45 minutes, and then he still had to walk. He found it right away though. Sharmilla opened the door.

She was a Muslim girl, who was in her full 'pardas' outfit, where everything is covered except the eyes; it was just like Suraya had described to Shane in school. She was originally from Zimbabwe, and her brother, Yunus, was with her.

They held the lease but shared with other people on a random basis. There was a girl from France and another boy from Wales. It was a five-bedroom flat, so they all had their own rooms. The kitchen, bathroom and lounge were communal ground.

She showed Shane to his room. It had a double bed, a small cupboard and a desk. Sharmilla told him that it was available for three weeks and that he could have it for 25 pounds a week. That was cheap.

He said that he would take it. Thinking about his previous night, even though he paid for a three-night stay, he told her that he would return that night.

It was just after 19h00 and she told him that she would probably be asleep but gave him her brother's number.

Shane went back to the lodge to collect his things. He was told that if he checked out that night, he would forfeit the amount he had paid. He did not mind and just packed up.

It was after 21h00 when Shane boarded the underground train again. Being as tired as he was, he assumed that he was on the right one, but he was not.

It went in a different direction. He was told that he should rather wait it out and let it circle back to where it was, as once you are lost, you make it worse by taking another one.

Back at his starting point after about an hour and a half later, he had the time to figure out the right route again and which tube to board. Upton Park was at around 23h30, but the worst part of the trip was still to come.

Shane had to pull his 20-kilogram suitcase along the road for about two kilometres, and he only got to the house at around 01h00. Sharmilla's brother, who was sleeping, was texted, and he awoke to let him in.

At peace, he slept that Sunday night. In the morning, he went back into the city to sort out the details with his teaching employment agency. All organised, he was told that he would be called when there was supply work and that he needed a UK bank account.

He went to a bank and opened an account. Everything was in place, and he toured around.

(I am touring around too much now, no? Well, here's picture proof of me on top of the open top of the touring bus.

Seems like I'm taking too long with this three-week visit. Thank you for joining me. ☺ *Let me give you a quick summary before I get a scolding from Shane.*

Those of you who have visited London may or may not agree. I was honestly way too lonely. As a place, London is divine. The people, though, are cautious and antisocial. Not just the weather is cold. For a socialite like me, I froze. Fine, I was living with nice people, but they were busy with their lives.

I toured by myself. I also did the walking tour and the boat cruise down River Thames. Green Park in the centre of the city is huge, a beauty of Mother Nature, within the oldest city of the world. At least She was with me. There was lots to see, but no one to see it with. ☹

When Sarisha got back to London at the end of that week, we met. She was London for me and showed me around over the weekend because she worked during the week. She took me everywhere. The time spent with her there is basically my London visit.

We were best friends touring around this huge city. I saw it all, I think? Enjoyed it as a holiday. The work was terrible though. I did supply work at a good pay rate a day at three different schools. Supply teachers are totally temporary.

The system they use there does not have relief teachers. If a teacher was not present, a supply teacher would be brought in. The learners knew I was temporary. To say that they were badly behaved would be an understatement.

The rights they had, in my opinion, are too liberal. There was no respect. I am not saying that their lack of corporal punishment is bad, but it is worse there. You cannot even raise your voice. It is almost like the children are in charge and you work under them. Why? Are we not shaping them? I am not being didactic but at least show some respect. I was not even greeted when I entered the classroom.

Then to get their attention was like a screaming match. Plus, I had pre-planned lessons to deliver. I was losing my personality. It was not all that I had expected.

Sadly, I had to return with my tail between my legs after only 3 weeks. Yes, I know that maybe I never got used to it, but I did not like it, and even though the payment was great, I suppose I do not let money dictate me.

I am very lucky to be saying that, I know, yes. I do have a loving family around me all the time, who support me willingly, in my fantasy, allowing me and my thoughts to grow individually.

I am sincerely forever grateful for that because, essentially, that is what gives my mind the freedom of thought that has shaped it.)

Chapter 9

Back In South Africa

Returning from London in just three weeks, Shane saw more prospects of following his dreams of stardom in Johannesburg.

Resh and Sanj lived there now, so he was given a room in their four-bedroom townhouse in Fourways, a suburb in the north of Johannesburg.

The free accommodation factor in Johannesburg suited his freelance acting lifestyle perfectly. His dreams were huge, but he had to start from the bottom again. The industry had forgotten him.

Not having any big auditions lined up, he went for auditions for extras on commercials, without any luck. He finally got some industrial theatre work through Laura, his old friend from Contractors, who had toured around the country with him in the beginning of '02.

She now had her own industrial theatre company, The Work Zone. The job he got was in an invisible theatre piece that the Walt Disney Company had designed for the Goldfields Mining Company. It was twice a week with a nice paycheque.

The rest of the time, Shane was busy with his normal wandering alone, waiting for auditions and lazing around at home. He helped another Gogo, the elderly maid who lived there, whom he loved as a mother, as much as he could.

It was lonely during the day because Resh and Sanj went off to work early in the morning. Shane also helped Gogo by babysitting his nephew Shival, who was not yet in school.

Nikita, Shane's niece, was at a primary school that was directly opposite the townhouse. In the afternoons, he would help her with her homework.

Shane also made close friends with the garden boys in the complex, Sam and Baathi. He would sometimes have lunch with them during their break and even drive them home after work to the nearby squatter settlement they lived in.

Finally, he got to see, first hand, what life was like for homeless people, who lived in shacks in a shantytown. It broke his heart to see the conditions of their life. He wanted so much to help them all, but he could not.

(I still have the same dream for the underprivileged. Again, as I said, a considerable portion of the sales of this book will be donated to charities.)

A few more months passed with Shane doing what he does best—'jolling!' *(Philanderous wanderings. LOL ☺)*

Before he had left for London earlier that year, another one of his potential partners, while he was busy with Marianne, was Pammy.

He had known her for a while and would visit her when he went back to Newcastle on some weekends. She loved being on the bike as well, and he would take her to the rallies.

She was also in love with him, but no flutter feeling for him though. As always. What was he looking for?

(Aye, even I don't know! No flutter at all. Care and concern but again, nothing. I told her how I think we have this stereotypical notion of what 'love' is, but for me, it fades.

Why can I not return love? I have received so many soppy notes about me being the 'love' of all their lives, even when it was casual. I felt care and concern but not the 'love' as described by our mass media understanding.

My original lengthy work had a few more casual affairs, me joining dating sites and so on, but I've removed them coz I am shortening it. Plus, I am removing the full stories of my next three engagements that I broke off as well coz, basically, the same old, same old. These relationships were a lot shorter—jump in, feel the water, even propose, but then jump out? I am a very sporadic dreamer.

No vanity again, but honestly, it seems like the same thing happens all the time. They fall in love, and I split up with them. I did do the concept of 'love' perfectly, but it faded because a relationship is actually work in a sense.

I want a balanced relationship with no one taking authority. 50% plus 50%, giving 100%. I refer to the imperatives as the 3 'C' equation. 1. Communication – straight talk breaks no friendship. 2. Compromise – give and take. 3. Compatibility – mutual understanding of each other, giving respect and receiving respect.

I dream?)

Chapter 10

The Book Begins

In the latter part of 2008, Shane did his usual wandering around. He was chilling in QwaQwa with his God and Goddess, in JHB with his sister and also went to Newcastle to the unoccupied Mandir.

He always felt strangely different to everyone else. Being in Newcastle, he was advised by his parents to go and 'open book', which is having one's destiny foretold or read from a Hindu astrological perspective.

The priest's accuracy amazed Shane, who was initially told that he has an edgy nature, as he attempts a lot but never completes anything—a true wasted talent at everything he does.

(This is because my entry into whatever I do is eager to try something new out. I, however, then somehow lose interest and pursue something else. The story of my life, I suppose? ☹)

What astonished Shane more was that this was merely determined from his birthdate of 12 March 1977, which the priest said was a 'bad day' astrologically. Shane had said nothing to him about his past or anything.

In his explanation, he used Shane's Piscean analogy. According to him, Shane was, apparently, like a fish that swims in one pool, and then growing tired of the pool before reaching the end of it, it finds another pool without an ending as well. And then has enough and finds another pool and so on.

He was always trying different pools, as he was always searching for an undefined or unknown destination. The search would apparently be endless.

After relating his story a bit to the priest, who then jokingly said that he should start writing a book as his life really was a story, a new thought entered Shane's mind. Around November 2008, he very excitedly actually jumped into the writing pool.

He dived in, but as always, he swam for a bit and stopped. He was stuck. Shane wanted to quit. The main solution came from another very personal note. At the end of 2008, he met with one of the friends he had gone to school with and had not seen in a long time; their friendship rekindled.

She was really successful now, happily married and running her own company, an advertising agency, in Johannesburg. Basically, she had made it, the clichéd big time, even though she was mediocre in school.

(As I was in a slump, I asked her how she had succeeded. She jokingly said that it was a 'secret' but that I could get it on DVD if I wanted. She burst out laughing when I went, "Huh?" She had to take a deep breath before she told me that it was a motivational DVD, 'The Secret', which had changed her and put her where she was.

I thought that I might as well give it a try. To be honest, it made a huge difference for me, as it pointed out a lot that one can do in order to change your personal destiny.

I especially liked the waking up every morning with a smile part. I do that every morning. Try to motivate yourself by saying, "Today will be a great day", and it will be. It will make a big difference. Also, the secret to what you get is all based on envisioning receiving what you want.

Dream, keep your hopes alive, look past the worst, as times do and will change. 'The brightest future will be based on a forgotten past. You can't get on with your life until you let go of your past failures and heartaches.')

Feeling inspired, Shane based himself at Resh's in Johannesburg once more, in his old room, and started writing. It seemed strange to him that as soon as he sat down to write, fully inspired, Natasha Bedingfield's song 'Unwritten' played on the radio.

(I took a deep breath and sang out the words: 'I am unwritten, can't read my mind, I'm undefined. I'm just beginning, the pen is in my hand. Ending unplanned.'

The ending inspired me the most: 'Today is where your book begins. The rest is still unwritten.')

As he was singing, Shane got the idea to combine the autobiography with a narrative in which he would play the leading character and also offer his own opinions, motivational advice, humorous comments, poems and plays, diary excerpts and quotes. Basically, join the reader *(in bracketed italics).*

At the end of February '09, however, while writing, Shane left Johannesburg, as Resh moved into a three-bedroom home due to tight financial circumstances. Shane then went back to his hometown, Newcastle. It was calm and peaceful.

The Mandir was empty and Shane was alone. He felt fated though—to be there to pursue his dream of completing his book. Just about two weeks in, on his thirty-second birthday, March 12, he was all alone. Here's an extract from his diary entry that he entered the next day:

(I never knew that my hometown would get so cold. Everyone is so busy with their own lives, which is probably why we will always miss the good old days, neh?

People were so much closer then. There was camaraderie and there was time for family. Now we have a closed-door policy, sticking within, which is actually understandable as far as safety is concerned. We seem to be privatising ourselves.

I am so busy with my book, though, that it doesn't really bother me. But yesterday was my birthday, and I was alone. My close cousin Sona is a cop. He came to see me, thankfully, at about lunchtime. He wished me well and then left. Then, later, at about 17h00, I got a call from my dear cousin Rishika, telling me that I had better be free, as she wanted to take me out.

She came to pick me up at about 18h00, and then we went to pick up Sona and his wife, Audrey. We decided to go to a popular local pub for the evening. At the table, Audrey took my breath away when she gave me random notes on astrology that she had done. I was shocked at the relevance adding to the card that Arthi gave me. Let me just share the most important parts.)

Pisces

Your studies, career and profession:

- 'This sign has also produced some of the best poets, comedians, philosophers and musicians.'
- 'The more sensitive ones may take up writing books!'
- Your faiths and beliefs:
- 'You believe in the supernatural. Mysticism, the occult and different thinking fascinate you.'
- 'You do your own research in this field.'
- 'You are gifted with a sixth sense.'
- 'The more philosophical Pisceans start their own cult or faith and attract a large number of followers.'

Chapter 11

New Thinking?

After a great night out for his birthday, Shane was back in his groove, bonding with his book in Newcastle. His mind was also going off into different directions.

Within his culture, losing someone has many ritual undertakings after they pass away. There is the funeral the next day, then a 3rd-day ceremony, then a 10th-day ceremony, a 13th-day ceremony, then 14th-day ceremony, then 6th-month ceremony and a 1st-year ceremony to finally let go of the soul.

(In my book, it's all mythologically based, according to the past history of living on the flat Earth, but it is kind of a family tradition that I do feel bound to participate in.)

At his Farm Pooah's, his father's late sister's 6th-month ceremony in April 2009, Shane received some more grounding in the simplest of ways.

(When I got there, there was a solo singer, setting up his equipment. I did not know who he was. To me, it seemed as if he was there just to sing the religious songs. Indian functions normally have live bands.

With all the normal commotion associated with functions going on, I think that just like me, no one from the crowd even knew who this man was. After some time, he was ready to start.

Here was a man, all by himself, sitting on a small low stool, with a microphone. His speakers were up, and he had a laptop in front of him. There were not more than 20 people inside the marquee. I was even speaking to my cousin when he started singing.

I actually thought that a CD was playing. I had heard that voice before, but I had no idea it was him. In shock, I turned around to see him holding the mike and singing live.

My fate led me to meeting the most amazing man—famous South African Indian singer Dharam Maharaj. He is a star, a celebrity but as humble as the proverbial 'man next door'. What a man!

Just like me, when he started, everyone's attention was caught. I went to sit in front of him on one of the seats in the empty front row. Being a trained performer myself, I was awestruck. His vocal range astounded me. It was totally natural. His talent took my breath away. The best part is that he is humble and simple, even though he is big in the industry. In-between songs, he actually spoke to people, trying to advise us about life, about being simple and true.

He explained his favourite Hindi religious song before he sang it. It is basically about how selfish we have become as a species. We deny a thirsty man water today but pour a sweetened milk offering into the ground. A man is alive but is not offered anything. Plus, it goes further. When the man dies, his loved ones do their best to make a sacrificial offering to him, denying other living people.

*Uncle Dharam said, "Cater for the needy, my dear people. They are **real**."*

Then he expounded further on the subject of how selfish we are. The example he used was that we take our children to toy shops and buy them toys for about R500, and yet when we are asked for a donation, we say we have no money.

We do, however, give money away if we are noted for it, don't we? We want to see our names in print, even carved on donor stones at temples and funeral parlours. "This is all because of so and so's generous donation."

What for? So our names are in spotlights? So people can one day say that so and so is and was a nice guy? I was more than touched. Uncle Dharam was speaking my own thoughts. I have never been more gripped.

He also mentioned that we should value our parents first before all the devotion to our Gods. Our parents are our Gods. I was saying that all the time. There is so much more to it. Let me just tell you about the song he sang 'Kabhi pyase koo pani'. I went to the internet and found this English explanation for it.

"This devotional song has a deep meaning. It says if we can't provide water to the thirsty, then offering even the elixir of life is of no use. It is better to help those in trouble than regretting it later.

If we don't respect and serve our parents, then visiting holy places and praying to God is of no use. When we go to holy gatherings and listen to enlightened words of saints, we realise that this precious human life is wasted if we don't have mercy and kindness in our hearts for the poor and needy.

Offering food to a hungry person is better than any alms or donation. Taking a bath in holy water only cleanses our body. But it is more essential to purify our soul. Even after deeply studying scriptures like Veda and Shastra, it is of no use to be called a scholar and knowledgeable if we don't believe in spreading its light in the lives of others, which gives the real satisfaction.

The actual pilgrimage lies in the feet of our parents. Loving, respecting and serving them is the only way of earning God's blessings and love."

How true is that? Mum and Dad were sitting with me. I remembered how, when I was little, we used to go and bow down to our grandparents and touch their feet, asking for a blessing, "Palah goo Aajah?" (Bless me, granddad.)

They would touch our heads and say, "Jeeteh raho beta." (Live long, son.) That has been forgotten. After the song, I got off the chair and bowed down for 'Palah goo' to Mum and Dad. Their blessings came naturally.

Uncle Dharam saw this and then sang a song about valuing parents. I was in the centre, with my arms around my Matha'jee and Pitha'jee. My tears were flowing, Mum's as well, and Dad wanted to, but he maintained his 'manly' composure. I am truly blessed.

During lunch, I told Uncle Dharam about Etherealism, my dream cult to unite humanity. An elderly citizen in his fifties totally agreed with me. He not only understood but agreed. I went further and told him a bit about evolution and so on.

He had not known but told me that I had sparked something in his mind. My deep question bamboozled him as well. I am agnostic and do believe that there had to have been a creator who made everything primarily.

I call this mysterious creator 'IT' because we actually cannot define if it's male or female or even one of the many names all religions have for their one.

What is clear is that we all have been indoctrinated to accept it as a male due to male power, 'machismo', them having ruled in the past. Sexism? 'God' could easily be a 'Goddess'? Hmmmnnn??

My question is if totally mutual in terms of sex or religion, IT created our Mother Earth and life, what created IT? And then ITs creator and so on? Infinite question!

Uncle Dharam agreed with me and said that we should rather focus on us and each other, as we'll never know. We must stay true and simple and that will lead us to ultimate happiness, respect our parents and elders and also to give the thirsty man water.

Our time passed so quickly, and I feel more than blessed to have met him—a truly blessed man. My mind was now, even more than ever, being shaped into a holistic way of thinking.

Another deep story happened when my dear cousin sister from Benoni, Melanie, who for a long time saw spirituality differently, made an appointment for me to see Beverleigh Esterhuizen, a parapsychologist.

That was a lengthy incident, but let me just give you the gist of it in the form of a letter to you.

A Visit to the Clairvoyant

To my dear reader,

At the risk of sounding too dramatic, my visit had me shuddering. This woman, a parapsychologist, knew me inside out. I merely gave her my birthdate, and from that, she knew my flighty nature, just like the Hindu priest.

In my mind, both of them linked my personality to the typical Piscean stereotype, a frivolous dreamer fish floating life away. Look, I do believe in astrology in a way but also think that people interpret generalised horoscopes for themselves. Also, we tend to embody our clichéd star sign personalities in a way.

I am what she told me, in a sense, yes, but that's the normal cliché. My initial, real, analytical mind was now being charged up. Upon being asked why I was there, I just gave her a brief history about how I see the world differently now, especially after my accident.

She said that the aura around me was very strong, which is probably why I had survived in the first place, defying medical realities. She went on to say that there was still a lot more for me to do and achieve before my time passed.

Out of the blue, she then said that she can sense that I have the spirit of my late disabled grandfather above me and that he is special just like me, hinting that maybe he was in a way responsible for colluding with the other spirits to return me to Mother Earth?

She said that he cannot walk, and aside from his disability, he is the same as me in a whole lot of aspects as well, mainly in his thinking and questioning of established belief patterns. She added that he is unconsciously guiding me.

(My nana, who was also a school teacher, was crippled and spent the latter part of his life in a wheelchair as he developed gangrene in his legs. He passed on when I was 12.

I guess I was too young to actually do much with him, but I once remember him coming to my house, about two kilometres away from his, pushing himself along on the tar. I can clearly remember my deep connection with him, as I gave up the soccer match I was playing with my friends in the park to push him back to his home.

In those 15 minutes, I bonded with my nana, especially when he said that I could push him a bit faster. He was like a child again, laughing. He said to me, "Thank you, Shane, you are special. I feel like a child being guided by a child."

I just smiled inside and took him home, then jogged back to kick the ball. I suppose we don't realise 'deep' moments until they happen to us. We think back and wonder about the past.)

My psychic reading astounded me with its relevancy. I was astounded at this woman's power. She even said to me that I am also powerful and that my aura was the most remarkable one she'd seen.

During the tarot card reading, I seemed to be analysed as the ruler, the King, with eager pending followers that I will find if I stayed true to my inner being.

Beverleigh advised me to invest in 'Doreen Virtue's Oracle Angel Cards' because I had an amazing power. This was probably why I survived and defied all odds.

Looking at my palms, she was more shocked. I was the first person she had ever seen with two totally different palms. Aside from my right hand's disability, the lines still remain.

The major thing she read was that my disabled right hand, which is now withered, has only one main lifeline or heartline. My left hand, the working hand, however, has two lifelines, as they have split. There is a shorter one at the bottom of the matching longer one.

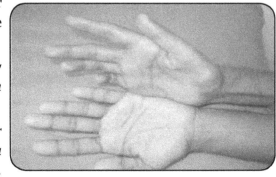

(Here's a picture of my hands. See the difference in the lines?

Now, look at your own. You probably have a matching pair, similar lines, but just going in different directions. There may be slight differences, but as distinct as mine?

A whole second lifeline on my 'alive' left hand! I now clearly remember the scared gypsy lady from Grahamstown. Remember?)

Beverleigh told me that I am a destined miracle, one of a kind. She said the main line at the top is the life that I lived, but it came to an end. Now the shorter one at the bottom is underneath the top one for a bit but eventually is free.

My mind spun. More than an hour had already passed. She led me out to Mum, who was waiting for me in the lounge. As Mum saw Beverleigh, her arms opened.

They embraced and she said to Mum, "You are special but never realise it, like your father. You have been gifted by having a gifted son who will do what you and your father were supposed to do."

My mum hugged me, and her tears brought tears to my eyes too. Beverleigh embraced us both. We felt the spiritual connection but had to part as Beverleigh's next visitor was there already and was waiting as my time had been extended.

Thank you, my dear reader.

Love & Light ☺

A few months after my consultation with Beverleigh, I had a small fire at home. There was a candle burning on a corner table, and an unexpected wind blew the curtain right to the candle. Luckily, just the curtain burnt up, and it was doused in time. Just that small fire, however, caused a lot of smoke, and the walls were black.

*Ethe*re*al*ism has, essentially, your creators—mum and dad—as the main God and Goddess, but also, you should worship your grandparents, your closest roots, as well. I, therefore, have pictures of them on my wall also. After wiping my nana's picture, some of the smoke damage was not coming out. Then I looked closer. Here is the picture.

Can you see a clear face on the top right? I do have this photo framed in my bedroom and am open to any tests to verify. I do know that I profess for us to work with our reality, but even that cannot be defined exactly, can it?

It is almost as if we are all trained by our family and society in a sense to follow a stereotypical version of normality. BUT what happens when you can see differently. Must you not...

See to Blind?

So, you don't clichédly 'toe the line'.
Your inner being is very different to everyone.
Being you then is hard to define.
You ponder, what needs to be done?

Must you be like all around you,
blind leading the blind?
But what do you do,
because you CAN see & know what to find?

Shutting your eyes and pretending is easy,
because the blind won't know you see.
To fit in, you do though, but that makes you your own enemy?
Or maybe guilt makes you proclaim that you can actually see?

Who will know though,
because they are all blind apparently?
Exactly! What if some, or even everyone around you, can see,
and you are just looking at things differently?

What exactly came first then, the egg or the hen?
Half full or half empty?
Do you encase your thoughts in a den?
Can you not declare what you personally see?

An answer to these questions, you will never know.
But your questioning it actually is a sign though!
Just that proclaims that you can really see,
because at least you question, and think differently.

But what are you looking for?
Because only you can open the door,
that will lead to an answer to find.
As long as you don't think that you are also blind.

My eyes were honestly being truly opened to a whole new side of reality. A few months after my visit with Beverleigh, I went to the acclaimed spiritual godfather in Johannesburg, Lionel Berman.

What a deep two hours! Lionel saw my gift as well and said that I must persist with my vision, giving me hope that I am special. He sees the 'real' me and described me as a 'Shaman Healer' with an indigo aura.

I am an Indigo Child. We are uniquely different thinkers. Also, apparently, with the planetary alignments shifting, many of today's children think differently also. Indigo?

The subject is too deep and intensive to go into. Can easily be researched online nowadays. We know so much more and can still educate ourselves even more.

The rest of my visit with Lionel was also deep and intensive, plus a bit too personal to even disclose here, quite ironically, in my biography. All positive though, rekindling my inspiration to dream of my goal for unity in humanity.

This was more enforced by Mr. Berman, a Caucasian man in his fifties, still living in his home in a rough, now mainly black, suburb of JHB—Hillbrow.

He said that he planted his roots there like 30 years ago, and there was no need to leave his home because things around it changed. Amazing. Plus, he did not even charge me. He said that he cannot charge me and presented me with his drawing of my portrait with the indigo aura around.

Personal thank you, Lionel, for also clarifying to me that we only judge based on established premises of what is good or bad. All this is based on personal interpretation or fixed behavioural patterns. We decide at the end of the day though. We can make our own choices.

Now, you can make a choice to go and compare the picture of the spirit above my nana with the Shaman Healer picture. I see a clear similarity. I had not even told Lionel about it. Also, quite strangely, he never drew my eyes as the quite obvious green that they are. Weird? Hmmmnnn? You be the judge of that. Let me next plant something imperative in your thinking mind before the whistle for this Second Half blows.

Chapter 12

Planting Something Fishy?

It was the last weekend of May 2009 and on one of his visits to check in on Shane, his close cousin Sona, who also fishes competitively, told Shane that they were planning a Saturday fishing trip. He and his elder brother, Naresh, whose nickname is 'Boyks', from who Shane bought Suzy, were taking their sons fishing.

Totally interested, Shane said, "Just you lot? Include me, please. I need to touch the ground again. The book I am writing is making my mind wander. I even talk to myself."

Sona laughed and told Shane that he will be fetched the next morning and squeezed into the back of the van (small pickup) with their two little boys.

In the morning, the van pulled up. It was just after 05h00. Shane was waiting for them. They got out and mounted his fishing rods with theirs on the rod rest of the van's canopy, the roofed enclosure.

His bag was thrown in as he jumped in the back of the van with the lighties. Aidan, Sona's son, was 11; Shaylin, Boyks' son, was five. They called him 'Shane Kaka' *(father's brother, pronounced 'car-car')*.

It was about half an hour's trip to the dam, but it felt quick in the van. Shane remembered how, when they were teenagers, they used to ride their BMX bikes there, which took much longer. He reminded Sona, who burst out laughing.

Those trips used to take around three hours to get there. That was with their supplies in their backpacks and smaller rods that were tied to the bikes. No boot required. They would fish, have a braai, swim, even play soccer, then make the long ride back home.

They both reminisced as they entered the gates that had just opened. Driving, Boyks headed to the left of the dam, the most favourable spot.

As the van was parked, Shane could, from the back window, see the water. There was a mist hovering over the water, like fog. It was almost winter, and a nip was in the air as the sun was just rising.

Shane jumped out, but the boys kept under wraps in the back. The men unpacked and took out just the rods and tackle. Most avid fishermen do that—get their lines in first and then set up.

The view of the sun rising took Shane's breath away, and he felt regret because he had left his fancy camera behind. He grabbed his phone that had a camera and snapped away relentlessly. Boyks jokingly said, "And now, mister cameraman! You came to take snaps or catch fish?"

Sona, who gave Shane the nickname 'Boozine' as he was growing up, blurted out, "Ja, Boozine, come, come! Now is the best time to throw your line in."

"I'm in no hurry my bhaias. The fish are in the water all the time," said Shane.

"That's why you need to put your rod in the water, you won't find them in the sky!" remarked Boyks.

Shane responded, "Then I'll be fly fishing!" They all laughed. Then he added, "I will get my rod in when I feel ready. You should know me by now. I appreciate nature first. And I work on my gut feeling. When I'm ready, I'm ready."

Looking over the water at the sun, which had risen, Shane's breath was taken away. He took a picture. Looking at the picture, it was like a hand lifting the sun from the water. The clouds formed the shape of a hand, with the globe of the sun in our Creator's palm.

(My blessed addition to the front cover pic. Close quickly and see...)

Shane smiled inside. The silent moment was enough to touch his heart as Sona called out to the boys, "Come, boys, it's warmer now. Come get your rods ready."

"Okay, Dad. We'll beat Shane Kaka," said Aidan.

Pointing at Shane, who was still sitting and looking at his photos, Shaylin taunted him, "Nah nah nah neh, we gonna beat you, we gonna beat you. Shane Kaka, watch out."

"I'm scared. What must I give you when you win?"

Shaylin answered, "Don't show me the Tickle Monster again, Shane Kaka."

This is because Shane would catch them, change his voice and tickle them till they said the name of who he became as he tickled them: The Tickle Monster. Only then would he stop.

(I love children's laughter. It warms your spirit. Try. Aim for their sides. The harder you tickle, the more they laugh. To make myself feel

young again, I slip into their realm, even speak the way they do, become a child. As we grow up, we lose our inner child, but it is still there and can be reached.)

Shane changed his voice into the Tickle Monster's, "But I will never leave you. I am the Tickle Monster. I do not make deals."

He hobbled towards Shaylin, who laughed and ran off. The monster caught him. Being only five years old, he was light. With the grass being wet, Shaylin could not be tickled there. The monster moaned as he picked him up and carried him on his shoulders like a bag.

Deeply, he roared, "I will take you back to my lair to do my job."

Aidan ran into them, clutching the monster's legs. "Don't worry, Shaylin, I will save you."

He locked his hands around the monster's leg and held it tight.

The Tickle Monster said to him, "You think you are going to stop me? I will hunt you down after I'm done with your partner."

"I will hold you back, monster!" shouted Aidan.

He folded himself around the monster's leg and sat on his foot. The monster laughed loud, "Try your best, my boy, you cannot stop me!" and he forced his way to the van, hobbling with Aidan attached.

He reached the van, and as he laid Shaylin on the mattress in the back, Aidan let go and ran off. The Tickle Monster bawled out, "Oh, run now, my boy, I will catch you when I'm done with your partner."

Shaylin was thrown onto the mattress. He yelped, "Arrgh!" The Tickle Monster attacked.

Aidan, now standing on the outskirts, screamed, "Say his name, Shaylin, say his name!"

Shaylin, however, was being tickled to death. He couldn't say a word.

The Tickle Monster gave him a break to breathe and asked, "What is my name?"

Still gasping, Shaylin answered, "Tickle Monster, Tickle Monster, Tickle Monster!"

The Tickle Monster replied as Shane, "Good boy. You are now free. Where's your partner?"

"Over there, over there, Tickle Monster, get him too, Tickle Monster, get him too!" replied Shaylin.

The monster turned around and ran after Aidan, who had already started sprinting away. The monster caught him near the park swing. With Aidan being slightly heavier and the monster's limp hand, he couldn't carry him back to the lair.

He noticed that there was no wet grass there. He threw Aidan on the ground and pounced. Shaylin had caught up from the back and ran into the monster, who just brushed him away and screamed, "You told me where he was, Shaylin. Now he pays!"

He attacked Aidan, who burst into laughter. After two minutes, he asked, "What is my name?"

Aidan answered, "Tickle Monster, Tickle Monster." The Tickle Monster stopped. Shane grabbed them both close and said, "Thank you, my boys, you brought Shane Kaka back."

They embraced and Aidan said, "Welcome back, Shane Kaka."

Shaylin added, "Ja, bye-bye, monster."

Aidan very seriously said, "Now let's go catch some fish. I told you, I'm gonna beat you."

"I accept the challenge, my boy. You, Shaylin, are you up for it?" asked Shane.

Shaylin laughed and again said, "Nah nah nah neh, I'm gonna show you, I'm gonna show you."

Shane replied, "Deal's on, my boys. Bye-bye, Tickle Monster." He held their hands, one on either side, and they went back to the van.

At the van, Sona said, "Come, come, boys, stop playing around. See, our rods are in already."

Shane and the boys started getting their rods ready, putting the reels on and threading the line through.

"Ja, come, boys. Get the rods ready," Boyks told them. "Shane, must I cast for you when yours is ready?"

"No thanks, Boyks, I think I'll manage," replied Shane.

"But with your hand, you won't throw far," said Boyks.

"Ja, true, Boyks. Give him, Boozine. You know he can cast far," agreed Sona.

"And you should know me," responded Shane.

"Oh, your 'water is water' theory," said Boyks.

"It makes no difference if it's near or far, water is water, fish are fish, and they are everywhere," responded Shane.

"Okay then, give me one rod to throw, and you take one. We'll see who wins," challenged Boyks.

"Challenge accepted, my bhaia. My rod's ready," said Shane.

Passing the other one to Boyks, he said, "Here is yours."

Shane took a rod stand and walked to the water's edge. Stabbing the stand into the ground, he rested the rod on the stand. Shaylin was already there with the 'mealie bomb', which was made from ground dried mealies, which would create a fake feeding area for the fish as the mealies were too fine to eat, but the bait was a sweet corn pip that they would find.

Shaylin told Shane, "Shane Kaka, Sona Kaka is casting my rod for me, and Aidan is big enough to cast his own, so I came to help you."

Seeing the boys as his sons also, Shane called them 'beta', the Hindi word for son, pronounced 'bear-tah'. Kneeling down, he said, "Thank you, my beta. I'll put the bomb on, and your special hands will put the sweet corn on the hook for me, right? We'll show your father."

"Okay, what must I do, what must I do, Kaka?" asked Shaylin excitedly.

Shane was almost done putting the bomb on the bomb holder. He made up the following myth, "Okay, you must rub your hands together three times. Then you take the sweet corn pip with your left hand, touch it on your forehead and pray for a big fish."

"Okay, Kaka, and for the other hook, because you got two?"

"Well, same thing, beta, but you must use your right hand. You know which one is your left hand and which one is your right hand, don't you?"

"Of course, Kaka, I'm five years old, duh!" giggled Shaylin.

Shane laughed, "Oh, sorry, sir!"

He handed the end of the line to Shaylin, who was rubbing his hands together.

Shane nodded his head as Shaylin continued by holding the pip in his left hand, and as he was touching it to his forehead, he asked, "Like this, Kaka?"

"Exactly, beta. Now bless it."

"I must say *Aum Bhoor Bawah Swaha*, Kaka?" *(The holy Hindu chant.)*

Shane thought deeply and responded, "Do you know what that prayer we say means, beta?"

"No, Kaka, tell me?"

"Well, I don't know either, beta. It's in Hindi, I think, which we don't speak properly any more. I was taught it, we all are. Now just say what's in your heart. Speak the way you do. IT will hear you!"

"Huh? Why you say 'it', Kaka?" asked Shaylin curiously.

"God, my boy. I don't give our Creator a name. I say IT as the creator could be male or female, and I don't like every religion's different name for the same thing. IT is one thing that controls Her, which is Mother Earth. But it is too deep to tell you about. Now, beta, get the hook ready."

"Okay, Kaka, that makes sense to me. But, coz in my playschool, the Muslim boys call God 'Allah' and the Christian boys, 'Jesus'. IT is the same thing, isn't it?"

"Yes, beta, I don't deny them. I make them one. That's why I call all the different Gods one thing – IT!"

"Okay, Kaka, but why we eat differently? My Muslim friends can't share my lunch because they say it's not ha... ha... haal..."

"Halal," said Shane.

"Ja, that, Kaka, what's that?"

"Well, to me, beta, Muslim religion makes the most rules. All religions make rules. Like how we can't eat beef, they can't eat pork. And all their other animals must be prayed for before killing them, making them halal."

"But why, they gonna kill it anyway, neh, Kaka?"

"Aha, that's why I can't see why," agreed Shane.

"Will my friends die if they don't eat ha... ha... halal, Kaka?"

"No, beta. Like how I don't die when I eat beef sometimes, even though I'm Hindu!"

"Then why, Kaka?"

"Well, beta, we make beliefs. Like we all believe in either heaven or hell, right. Now how you live will determine where you end up when you die."

"But, Kaka, my friend Johnny, the white boy, eats pork and beef. Is he gonna go to hell?"

Shane was stumped. "Well, beta, I can't judge that, no one can coz we don't even know whether heaven or hell is there or not."

"But, Kaka, Daddy says Ursula is there in heaven?"

"She must be, beta, if there is one. Now come on, bless the mealie."

"Okay, Kaka, but one more question, please?"

"Okay, beta, last one."

"Why the Muslim boys got different toilets and sit like that, Kaka?"

Shane did not say what he wanted to about the quantity of faeces and just smiled and said that it was easier for them to make poo.

"But we all the same, neh? Is it easier to make poo like that, Kaka?"

"If you want to try, you can, beta. Nothing's stopping you."

"But I'm scared, Kaka. I'll fall down on our pans. They got one down on the floor already. There's a hole in the ground, Kaka."

"I know, beta. Okay, 'nuff now. Come, let's get the bait ready. See, your daddy got his reels and tackle on. He's grinding some more mealie bomb. Let's put ours in first".

"Okay, Kaka. But I understand you. You make sense. To me, we are all same, right? I can kick the ball like Zubair and Jack and Temba, can't I?"

"Aha, beta, even better. Now, for the last time, hurry and bait the hook."

Shane's heart smiled. Here was a little kid who just touched on the heart of the matter by understanding his theory of humans being the same but creating differences, all stemming from different religions after evolving into different races, each with their own cultural mythology and traditions.

As Shaylin was baiting, he said, "I'm gonna ask IT to bless this hook, Kaka."

"Aha, my boy, IT is one! Now get the rod ready, we'll ask IT for ITs blessing to ask Mother Earth to give us one of Her big fish!"

Shaylin looked a bit confused but held the mealie pip on his forehead, closed his eyes and said, "Okay, IT, tell Her to give my Kaka big fish!" Then he laughed as he added, "And make him beat my daddy!"

Shane burst out laughing as he glanced over to see if Boyks, who was still busy baiting the other rod, had heard. Luckily, his head was still down.

Shane winked at Shaylin and said, "For this other one, don't say it so loud, beta. Your daddy will hear you. IT can hear you when you pray to IT, even in your head, without saying anything."

"But, Kaka, why you don't do this one so IT can hear you also?"

"You are right, beta. IT needs to inform Mother Earth, and I am a womaniser."

"Huh, Kaka? A what?"

"Never mind, my beta, you're still a lightie. Give me the pip."

Shane did the necessary, then took off his shoes, socks and jeans and walked off in his underpants to wade in and cast. He stopped on the shoreline. Facing the water, which he knew was cold, he held the rod on his shoulder and then paused and admired the beauty of the source of life itself.

From the back, Boyks asked him, "And now, Boozine? You checking for fish to warm you up so you know where to cast? I'm ready, but I'll give you time."

Sona added, "Just like how you check for 'fish' in the nightclubs to warm you up? But like there, there's a closing time here also."

They burst out laughing. Aidan asked Sona, "Dad, the fishing club closes?"

Sona smiled and said, "Ja, my boy, the fishing clubs."

Aidan laughed and screamed down at Shane, "Ja, Kaka, you better check for the fish coz I'm gonna beat you anyway."

Shaylin told Aidan, "He will beat you coz I helped him."

Aidan then said to Boyks, "He'll see, neh, Boyks Kaka, you'll beat him, neh?"

"That's right, my boy. I'll show him," agreed Boyks.

Shane smiled, stepped into the cold water that he psyched himself up for to feel as warm, waded in for a bit, then stopped knee-deep and cast the rod in. Given his disability, it didn't go far.

Boyks laughed and said to him, "Oh, are you aiming for the small fish? I'll show you where the big ones are."

Shane was still in the water, busy putting on his 'policeman', which is basically a marker on the line so that if it moves, one knows that there is a fish nibbling away.

Boyks, who was wearing his waders, the waterproof pants and boots combination, walked past him and said, "I'll show you where the big ones are, my boy. I'll show you."

As he stepped into the water, Shane responded, "All the best, my bru, but remember, fish are everywhere in the water."

Boyks chuckled, "And everywhere in the clubs also."

Shane laughed, "But some are already caught?"

Sona laughed from the back, "Or about to be caught by you, mister 'fish-her-man.'"

Boyks smiled as Shaylin added, "Ja, Sona Kaka, they are everywhere for Shane Kaka because I asked Her to give him one fish."

"Huh, I hope you never tell my son about the other 'fish' yet, Boozine?" Boyks said.

Ruffling Shaylin's confused head, Shane laughed and said, "Nah, he's too small for that."

Sona joked, "He's got the blood though, the Praag blood, our father's blood!"

Shane and Sona laughed. Also smiling, Boyks said, "Too young for that fish. Now, I'll show him a 'real' big fish!"

Wading in a lot more than Shane, Boyks sent the line flying deep into the water, almost double Shane's cast. When the line hit the water, he boasted and said, "Now that's how you do it, Boozine. That's how."

Shane smiled and responded, "Nice cast, Boyks, but like I said, we'll see."

Climbing back onto solid ground, Boyks laughed and passed him the rod to put next to his on the double rod rest stand and put a policeman.

With all the rods in the water, they set up the site. It was still chilly, so Shane put his jeans and shoes back on. Sona took out the 'Skottle', which is a big flat pan that goes under a gas burner, and they all helped prepare breakfast, fried sausages and eggs. It was ready in under 20 minutes. They were all hungry and ready to eat, especially the young boys, as the aroma was making their tummies grumble.

Their chairs were together around the skottle pan. Sona passed them the bread rolls, and they tucked in. Shane was about to take his first bite when his mind wandered. He looked out at the dam, glancing at his rod. He kept his eyes on the beautiful view. As the bread roll reached his mouth, it got thrown to the ground.

No, it was not the taste at all. Shane ran to the rod that he had just cast in. The policeman was in the air, signalling a fish on the hook. He reached it and, with a tug, lifted it up.

The fish was on. He turned back and said to Boyks, who had just stood up, "See, I told you the fish are everywhere."

"I also know that," Boyks replied. Jokingly, he added, "The small ones!" Then he asked, "You need the landing net to scoop him out, Boozine?"

Shane, feeling the small tug, said, "Nah, bru, a small one."

He carried on reeling in. Luckily, it was the self-casting reel that his disabled hand could work as he jammed it against the handle that he turned. It was a short reel in, and in no time, he saw the fish. It was almost home.

Then, when it was almost on the shore, the hook suddenly came out. The fish was free to 'turn tail' and swim away. Shane threw his rod down and lunged forward to go in the water to grab it. As the shore was wet, he slipped and slid to the ground, ending up flat in the water.

Everyone burst out laughing. He was on his side in the water and could still see the fish. Now his mind wondered. That fish was free to turn tail and head off into freedom, to put itself into reverse gear. However, it was almost as if it wanted to swim to shore because it was still facing the shore.

Shane put his hand behind it and scooped it onto land. The crew were rolling on the floor with laughter as Shane stood up. His shoes and jeans were wet.

Laughing, Boyks asked, "All that for a small one?"

Walking to pick it up, Shaylin said, "It's still a fish, Dad. I blessed the bait. That's why it came."

As Shane sat and took off his shoes, he said to Shaylin, "Mother Earth sent it as blessings, my beta."

"Sharp, Kaka. I also just told IT thank you in my mind," Shaylin responded.

Shane smiled, ruffled Shaylin's hair and said, "Thank you, beta, thank you. That did the trick."

Aidan already had the keep net there. They put the fish in and took it to Sona, who had his waders on and carried it into the water.

Standing on the shore, staring at a beautiful morning sun over the water, Shane froze the moment. He took in a deep breath of the fresh air, looked up at the heavens through the now silver-lined clouds.

He closed his eyes and, in his mind, said, "Thank you, IT. I not only caught a fish that you told Her to send to me but also my totalitarian ideology caught Shaylin in a sense."

With a smile on his face, he opened his eyes and Boyks, with his waders on, said to Shane, "Welcome back. What you asking for? Dry pants?" He laughed. "See, my jeans are in the van. Wear them and hang yours to dry."

"Thanks, Boyks."

Walking with bare feet, which he loved to do as he felt connected to Her, Shane headed to the van. He put the jeans on and went back to the guys for breakfast. He was surprised that even the boys were waiting for him.

Shaylin saw his surprised face and said, "I told them, Kaka. They can wait for you. You got the biggest fish so far. Because I helped you, isn't?"

Shane responded, "Thank you, beta, but you know who you asked to give it to me, neh?"

A big smile was on Shane's face as Shaylin nodded his head. All together, they congratulated him and patted him on the shoulder as they tucked in.

The rest of the morning was quiet. No more fish were biting. There was also a game farm there. Like with most of his past fishing trips, even with his father, Shane would always want to see the surroundings and go on a hike. He asked the boys to join him. Aidan said no because he had come to fish. Shaylin, however, excitedly agreed.

They started on their way and were not going too far, as Shaylin was still small. Along the way, they saw three zebras.

Shaylin wanted to touch them, but as he got closer, they ran off. Shane quickly took a photo.

The hikers returned to the site. Shane showed the fishermen the photo, and they were amazed at how close Shaylin actually got to the zebras.

Shane then joked and asked, "No fish, boys? I took the only one." Extending the 'big boy' humour, he added, "The 'fish' come running to me."

Sona laughed and said, "Small, loose ones, ja, but you have to wait for the big, proper ones!"

Boyks, also laughing, added, "You have too many loose ones, Boozine. I know your big one got away a long time ago."

"And how? I never caught her in the first place, and even though I wanted to catch her, and she wanted to be caught, I couldn't."

"But there are a whole lot more fish in the sea though. And you enjoy casting," Sona told him.

"Aha, but like earlier, even when I catch fish and want to set them free, they still want me to put them in my bag and take them home," laughed Shane.

"They must be wanting that, Boozine, because you are the 'dada', the king. I don't think any fish will want to leave you because even when you say you don't want to catch them, they still want to be caught," Sona replied.

"But why, Sona? I say it straight, bungee, no strings, just casual."

"Not strings, my bru, blood. You got the Praag blood, my boy," said Sona. Then he smiled and stuck his fist out, like they always did, and said, "Touch my blood!"

Shane punched his fist out to touch Sona's and said, "Blood is thicker than water." They performed the same ritual with Boyks and then got the plates out to have lunch.

By now, they were hungry. They dished up salads that Audrey, Sona's wife, had made for them onto their plates and got the meat from the braai stand. Being professional fishermen, Sona and Boyks also had alarm detectors on their lines to signify a bite.

They put their heads down and were just about to tuck into their food when Sona's alarm siren went off. "I told you, Boozine, wait for the big fish."

"Mine was having breakfast with us, now you have a bigger lunch date," joked Shane.

As he walked up, Sona laughed, "My club has an age restriction. You have to be big before you are allowed in!"

He put his plate down and calmly strolled to his rod. The fish was already hooked on the line.

Before he reached the rod, his tone turned deeper and he turned and said to Shane, "Like in your life, my lightie, just wait for the big fish. It will come."

He walked up to the rod, lifted it off the high rod rest stand and started reeling in. Shane had his shorts on, and with the water now being warmer, he strolled in with the landing net and landed a carp of about five kilograms. It struggled as it went into the keep net.

They had their lunch, and with their fish in the bag, Shane and Sona had broken their ducks, the term used for catching a fish, similar to the term in cricket for scoring a run, and had lost interest in being patient on a very dead day. No more fish.

They both loved to have a jol. Sona's sense of humour and spirit were just like Shane's. They shared jokes and danced to music from the car radio. The afternoon passed.

As they packed up, Boyks came to Shane and said to him, "Well, my lightie, you won the bet. Well done. You caught the biggest fish."

He extended his hand for a gentleman's shake. Shane had tears in his eyes. He took Boyks' hand and held him close.

(Those tears came because Boyks' story is longer, way longer. I will give you a brief rundown for you to understand my tears. Boyks lost his wife, Yogi Bhani, and his 16-year-old daughter, Ursula, in a car accident about three-and-a-half years before the fishing trip. ☹

His whole family were in the van together, going on a family trip to Durban. The children were in the back, laying down on a mattress, under the canopy.

At an intersection just on the outskirts of Newcastle, there was a biltong stall, and Boyks wanted to stop and get some biltong. He waited for a truck going in the opposite direction to pass before he turned right to cross over the road and go to the biltong stall.

But a car crashed into him from the back, sending his van into the opposite lane, where they were hit by the truck. Horrendous accident. He lost his lovely wife, Yogi Bhani, and the sweetest teenage girl ever, Ursula, on the spot. ☹☹☹

Shaylin was not so badly hurt as his little body was thrown around. But Boyks was in the hospital for a long period. Luckily, he survived, to parent his child.)

Shane took a deep breath and whispered into Boyks' ear, "You had a bigger fish, my bhaia. Much bigger. She was the biggest. She took her copy with her, but she left you a special copy of yourself to look after."

With tears welling up, Boyks said, "And you didn't go where they are because your job here is not done yet. Today you showed me that you can affect the way youngsters think. Shaylin told me, when I took him to the toilet, about how you think, how God is one and what you taught him."

Pulling away to look Boyks in the eyes, Shane said, "That is my ultimate goal, Boyks. The young minds. To set them free to think differently and to show them that we are all the same."

"Exactly, our minds are too old, too trained. We cannot change. Some may. I will change my thinking but not my beliefs. I was born like this. I am a Hindu; I will do what my father did."

"Me too, Boyks, but my father did what his father did and his father's father's father and so on. Praag blood I am, but I see the world so differently. I am not objecting to Hinduism or any other religion or even asking you not to believe what you have been taught and change."

Boyks asked, "What must change then, Shane?"

As he asked this, Sona came to them.

"And now? What's happening, what's happening? We packed and ready? Just the keep net is in the water. You got shorts, so go for it, Shane," Sona said. Jokingly, he added, "You think you can carry my big carp?"

"Just now, Sona. Just telling Boyks about my changed view."

"Oh? I follow my father, but it does make sense to me. What you call that thing?" asked Sona.

"Well, I can see myself coming here again for a kind of a sub-cult that I am making to unite us all. It's called Ethe**real**ism. I want to save Her and, therefore, us. That's what needs to change."

Totally interested, Boyks said to Shane, "Tell me about it in the van. Go get the net. The gates are going to close just now."

"True, Boyks, it's a long story. Let me get the net."

As he walked off, Shane smiled at Sona, who was making sure that the rods were tied properly.

"Boozine, you have a funny look on your face. What's up?" asked Sona.

"I will let you know just now. Let me go bring the net. Load the boys and go wait inside for me. I'll put the net here," Shane responded, pointing to a spot in the back of the van.

"Okay. don't be long. I know, in a way, what you're going to do."

Shane entered the water. He reached inside the bag, pulled out his little fish, touched his third eye, his forehead, with it, then looked up and said, "Thank you, IT, for telling Her to send this fish to me. You wanted Shaylin to understand you, and he now does. I am returning it to you because, as my brother says, I would rather aim for bigger fish."

Even in this serious time, Shane joked to himself, *"And you'll come and catch her when she's grown and ready for your 'pan', not so? I know you. You are naughty, naughty."*

As he put the fish in the water, he said to it seriously, "You wanted to ground yourself for me, but like the many other fish that wanted to be grounded by me, I set them free. I ground myself in water. I'm in constant motion. Swim free, my dear."

He smiled and breathed in as he released it into the water. It seemed to pause at first but then darted off. Shane watched it swim away into the water this time.

There was a calm atmosphere around as the sun behind him was setting and the water turned red. He turned around, picked up the net with the big carp in it and headed towards the sunset.

As he placed the net into the van, Sona, who wasn't even watching him, said, "Thank you for my fish. Yours wasn't big enough for the pan."

Shane responded, "Ja, Sona, I want bigger."

"Nice, Shane, that means we are even now, neh? You got no fish to take home?" retorted Boyks.

Sadly, Shane said, "Well, I got no home also, Boyks. I just wander around. I had a few homes that I threw away with the fish as well."

"Don't stress, my lightie," said Sona sympathetically.

"Ja, Shane, when you dock, you dock, but you can never say when a ship will have a storm. Look at me!" Boyks told him.

"Ja, I know, Boyks, but at least you have a ship and a young sailor. I got nothing,"

"That's because there's nothing wrong with searching for the big fish to give you a sailor, Shane," Sona told him.

"And finding and trying a few small boats on the way," added Boyks.

Shane smiled, "Because you will only keep what's big enough for your bag!"

They all laughed as the van headed off.

Boykssaid, "Okay, Shane, tell me more about your Ethe**real**ism thing."

"I will sometime, Booiks, but this is not the time or the place," Shane said.

Boyks smiled and said, "Okay, my lightie, when you ready. Plus, you must be tired. You had a big ou to reel in."

"Had to be, to beat you. I had to make sure you know. Taking you on is the bravest thing I did."

"You got that right, but I accept defeat. You were lucky this time, but the next time, I'll show you," Boyks replied.

Sona scolded, "Quieten down, boys. The lighties are sleeping."

"Oh, sorry, Sona. Your lightie is lucky he's sleeping, otherwise the Tickle Monster would have come to say hello," Shane said.

"Aha, and lucky Shaylin bought him off by working for him," Boyks replied.

"Also, he listened to me, and even though he looked confused, I made sense to him. This story has inspired me and is definitely going in my book."

"Do what's in your heart, my boy. You came back for a reason, and even if we and lots of people will disagree, the essence of what you are saying makes sense," added Boyks.

"All the best, my lightie. Just get us a bigger van coz your'll are stuck in the back there. All the things as well," said Sona.

"Ja, I will, bro. I'm dreaming big."

As he said those words, Shane reminisced about the day and felt more at peace about his ability to change young mindsets to a greater understanding of this existence.

(Sadly, dear grown reader, essentially, we all follow what we are indoctrinated into following and not open our own minds to think. On YouTube, check out 'Get them while they are young' by Seth Andrews.

Think? Open up your mind for the Extra Time?)

BOOK 3
EXTRA TIME

Chapter 1

First Half – Short Stories

*(The match has set us up, yes, to complete the Extra Time. Because of this book being shortened from what it was, I have only chosen important sculpting stories of my thinking in the First Half, then Ethe**real**ism in the Second.)*

Marking the Completion

In the early hours of 11 July 2009, Shane finally completed the lengthy initial draft version of *'Crash Landing, Born Again For?'* It had Ethe**real**ism as its final chapter, and he had made a vow to himself to get a tattoo of the word Ethe**real**ism.

(Here is that initial ending of the lengthy 1ˢᵗ draft of Crash Landing:

My dear reader, you must make the best of your life while you are here, as you never know when you are going to leave Her. But do your best to make Her a better place.

I am hoping I've done my best—257,689 words with just one finger of my now main left hand. I believed I could finish and did. Yay! ☺ *It is now 04h45. I have made it. Just now, I will tattoo 'Ethe**real**ism' on my blessed hand. It will establish a kind of emblem with* *a Scientology 'S' and have the five other dominant religions encircling the bottom.*

That's the plan, as the dominant religions have shaped the majority of us, but I sincerely believe it is now time to reassess and understand our existence. This means that kind of like Scientology, which comes from a more intellectual perspective, we should use knowledge to combine them all into one.

Basically, see all humans as the same, regardless of race and religious differences. This is because the religions we follow were formulated a long time back—when there was no knowledge of Earth's true shape, solar rotation, evolution or even dinosaurs, etc.

*Knowledge has been gained. Think on it? If the spark has been lit in your mind, burn an explosion into the way **you** see life and the world.*

Please take into consideration that you are living your own life, so you get to decide on everything personally. From the book 'The Secret': 'The way you think creates your reality!'

*I was born again to live and, more importantly, to think. Let me part with a well-wish phrase that I have rooted in my belief about us being one. It can be seen as Ethe**real**ism's holy phrase. It is '**Love & Light**', which, at the end of the day, is all we need.*

The sun provides 'light', which to me equals life. From there, we grow 'love'. Anyway, I am done. You take care, and I wish you all the best.

Love & Light ☺

Shane had made a vow to himself that after completing his original 257,689-word book, with Ethe**real**ism as the final chapter, he would go to Dee, a tattoo artist in Newcastle.

He had been there before, planned the design and told Dee to give him a few weeks and he would make an appointment, when his book was completed, to come and have it done. The time had come to mark the completion.

A few months earlier, in May, at a biker rally, he had completed the first part of his tattoo design. It was Mother Earth and the moon inside the sun, surrounded by stars, our essence.

Shane's God and Goddess were there in Newcastle for the weekend. Sheila made him a cup of tea and then gave him R300, hers and Manny's contribution and half the cost of his R600 tattoo. Shane never asked for it at all, but she forced the money into his hand.

(It was more than the unwanted financial gain that brought tears to my eyes when she gave it to me. She held me tightly and said, "Your mission is complete, beta. This half is for you because this world needs your Etherealism, combining all beliefs into one to focus on Mother Earth. So, Dad and I will be blessed for having some part in it.")

Shane left, feeling content and complete. He rushed to get to Dee and made it just in time. In that state, he was all abuzz as Dee stuck his template onto his forearm.

(ETHREALISM finally found its place of belonging in capital letters on my forearm. Just seeing it thrilled me. I put a pillow behind my head and lay back with an extended arm.

Yes, there are those who have noticed. It was a mess-up as the word was spelt wrong. The 'E' after the 'H' was missing. Dee tried to correct it by making the H smaller and putting the E in. The HER in ETHEREALISM is different, putting the focus on HER— Mother Earth.

I never even realised until now that She was in Etherealism. Maybe this was fated guidance towards HER? My mind was at ease. I, personally, like making a plan to fix something in a sense. We can correct anything in our lives.

Before I edited and cut this book, I had a long story about a stray dog that I called 'Boy' in Newcastle. Boy used to visit me every day in my solitude while writing in 2009.

In my compassionate search for inner peace, I used to feed him as much as I could. We developed a great bond.

I see every living thing as part of Her plan for this Earth. The lengthy part of my story with Boy is that even in small-town Newcastle, my appeals from butcheries to donate some off-cuts to help feed him were not granted. They wanted to charge me for scrap bones.

I told you, we live in an extremely capitalist society. Anyway, let me at least give Boy some of the spotlight now. He was designed for it because, honestly, he knew me, and I knew him. His internal pain and suffering, all gone with a wag of his tail. If only life was that simple?

Let me end this story with a picture of him I took while he braced my hand and gazed with his gorgeous eyes at my new ETHEREALISM tattoo, almost intentionally, bracing my arm, as a saviour for him as well? That is my opinion, but you be the judge?)

Running to an Inner Journey

(I was in Newcastle in August 2009. My school friend Rashika is an accomplished runner. She now lives in JHB but was down for a run. Knowing I love to run casually, she invited me to join her for a cross country run in the mountains, organised by professional runners.)

Early on Saturday morning, as Shane reached Rashika's house, she was already waiting outside for him. They had not seen each other for a good few years. Her mum was also outside to bid them well. Shane got out of the car and paid his respects. Rashika's mum told him that he must join them for lunch afterwards.

Shane accepted as he and Rashika headed off. She was all excited and ready for the challenge. Their chatter started—mainly from Shane, who had so much to tell as his life had changed in such a big way ever since he started his book.

Rashika was now a school teacher in Johannesburg. She was married but had divorced her cheating husband a few years back. She had no children, and her focus was now on her love for running. She had a rough idea as to how to get to the area.

They did not know it was a gravel road for at least 30 kilometres though. But even on the gravel road, Shane's spirit never died. He gave two hitchhikers, Sipho and Thabo, young gentlemen, a lift; they readily joined in his conversation with Rashika.

It was mainly about a new understanding of the world, and Shane's deeper perspective regarding uniting the different races and religions. He was totally understood again, so his hope and faith in his dream were reinforced.

After a bumpy road for about an hour, they finally reached their destination. It was in the mountains, a cross country race. The beauty of Her all around him, with Her fresh morning breath, had Shane on a high already.

The guys got out and thanked him for the ride, telling him that he was a real nice guy. Shane smiled and wished them well. Then Rashika took him to register for the race. He bottled up his little grievance of having to pay an entry fee of R65 just to run. The logic behind it escaped the principle. It is, however, a financial world.

The three events each had a different distance to cover. Rashika was doing the 24-kilometre event. Shane was brave and decided to try the 12-kilometre race instead of the five-kilometre one that he usually covers on his social jogs.

He set himself a goal. Rashika started at exactly 08h00, and his race was due to start half an hour afterwards. He walked around, feeling a bit out of place. They were all professional runners there. They were fully kitted in their running gear.

Being a bit chilly Shane just wore his tracksuit pants with a long-sleeved T-shirt. Even though he looked shabby compared to the other professional runners, his spirit was still high. He was wearing the race number he had paid for, so there was no need to feel out of place.

Excitedly, he did his stretches and warm-ups, trying to look professional, not knowing that the most moving moment in his life was about to begin. The starter gave them instructions on the route.

Thinking that it was just on a gravel road, Shane assumed that it would not be too challenging. Almost foolishly, he thought that it was just going to be an outdoor run in the freshest air he had ever inhaled. He could not do it in silence though, and as he would normally do when running, he put his MP3 player on.

The starting gun fired. Shane was in the middle of the pack of 167 entrants. It was a thrill to run on the gravel road with people around him. His love for running actually came from running alone around the complex in Johannesburg, where he had stayed with Resh the previous year.

Now there were people with him. He stayed with the bunch as they ascended towards the mountain. It was a bit of an uphill climb. After the first two kilometres, the gravel road changed to a jungle pathway.

Shane was awestruck. It now became a hiking trail up a steep mountainside. He had not expected this. It was true beauty for him to love, being out in the open, with Her. Then he headed into the forest.

It was turning out to be just like in his favourite movies with jungle scenes, following a pathway and jumping over fallen trees. It was a real adventure as the route climbed. There were even streams to be crossed on rocky pathways. It was beautiful. Even though Shane slipped on a few rocks as his shoes were now wet, he pushed on.

There were refreshments at the four-kilometre mark. Shane stopped to have some water and looked down the mountain to see the starting point off in the distance. He was in his heaven. The air was so pure, and realising that his chest was heavy, from the very casual smoking he never thought would affect him, he decided to quit for good.

As he was leaving, he noticed a few paper cups on the ground. How could they destroy Her by littering? She was so beautiful. He actually picked them up and ran back to the table to throw them away in the refuse box.

The attendants were amazed. He just smiled and proceeded. The route got a bit trickier. It was going through the mountain on a pass called Donkey's Pass. He was now mainly in solitude as the packs had split up.

It was thrilling. There were uphill climbs, even slightly dipping valleys. At one stage, Shane bumped into five cows on the path. It was really touching to be sharing the path with animals.

Even a warthog ran across his path. On the level ground, Shane jogged and actually ventured off the path to climb onto rocks to get a better viewpoint of the scenery as he caught his breath. He could not believe Her beauty, which he now looked at differently.

On the run, Shane had stumbled a few times, touching the ground, but he got up each time and dusted himself off to proceed. He did have the company of some fellow runners, who he spoke to but passed. The jungle path was a single lane, so there were quite a few stop-and-walk moments.

Shane's blabbermouth took over as he chatted away. He removed one headphone and chatted to a few runners. When he actually told some of them a bit of his story, that it was his first time taking part in such a race and that he had a metal pin in his leg after his accident, he was praised.

His drive increased. At the next stop, he asked how far he had gone and how much further he had to go. He was more than shocked when he was told that he had already covered eight kilometres and the last four were basically down a gravel road.

As Shane ran on, a slight drizzle started. He stuck his arms out. Yes, it was a bit chilly, but the calm, cool serenity of the raindrops on his face had Shane more than ecstatic. The road made a few steep descents on gravel.

It was tricky, but Shane still pushed it. His late stamina that he always had luckily kicked in, and he ran on. There was a runner a few metres ahead of him, and he stuck with him. The road branched, but he followed the man.

They approached the finishing line, but their trail was going around it. They somehow passed the entry point from the outside. On entering from the back, they were redirected. Shane just smiled because he had completed maybe a few hundred metres more than the 12 kilometres.

He reached the end and clocked in. It had taken him one hour and 35 minutes. He was placed 51st out of 167. He was ecstatic.

The weather changed for the worst, with the drizzle becoming a downpour. Everyone headed indoors.

The prize-giving for the 12-kilometre race started, but the woman's voice was too soft to compete with the rain on the tin roof. Shane volunteered his loud voice and went to announce the winners. Amazingly, the fastest time was less than an hour.

Being the Master of Ceremonies reminded Shane of his talent for public speaking. It was great for him. Afterwards, the woman in charge saw the scar on his throat and his disabled hand.

She asked him what had happened. After his quick explanation of his situation, she actually handed him another medal for his achievement of overcoming his disability by running with a metal pin in his leg and invited Shane to join them in the future.

It was a great experience with a great group of people. Running clubs were just like his biking clubs. Human companionship is the basic essence of our existence.

Time was passed by hanging around and making new friends in a very social environment as he waited for Rashika to arrive. About half an hour later, she did. She finished the 24-kilometre run amongst the top 20 women.

Shane took his hat off to her because it was still pouring rain. Rashika actually needed to warm up at the fire to dry off. Shane, even on the way home, could not thank her enough for opening his eyes to Her.

Rashika has the same way of thinking as Shane in terms of valuing Her first. She teaches biology. She told him that his book is urgent. She agreed with everything and said that it is more needed regarding indoctrinated religious belief patterns.

This discussion was continued when they had lunch with her parents, the most simple and loving people, just like Shane's Lords. They also saw his points regarding creating religion. In fact, a few years back, even though 'trained' South African Hindus, they themselves stopped some of the ritual practices that did not make sense to them.

Her father said to Shane that it is about time these opinions are voiced. He was doing the right thing by asking, addressing and having the guts to voice his opinion in his book. He added that it is urgently needed, especially in today's times.

(I felt like heaven after I left. Here were total strangers in a sense, elderly, trained in the dark, but they saw light to my new approach. My confidence in Etherealism has grown so much more in one day, and my love for Her has grown.

Three days afterwards, as I started writing, I realised that it was such a beautiful morning. Rashika is still in Newcastle as well and wants to do something. Pay her back, Shane, and take her for a ride on Yummy?

Yes, sir, coz Rashika wants to experience a bike ride. I will take her to the waterfall and go on a hike. I now value Her more than I did before, all because of Rashika taking me on the run.

Pay her back, have a picnic and come back to me. *You are so right, Shane, life is to be lived.* Not too much though, don't lose focus on me. *How can you even think it?* Just that I know you from before, you do so much and always give your all but then take it away. *I'll never lose focus on you, my salvation, my book.* Great, now go and tell me and our lovely readers when you get back.

What a trip! The ride there, cruising in the fresh air—heavenly. Reaching the waterfall, it was locked though. Here is an example of Her in one of Her most beautiful places, and it is not accessible. People's visions have changed.

It is a fully equipped park that once had a heyday, especially in the apartheid days for the Whites, but now hardly anyone even knows of its existence. Our focus has dramatically shifted, not so?

I hid Yummy behind some rocks, and Rashika and I climbed over the fence. The park was empty. There are lovely picnic areas, but they are abandoned. We hid our helmets, jackets and picnic backpack in a big hole, covered it with dried grass and made a smaller bag to carry on the hike.

With a lovely breath of fresh air, we started the hike down. I was always in love with Her, but now it is a whole lot more. Rashika and I hiked down the mountain and away from the falls. Along the river, I never saw Her like this before.

Memories of childhood summers spent there filled me. We actually used to dive into the deep pool from cliffs. Being here again, I was tempted to at least be in the water. It was cold, but I felt like being naturally me for at least a minute, 'au natural'. I decided to skinny dip.

Surprisingly, Rashika felt the same way. I waded in, dived down and came back up freezing as she was wading in. It was too cold for her to wet her long hair though, so she went in waist-high. I had carried a towel in my backpack, so we dried off and lay naked on the rocks.

I felt so free and alive, surrounded by and feeling Her. My most blissful place to be in. Yet, I do know of Her current dilemma, caused by us, and sorrow filled me inside. I know, sincerely, that I am merely a speck of dust, on a grain of sand, if we equate Her to a desert, but I cry for our...

Mother Earth

The most essential part,
of our existence,
a complex work of art,
who bears our persistence.

She takes whatever we throw,
hurting her immensely,
but most of us do not know,
that our simple actions are costly.

We are now paying the price,
because She's wearing thin,
by gambling with the dice,
we've committed a sin.

Is it too late for change,
sending our focus to Her,
breaking our indoctrinated range,
*and focusing on our **real** Mother?*

(I felt like I was in her womb. Only me, naked on her. What a pity I had to return to our indoctrinated normality though. I always wondered why there was a thrill in a nudist colony. Now I know. You feel so connected, naked out in the open. I must attend one sometime when I'm done and can find one. To just be me, totally natural, was such a thrill for me. ☺

The thrill had to end, so after drying off, clothes went back on for the hike back up. On the way back up to the campsite, I found my stone to keep and one to present to Rashika as a thank you. We got up, took our bags out of the hole and laid out the picnic blanket she had brought.

We sat down, and I gave her the rock to remember the trip by. She accepted it and had something of her own planned. I was touched like never before. Here's what she did. She gave me a Chappies bubble gum and the first handwritten note of five.

1. *Learn to appreciate the simple things in life! Each experience is meant to be an opportunity for you to appreciate the little things that life has to offer. Live the moment and you will be amazed at how each experience enriches your life. Like this Chappie. Enjoy and savour it. Simple.*

2. *People enter your life for a reason. Each person that you meet is meant to teach you something that will improve your life. Give people a chance to share their wisdom with you. You, in turn, will give them all your wisdom. But remember to always let them talk and you listen. You will be amazed at the results.*

3. *Ego, pride and self-praise are obstacles in your path to make you grow into what the Creator intended you to be. When you put others first, you put yourself into their situation and forget about you. All energy should be given to them, and talking about you is fruitless as the Creator knows already and that's all that matters. Be simple, be yourself naturally.*

4. *Love is moments in your life that you feel good about yourself. Love unconditionally and love without selfish want and need. Love to give, but not to hurt. Love is true, honest, respectful, moral and consistent. Don't stop loving, no matter what your circumstances may be.*

5. *(When opening the last note, there were little mustard seeds folded inside it.)*

Each little mustard seed grows into a huge tree that produces tonnes of seeds. Nurture what nature has provided. Shane, take care of yourself, and although you are a little seed, and your spawning may be difficult, it is your purpose. You need to be the seed that grows and spreads your knowledge to other seeds. You are the seed that needs to grow into the tree of giving. You will be that, I know it.

.........☺ ☺ ☺.........

With gentle, happy tears, I took her in my arms and held my spiritual soul sister close for reinforcing to me that I had another seed to plant, to reveal some truth to the world to open their eyes? Educate your mind yourself? That is the only way you will get the...)

Truth Revealed

On the 11ᵗʰ of September 2009, I watched 'Zeitgeist'. The remastered version first, then the Addendum. The coincidence with the date? Unreal.

*I am also jittery because the principles behind it match my views totally. Like Ethe****real****ism, it wants to change and help the world. It shows us that we are all the same, following mythology and what we are taught and told.*

It is too hard to describe, and I'm sure they would not mind if I give you some information about the movement from their web page. Their goals are so much deeper though. Please get access to it. Surf on the net for 'Zeitgeist Movement' and watch the digitally remastered version.

Get access to it. Your life and our lives will change, the needed change for our survival. We are all fooling ourselves into believing that our system works. It does not, full stop. Zeitgeist has brought up ideas and thoughts that I've always had. Now is the time. Here's a brief snippet.

"The Zeitgeist Movement is not a political movement. It does not recognise nations, governments, races, religions, creeds or class. Our understandings conclude that these are false, outdated distinctions which are far from positive factors for true collective human growth and potential. Their basis is in power division and stratification, not unity and equality, which is our goal.

While it is important to understand that everything in life is a natural progression, we must also acknowledge the reality that the human species has the ability to drastically slow and paralyse progress, through social structures which are out of date, dogmatic, and hence out of line with nature itself. The world you see today, full of war, corruption, elitism, pollution, poverty, epidemic disease, human rights abuses, inequality and crime, is the result of this paralysis.

This movement is about awareness, in advocacy of a fluid evolutionary progress, which is personal, social, technological and spiritual. It recognises that the human species is on a natural path for unification, derived from a communal acknowledgement of fundamental and near empirical understandings of how nature works, and how we as humans fit into/are a part of this universal unfolding we call life.

While this path does exist, it is unfortunately hindered and not recognised by the great majority of humans, who continue to perpetuate outdated and hence degenerative modes of conduct and association. It is this intellectual irrelevancy which the Zeitgeist Movement hopes to overcome through education and social action.

The goal is to revise our world society in accord with present-day knowledge on all levels, not only creating awareness of social and technological possibilities many have been conditioned to think impossible or against 'human nature', but also to provide a means to overcome those elements in society which perpetuate these outdated systems."

There is a lot more on the Zeitgeist site. Especially opinions on the 9/11 disaster! I just wanted to give you a glimpse. Let me just keep my faith that this book is going to be out there to educate the masses. I personally do not want much from it financially.

In fact, after watching Zeitgeist, I blame money, capital and finance, just as they do, for our destruction.

Please get hold of it, watch and maybe change? I'll leave with a quote by the late Mohandas Gandhi from their home page:

"We must become the change we want to see in the world".

So true for all of us, Mahatma. We all need to reach a point in our minds when we can say that our...

Life is Healed!

With a complete book, in the beginning of 2010, Shane was relentlessly pursuing the media. It was a really mammoth task as he had no direct leads. The 'Soul Magazine' did a short article, followed by the 'Eastern Tribune'.

These stories proved to Shane that he was newsworthy, and even though bigger publications were hard to get at, he never lost faith.

This self-belief was highlighted when a lady, bearing the same name as his mum, contacted Shane after she saw a brief article about him having a 'novel life' and his dream to find a publisher in the 'Post Newspaper'.

It was the first weekend in May 2010. Shane was in JHB and met Aunty Sheila at a shopping mall on a Sunday, after his morning biker breakfast run.

He was still in his full biker attire. Sheila was more than awestruck by his story and referred him to the woman who had changed her life, Ashika Singh. Ashika is the Louise Hay 'Heal your Life' workshop coordinator in Johannesburg.

(I never knew who Louise was but was sincerely awestruck when I researched. As a brief outline, the following information was taken from Louise Hay's website.)

"Louise was able to put her philosophies into practice when she was diagnosed with cancer. She considered the alternatives to surgery and drugs, and instead developed an intensive programme of affirmations, visualisation, nutritional cleansing and psychotherapy. Within six months, she was completely healed of cancer."

(The principle and premise behind this is that the body can heal itself, but it just needs to be trained to. External medication that we rely on does help, yes, but is not really needed to the extent that we use it. The mind is more powerful than we can even imagine.)

After his research, Shane called Ashika, who was very busy with her bookings but set a meeting for him in a few weeks' time. Ashika told him that she was intrigued by the little of his story that he told her, so she kept a whole Thursday afternoon, from 2 pm onwards, free, just for him.

On that day, Shane was happily on his way to his destiny, but then the car's accelerator cable snapped. He was stuck alone on the road, and even with a few people around him, they actually wanted to charge him to help push his car off the road!

Luckily for Shane, one friendly pedestrian helped him push it to a garage that was close by. There, the mechanic said that they would be fine to do the labour, but they did not have a cable. Shane was told that it could be done the following day when they were free to go and get a cable.

Knowing that he had to go and meet Ashika on this particular day, because the chances of her being free for him again were very slim, Shane took a taxi to a motor car spares shop to go and see.

Unfortunately, the shop he went to did not have a cable either. Luckily, there was another shop which specialised in Volkswagen parts up the road. Shane decided to take a jog there.

It was a paved road, and while jogging along, he tripped over a brick that was sticking out from the rest. This was because Shane's right leg does not really lift that far off the ground. He fell, banging his left knee on the hard pavement, as he tried to catch himself.

The pain was agonising. His knee swelled up immediately, but he got up, dusted himself off and walked fast to the shop. They had the cable.

Shane got the cable and took a taxi back to the garage. Luckily, they had already removed the old one. It took about 20 minutes to fit the new one.

It was now almost 17h00. Shane had called Ashika when he had broken down and was told to call her when and if he could still make it. Expecting her working day to be almost over, he phoned her nonetheless.

She told him that he was destined to see her that day and that three hours would not change destiny, so he must still come over. Shane's heart smiled as he drove to her home. He could already sense her divinity.

Getting there, he limped into the yard. Ashika was standing at the door and laughed when he explained his afternoon. He joked that even though his car tried to break down, as well as his leg, he had still made it. She smiled and said that he must be quiet, gave him a warm hug and took him in.

(Honestly, you never would have said that this was our first meeting.)

Shane limped in and followed her down to the lounge. Ashika told him to sit and show her his leg. Shane lifted the leg of his jeans up and sat down, sticking his injured leg out straight. Kneeling down, Ashika looked at it and told Shane to close his eyes. She then performed Reiki, a healing massage from the East that works through the transmission of energy.

(It is hard to describe, but honestly, my pain went away. Her power is immense.)

Shane was amazed at the relief in his leg in just five minutes. He could not believe it. They then sat and spoke for a long time. Shane spoke about himself and his vision. Ashika was more amazed by his different hands. He told her that he thought differently as well and showed her his tattoo.

Ashika told him that she sees the world in the same way as he does and that we are all truly the same with created differences. She advised him that he needed to combine his dreams and thoughts into one.

She then took out her Angel Cards to do a reading for him and passed him the box to remove the cards himself.

(When I took the cards out, there was R200 in the box. After I told her, she said that I was destined to find the money, and because I needed it more, especially after being more broke due to the car breakdown, it was mine to use. She told me that it is blessed money and even though I see money as the basis of humanity's destruction, I should take it and use it in a beneficial way.

I felt so warm inside as I took it, in a way that even I, Mr. Verbosity, can't describe. ☺)

The basis of Shane's reading was again deep, but it was also that he needs to find inner peace in his uniqueness and keep his faith and dreams alive.

(I suppose that is why we have grown as a species? We can dream, grow and prosper. It all starts with a dream. Guess what mine is? Yes, yes, to make my heart flutter again but, more importantly, also to get this book that I slave away on out to the world and get people thinking, in a small way, helping Her.)

After the reading, Shane thanked Ashika profusely, as deep inside he knew that his dream would come true. Then her daughter Tash, 11, came in with her brother Ashvir, 20. Because Ashika is originally from Pietermaritzburg, Shane actually taught at the school when Ashvir was there.

Even though Shane did not actually teach him, Ashvir remembered him. Tash had just met Shane, but she brought him a cup of tea and biscuits, and they chatted like the best of friends from a long while back.

(Aha, you guessed right, Tash is an Indigo Child as well.)

Ashika patiently sat and watched them. Tash is also extremely interested in different thought patterns and spirituality. This comes from her mum being a 'Unitarian Universalistic', on the exact same lines of Etherealism, uniting everything.

Shane glanced at the clock. It was now close to 19h30. He had been there for more than two hours and decided that he could not overstay his welcome. As he parted, Ashika told him that she would have a 'Heal your Life' workshop, like the one Sheila had been to, in two weeks from then and offered him a spot because there was one place free.

She told him that he could join and confirm on her web page: www.ashikasingh.com. As soon as he was home, Shane went online to the site. The workshop was for the entire weekend, Saturday and Sunday.

Expecting it to be an expensive, out of his price bracket workshop, Shane was shocked that even though Ashika did have a set fee of R1,800, it was stated that paying this amount was purely optional. In order to confirm though, one just needed to put a minimum of R150 into her bank account as proof of attendance.

That same evening, when Mr. Praag called him to see if he was okay, Shane told him all about his experience. Sensing his son's excitement, Mr. Praag said that he would do the deposit to confirm for Shane and asked for Ashika's banking details.

With his place confirmed, a few days passed as Shane figured out what to do with his blessed R200 from his soul sister. He knew that it would have to go to needy children, as nothing compares to a child's smile and thank you.

An idea came to Shane to go to the supermarket down the road, buy a few loaves of bread and some polony (meatloaf) and take it to the first children's home he could find.

At the till, Shane asked a lovely lady who worked there where he could go. She suggested he go to the Thembalethu Community Creche in the Alexandra Township, which was not far away.

Shane decided to also buy a few tins of beans and some maize meal. With 10 loaves of bread, sliced polony, three tins of baked beans and a sack of maize meal coming to R190, Shane kept the R10 safely as he knew what to do with it.

The Alexandra Township, under the apartheid regime, was the area for the Black workers, neighbouring Sandton, the commercial hub house of Johannesburg. This was Shane's first time actually going into the 'Bronx'. Aside from the sadness he felt to see his fellow human beings' living standards, he was not afraid at all, even though he'd been warned several times to not go there.

He reached the creche and was directed to go to the back first to meet with the lady in charge, Ma Ethel. As he carried the food up the driveway, Shane could see lots of children staring out of the window at him, wondering who this stranger was.

Ma Ethel opened the door and was very grateful for his small offering. They really needed it. Shane left the things in the kitchen and was taken to the children as he had requested.

The majority were toddlers, between two and five years old, at least about 40 of them. Shane smiled at the children, who really warmed to him.

(I'm lucky, I guess? Honestly, most children warm to me. As mentioned, it may be because I'm an Indigo Child.)

He played with them all and taught them his special cool hand sign, where you make your thumb and finger like a seven and say, "Olla seh-vehn."

(I love being with children. Even though I was only there for half an hour, they were sad when I had to leave.)

Driving back to Resh in total bliss, Shane was now even more excited to have the life healing workshop with Ashika and see her again, his main reasoning for feeling the way he did. There was just a week to go.

The day came and Shane rode Yummy there early on a cold Saturday morning. He would be one of a group of eight, but he was there first. Ashika welcomed him in. She gave him a corner in the lounge where he could leave his biker gear and bag. Because the instructions on Ashika's site asked for comfortable clothes, Shane had a nice warm tracksuit on under his biker gear. He changed, and Tash, who was helping her mum to set up, brought him a cup of hot tea.

Shane's heart smiled as he thanked her for it and sat down to fill out his form. The others started coming in, and everyone introduced themselves and filled out their forms. The group of eight consisted of three men and five women. Shane already felt comfortable because they all were sociable and were bonding.

(As humans, we all do that, I suppose, needing a mutual sense of camaraderie. Here, I was with a group of people who needed their lives healed, just like me. We never knew each other at all but were so close.

You know, honestly, I now suddenly feel as if my full disclosure will give the details of the workshop away and kind of intrude? This not only on their privacy but will also compromise the underlying principles and teachings of Louise Hay's work.

I sincerely know she would not mind, but I cannot intrude. I will, therefore, not give too much away about the tools used to sculpt me into the happy person I now am, regardless of all the pain of feeling lost in this world.

Let me just say that it was intensive beyond my wildest imagination. It was too deep to cover, even if I wanted to, and what's more, there was not a time limit. Ashika believes that time limits restrict true destiny. How true?

On that Saturday, even though we were not really totally tired, we forced ourselves to stop at about 02h00! We had spent at least 16 hours with our intense healing of each other and were still not yet done with the first day's programme.

As I told you, it was intensive. One thing to mention before I go on to the next day, and my results from it, is that I misplaced my Yummy's keys. When it was time for me to leave with everyone, I could not find them.

While I was searching, feeling stressed, my soul sister came and said to me that we need a greater understanding of why we experience the things that we do. We need to read a deeper meaning into events, as everything happens for a reason.

She said that the forces are telling me that they do not want me to ride this late at night with a drained mind and that with a fresh mind in the morning, I would find my keys.

Luckily, I always carry a small overnight bag with me wherever I go; Ashika said I could sleep on the sofa. I truly felt her love. I mean, here was a single mother with two children allowing a stranger to spend the night in the lounge. I knew then that she saw me as special.

I laid my weary head down and had a blissful four hours of deep sleep and awoke automatically at 06h30. I peeped out of the window, like I always do, to praise Her for a beautiful morning, and Ashika was already outside, meditating within Her.

With her eyes closed within Mother Nature, I'll never forget the look of inner peace on Ashika's face, surrounded by a unique aura. To me, personally, it was glowing like a sunrise. Here was a woman who came from troubling times but had found inner peace.

I smiled as I awoke, stretched my body and went to shower. I came back into the room and Ashika was there with a smile and radiance I cannot even describe. You take over, Shane, we need some direct speech and normal font now.)

"Good morning, my brother."

Shane smiled and said, "Good morning, my soul sister. You are truly divine, from another realm. I saw you outside earlier. You are showing me that there is so much more to this world."

"Yes, Shane, we just need inner peace, by wishing ourselves a good morning and knowing that it is our lives and we are in charge."

"I know that, Ash, and just by staring at you in your blissful state earlier, I do know that it can be reached."

"That's right, Shane, you will find yours. You are in control of your life. Remember yesterday I taught you about internal meditation. You have mind control to send your thought to whatever you are thinking about."

"You are right. I have physical as well as mental control over myself, and this is my life."

As she pressed play on her CD player, Ashika stated, "Great, now let's clear up and get this place ready. You can look around for your keys as well, and afterwards, you can have some tea and toast for breakfast."

With blissful music in the background, Shane made his bed and cleared up. Ashika was almost dancing as she picked up the cushions and tidied up. Her ability to see the best in life and in others during anything touched Shane immensely.

While having his breakfast, Shane gave his keys some thought. His mind directed him to check-in his biker boots, and there they were. He had not looked there in his search. They must have fallen inside somehow. Were his angels guiding him to spend the night there?

By 08h30, the others had arrived, and the session continued. In Shane's cleansing process, he was given more hope and faith in his story and getting it out there.

(I was excited again, and the healing really helped me. On Sunday, we finished at about 17h00. The most important thing we all learnt is self-love, the simple mirror trick. Readers, we can all so easily do it. Every time you see yourself in a mirror, compliment yourself.

Say, "Hey, good looking. I love you."

This is probably the simplest thing my soul sister gave me—self-love. We all strive for that, don't we, and believe me, even with the world as corrupt and painful as it is, we can all meditate in order to find inner peace, BECAUSE we CAN!

Inner meditation is so easy. We just need to realise that we are in charge of our own lives, the physical as well as the mental. The mind's power is immense. Utilise it. Affirm to yourself that you can control it and shape the outcome of whatever occurs by your own judgement.)

Shane's group now had a greater understanding of this life as they all helped each other and were the best of friends, just for those two days. Even though knowing that their paths would probably not cross again, they had helped and assisted each other and got the same in return. Perhaps this is the principle behind human lives.

Everyone parted, and Shane was the last. He wanted time alone with his soul sister. As a parting gift, even though Ashika never asked for the payment, he took her angel card box and put in the remaining R10 from the initial R200, plus R10 of his own as a symbolic contribution for someone else to find.

Then, as the R10 was symbolic, he presented his soul sister with just a R10 note as a thank you for opening his eyes to his healed life. Ashika understood the symbolism, so she took it, even though she did not really want it.

Shane's inner bliss and peace bubbled up inside him as he rode back to his real sister's place.

(Thank you, Ashika, my soul sister, for introducing me to Louise Hay and changing my whole outlook on life. Inner bliss is what we all seek, and I learnt the easiest way to find it—Self-Love. ☺)

Okay, in my longer version, I had a story entitled 'A Child's Smile', what I value the most in life—to make others smile, especially children. ☺

I returned to the Thembalethu Community Creche with some biker friends, and I had asked for the kids to put on like a show—dancing and singing—after which I rewarded them with toys, kind of showing them that to receive things, it's better as a reward that you kind of 'worked for'.

Quick Free Kicks to Fill You In

(I have just realised that my chopping and changing of this new book's soccer match, just to get to the deeper essence of it, has left out a few chronological details that you need because there were two older, longer versions of this book also. Sorry. Let me shoot a few detailed, relevant spot kicks to fill you in quickly before the Second Half of Extra Time.)

- In the latter part of 2009, I completed my book on my laptop, close to 1000 pages. I had no idea about publishers or anything in the industry, but I had a dream. I sent a few emails around to various publishers but no luck. My zeal took over, and I decided to privately publish. But I would need an editor. So, I Googled and found one in Cape Town. My loving father gave me the required R10 000 that she charged to edit. I emailed my full book. After the 3 months that she had asked for, nothing was returned. I emailed her. No response. Tried her phone number – line cut. I fell for a scam. But my dream never died.

- At the XXX Memorial Biker Rally in September 2010, in Greytown, Kwa-Zulu Natal, I made friends with a remarkable young biker from Durban, Sherwynn. We truly clicked because he was an Indigo Child as well. Strangely, at just 20 years old, at 10 minutes to 10h00 on the 10th of October, the 10th month in 2010, Sherwynn passed on. He was doing what he loved, riding on a Sunday morning breakfast run, but had an accident. Sherwynn was really a true spirited biker who I met just once, but it was like we were friends forever. I couldn't make it to his funeral, but I can actually imagine, like Moonshine, the smile on his face.

As I've mentioned countless times before, our time here on Earth is 'fore written'. Enjoy your life to the most as it will pass, but you rather have a smile on your face and see everything you do as fated to happen?

Me meeting Sherwynn was fated for this book also because he referred me to his sister, Tash, who runs 'Widowed S.A' for single women. It is a support group for widowed ladies like her. Even though in a steady relationship already, she could maybe help me find a soulmate.

Sadly for me, it was not a dating site, just a support group, but I maintained contact with Tash via email. In the first week of December, in her usual everyday motivational emails, Tash mentioned that her book would be privately published and released the following year and that she would like some support.

I enquired as to who published it for her and she referred me to Reach Publishers in Durban, who I contacted. A deal was struck, and I then thought that it was logical to shorten my lengthy biography, which was actually complete, as it was a bit too long because the first draft had Ethe**real**ism in it as well.

Reach Publishers gives authors full copyright of their work. All they do is charge to make the digital draft of it. Authors then themselves handle printing costs, marketing and selling.

From around 1000 pages of the original version, it was halved – less paper, cheaper, PLUS fewer trees cut. Ethe**real**ism was removed, and Mr. Praag and Resh joined together and readily covered the cost of making 500 copies of 'Crash Landing's' first publication released in April 2012.

- Newcastle held its annual Winter Festival, where I was given a complimentary stall in the section for disabled people to sell my book. Sadly though, at that stage, the two main newspapers didn't want to cover me launching and selling my book. The main one was still run by racist Afrikaners, and worse still, another smaller one, with a Muslim editor lady. I even wrote and submitted articles for them as a third person so that they could even add their names to it, but no. My story didn't interest them. The Muslim lady was worse because she angrily said to me that anybody can type and self-publish a 500-page book.

My disability and story were not news for them. Again, no vanity, but I do know that I have something, as the people who were given my draft to reference on the back cover state.

Yet, I am where there is the worst form of 'racism' in a sense, with people of your own race? Remember me losing my flutter that I was not even allowed to get?

Straight talk again: Indians are different. Biggest difference is between the two set religions, Muslim and Hindu, yes, but to kind of make it worse in a sense, Hinduism in this country has three distinct different language groups—Hindi, Tamil and Guajarati.

Even though we're under the same religion, there is still this huge segregation in a sense. Expecting a fellow Hindu Indian from another language group to stand by you, even marriage in some cases, is tough. Why, people, why?

Amongst the Africans of this country also. Segregation between the many different languages and tribes—Zulu, Xhosa, Sotho, etc. I also have to specify even the Caucasians (Whites). There is mainly the Afrikaner of Dutch descent and the English.

Hectic, hey? Let me term this discrepancy amongst Homo sapiens who match biologically but differ culturally 'Culturalism'. Kind of like from a different kingdom in the Dark Ages?

Turn on the light above your head and think! Do you know that chimpanzees, our closest genetic match, are extremely territorial and even fight for trees with their own species from a different group? In fact, all the monkey and ape species battle amongst themselves. Are we just being true to our hominid roots then and behaving like the clichéd 'monkeys'?

Come on, people! Have we not moved past our own past of conquering and destroying neighbouring lands? Ravaging our own neighbours goes beyond racism, as you share the same race but not the same tribe or clan – different blood? Different kingdom, not DIRECTLY from your own bloodline, so take over it?

What makes it worse in a sense is that regarding blood, a point I love to make is that if you need a blood transfusion, the essence of life, it does not specify culture or tribal group. More importantly, not even racial separation! It just matches the type. That means we have the same fluid inside, but on the outside, racially and, more painfully, culturally, we differ???

Maybe we need to get like permanent x-ray contact lenses fitted to us? That way, all we will see is the same colour skeletons on everyone and realise that we are basically, all the same? Hmmmnnn?

Okay, scientific autopsy people, I agree. Racial skeletons do slightly differ, yes. There is a minor cranium thickness difference. Also, sexually, the cervix bone amongst females differs to males, but we're using x-ray contact lenses, not microscopes. LOL ☺

Yes, I do like to smile and laugh at things, but think? Sexually? Hmmmnnn? Any human race male and female can copulate and make a child. Mixed BREED is mixed RACE, but NOT mixed SPECIES!!!

WTF? What more do we need to see that we are all the same even when we F###! I have put my 'thing' into many other racial women's 'thing'. Worked. Don't believe me? Try IF you can. Make love, not war. ☺

Jokes aside. I racially battled for publicity, in my hometown and my country. I sent various copies to the national press with a brief synopsis. Even the radio stations. Almost seems like the media are still controlled? Either that, OR they are too busy?

*At least I had a television interview in August 2012 on the Magazine Show – 'SABC 3 Talk'. I did like a 15-minute interview, getting to the essence of my vision of Ethe**real**ism – uniting us, but the pre-recorded interview was shortened. They cut off me showing my hands as well as my nana's photo.*

At least I explained my main point about us all being the same and how the collective faith from everyone who knew me, family, friends, the school etc., regardless of different race or religion, prayed for me.

If you want to view it, look for 'shane manilal' on YouTube: https://www.youtube.com/watch?v=FWkHjvePHhc&t=5s

Why can't we just see people as people? Our spherical planet is not the same as the flat one (sic) of our ancestors, who created an understanding of it and indoctrinated us with it.

Yes, you can proudly retain your inherent culture and religion, but you have to think bigger. We need to help each other out regardless of race or religion.

I felt this the most after the Mayor of Newcastle at that time, Mr. Afzul Rehman, a staunch Muslim man, who I met at the Newcastle Winter Festival, had a long chat with me about our flailing system that portrays differences. He was very taken aback by me and suggested I become a speaker at schools.

Direct School Reference Notes

"At the onset, please excuse my informal stance, but all I can say is 'WOW'! Shane Manilal has come to my school, St. Oswalds Secondary, to merely just explore his newfound passion, to be a motivational speaker and Life Coach. He has such an amazing story, had my learners gripped to focus." – **Mr. Dhewlall, Vice Principal**

"I am a Life Orientation educator at Lincoln Heights Secondary School in Newcastle. Shane Manilal, who I actually taught, has just taken my breath away again. I sensed that he was special from the onset, but not to this extent" – **Mrs. T. Pillay**

"What a man? Unbelievable spirit. What a story! And my learners were more than just spellbound. He freely came, upon request, and had them gripped and motivated them so much to change for the better." – **Mr. I Moosa Principal, Dannhauser Secondary**

What strengthened my self-belief even more was the reference letter from the mayor...

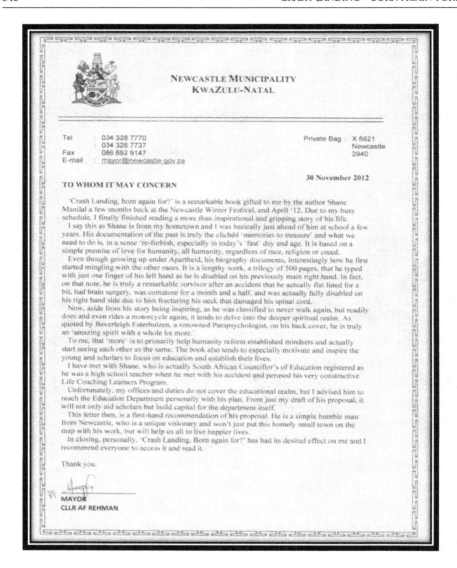

But even a letter from my hometown mayor never really got me any attention, even from the Provincial Department of Education I tried.

I suppose, again, it's a 'who you know world' to do business, and I knew no one. Plus, I am not a businessman, so me actually attempting to sell my book was challenging.

I gave quite a few away on credit, where people signed that they will pay, but they didn't. Come to think of it, around 100 people owe me money. But I believe in the concept of karma; everything we do is being seen in a sense, and we will pay in some way or the other. So, I just trust everyone, smile and carry on. I don't like to stress.

Why is it that if out of 10 things that happen, nine are good and one is bad, our minds focus on that ONE bad? Hmmmnnn?

We control our minds, so why don't we? Song in my head now: 'Always look on the bright side of life!'?

Bright side for me was I did, however, make some money, some of which I donated to Nil Desperandum, a society for the disabled, and also gave some to Eternal City – Child Support Group.

Oh, because I self-published with the cheapest printer I could find, around 80 books were really badly printed, so I had to return them. Plus, I also gave quite a few away to the media. So, I only actually made money from around 300.

The majority of the money I made, I spent to refurbish the Mandir, kind of like payback in a sense.

Like I did in JHB a long time ago, I got people off the streets who needed work. Ma'vusani (Ma-voo-sah-knee) & Shakes are still my friends. We repainted, removed old fitted carpets from the lounge and dining room and put tiles. Also, I retiled the kitchen.

Mum and Dad didn't even know I was doing anything. Everything was finished on time for Friday evening before they could arrive from QwaQwa. It was dark.

I closed the yard's gate and put a note on it that asked them to not use the kitchen entrance, like they usually do, but to come through the lounge door.

On the burglar gate at the lounge door, there was a huge bouquet with a note that said, "Ma'jee, you love flowers, so come and put them in the empty vase on the dining-room table of your 'new' Mandir."

I was standing behind the closed door and heard Mum say to Dad, "Aye, I wonder what this mad Shaynah is up to now."

Dad opened the door and put the light switch on. The look on his face, even I cannot describe. Mum stepped in from behind him; she got *weak in her knees. She sat down on the sofa and could not move. After I gave her some sugar water, she stood up on the new tiles and walked to the dining-room table and put her flowers in the vase.*

Ajesh, my sign writer friend, kindly made me a complimentary sign for the glass top dining-room table.

It read:

The Praag Mandir
Est. 1977 by Manilal & Sheila

Now Dad, being the busybody that he is, was already walking around in admiration with a huge smile on his face. To do that, my weak-kneed mum actually needed to lean on me as I escorted her around. Her happy tears then now bring tears to my eyes as I type.

At their closed room door, Ajesh made me a sign: 'Manilal aur Sheila ka pyar ki ashram' (Manilal and Sheila's love temple). When the door opened, Mum was even weaker in her knees because I changed the colour from white to soft pink to warm it up.

My God and Goddesses' hearts were more than warm. I have honestly never felt more valued as a son than when I helped and surprised my parents.

The next morning, I showed them how I blessed the Mandir with a big 'Love & Light' copper sign with copper butterflies that I mounted on the wall.

I also christened it, in a sense, with a sign reading the same as the one on the table inside, entitling our home.

More importantly, a wire sign was stuck outside the lounge door.

JOY, KIND, HOPE, LOVE, PEACE

Five simple words that actually sum everything up. The most inner peace I find mainly when riding Yummy, who I sadly didn't have when my book was 1st released in May 2012 because I had my 5th accident riding in JHB traffic. I already gave you a full account of my 1st crash with Yummy at the race track.

There was another smaller one at the race track and two major ones on the road due to inconsiderate drivers. With three going through insurance for repairs, Dad, who was paying the premiums, was written off their books due to numerous claims.

Mr. Praag had just started a new policy with a new company, and within that same month, I had my 5th crash. Prior to Dad taking the policy, I had to have Yummy assessed and I pointed out my disability and adaptation to the bike. They approved it. But now when Dad made the claim, they turned it down because, apparently, the bike did not have a disabled rider sticker on it.

To repair Yummy privately, I saved for a few months, and now money from book sales covered up the rest. It took just over a year to support the 'backyard mechanics', but I was back on my 'as good as new' baby.

Ma'vusani, the guy who helped me to paint the Mandir, needed a place to stay. With there being an empty shack abode at the back, which was previously used as a maid's quarters, I gave him a place to live.

No rent, just help out here and there. I needed someone around as well, especially with my disabled hand. Some odd jobs, 'handyman's' work, no pun intended.

I used to pack Yummy up with my usual rally gear—tent, air mattress, sleeping bag and clothes. An Insane Shane weird idea came to mind. Hmmmnnn?

Maybe Ma'vusani could help me to pack my bookstall, plus 15 of my 500-page books along with my usual rally camping gear on Yummy? So then, I can ride to the biker rally again, not like my few car trips before. Hmmmnnn?

But so much on the back of a superbike? Possible?

'To believe I CAN' is everything? So, proof in the pudding: I made some press coverage as the featured story of the 'BikeSA Magazine', February 2013.

I suppose, primarily, biker camaraderie does stem from the fact that when riding, with your full gear on, from a distance, you cannot notice race or religion, so there is unity. But it does go further. Some do express a true brotherhood, like Simon Fourie, the owner of BikeSA, who also understood my plight and vision. He allowed me the grace of complimentary stalls at any of the BikeSA rallies.

Just a pity to me that even when scanning and emailing this magazine story to other media as well, I was ignored. I even included the link to my TV interview, but I suppose, at the end of the day, the media decides for themselves what they are going to make news of or not, hey? Again, it's also a 'who you know world' we live in, sadly. If only I knew of another biker in the media. Hmmmnnn?

In June 2013, at the BikeSA Paradise Rally, I had my stall and sold a few books. More importantly, I made friends with an Afrikaner lady, probably in her fifties, Colleen. Her email a few months later really uplifted my dying spirits.

<colleenstokes92@gmail.com> wrote:

Hello Shane. What a book you wrote? Never have I been so inspired before! This is a brilliant reading experience from the first page till the last! I've just finished reading it for a second time!

What a moving and changing experience? As the man behind the book, you are truly unique. You taught me a very valuable lesson the day I purchased this book – to have faith in humanity! You just met me at the rally, a total stranger, yet you let me take your book home on a Saturday, saying it's okay IF I pay you on Monday!

I will never forget your words – 'IF I don't pay, everything is being watched in some way, and judgement does happen.' You laughed and joked about Karma.

This is so true to how you profess in the book also! To keep your dreams alive and to always stay true and believe in what you do, even if let down, above knows, so don't change you.

And, importantly, that everything has its own destined time, and will happen only when it is meant to happen.

I'm honoured to own Crash Landing, and it was more an honour to meet you, Shane! If only there were more thinkers like you, who actually focus on Mother Nature, our world will be a better place.

Best wishes,

Colleen Yvette Stokes

(Well, dear reader, what are your thoughts about Destiny and Universal Timing? Fate? I do believe that everything happens for a reason and that destiny is decided and set. Like me meeting Colleen, her buying and reading my book and then the much-appreciated complimentary email. Could I have predicted it?

Can we predict and control anything? Just had a strange thought that, maybe, our controller actually controls what we think we have full control of, even putting that thought into our minds!

Take suicide as an example. We take our own lives. BUT what IF it was designed to be like that? Because, at the end of the day, there's no intervention from 'above' to stop it from happening.

We honestly have no control over what we do. OR do we? I now want to scratch my nose. I just did. So, was it my thought to scratch my nose OR did something put that thought into me initially, to scratch my nose? Hmmmnnn?

Surely there is a grand design to everything and even a designer in the first place? My thoughts, me writing this book, not giving up in getting it out there, you getting it and reading it. Maybe even thinking a bit differently now? Surely planned as part of a grand design.

By who exactly though? We will never know, but SURELY there is a Creator! Now, atheists are re-questioning their 'coming from nothing' theory because maybe our Creator designed them to be like that and believe that there is no Creator? Or like Richard Dawkins, atheism's godfather, said in a TV interview, aliens made the universe because it HAS to be something that MADE everything!

Personally, I dream and just let things happen. Yes, I do kind of play some sort of control in a sense, BUT I'm accepting of that old cliché 'What's meant to be, will be!' One last point on this tangent follows.

The Creator made the universe, stars, suns, planets, moons, water, then life on Earth and gave it the power to evolve and adapt, right? That also means that disease was also made for specific reasons, right? BUT then, that means that doctors and pharmacists are going against the Creator's plans?

Nowadays, it is part of the norm to go to them for relief, but a few centuries ago, doctors were actually banned from going 'against God's wishes', especially in the Christian community. To help us now, if not Jehovah's Witnesses, we can go to doctors and take medication and go against God's plan?

BUT why do we even have people with brains to become doctors and pharmacists in the first place then? Did God give us the brains to go against the original plans?

Me? Back from the dead. Having brain surgery, comatose and under heavy medical care? Kind of lucky I did, or am I? I have strange and different thoughts. Was I destined to come back? What for?

To write a book and spark your mind more into another crazier thoughts? Why do we brush our teeth? Is there not a reason for the plaque and oral build-up? No animals brush their teeth! But they survive.

Hmmmnnn? Even bathing with soap. Are we not natural beings and forcefully removing what was planned to be there?

Here is the bigger 'BUT'. Get your mind ready. Okay, accept that disease was made to kind of control population, BUT why then give us the intellect to go against the initial plan and become doctors and healers? Or simply, manufacture toothpaste and soap? Antiperspirant deodorant? Is perspiration not natural?

Think on it. We go against the design. Humans control the world in a sense. Okay, not living forever, but a whole lot longer than thousands of years before, as we fill up the Earth with ourselves!

According to statistics, in the year 1935, there were around 2 billion people on Earth. In 2015, around SEVEN BILLION! We are making more children but also living longer. We perform heart surgeries? Organ transplant? Blood transfusion?

In reality, yes, we do have control, technologically as well, BUT we have honestly ascended to a level where self-depletion is looming with our destruction of our own planet!

All of the harm to Mother Earth comes from human progress. Straight fact.

To use an analogy, our progression is bad for Her. We are draining Her blood (oil), eating away at Her body (mining) and suffocating Her (air pollution), weakening her sunglasses (ozone layer).

Our Mother is being destroyed by us! Words of Michael Jackson's 'Earth Song' ring in my head now. Actually sad but reality music video. Access:

https://www.youtube.com/watch?v=XAi3VTSdTxU

Heartbreaking, what we are doing to our Mother, but we have to live and survive and obey and fit in with everyone. I spray my deodorant, after an electrically heated shower, drive my car as well. We are all part of the puzzle.

Hectic, hey? BUT maybe that was the Creator's plan all along? Our Creator let life happen in water and millions of years of evolution until the dinosaurs. Then IT plans for the whole of the Earth to die out, with that apparent meteor strike that shifted the Earth's axis slightly, but mainly only dinosaurs die out.

Life remains and the Creator possibly thinks that it will be too much of a mission to reload the gun to fire another meteor at Earth because IT knows that in a few million years' time, Homo sapiens would evolve and actually do the job and destroy Earth themselves, mainly in their last 150 years of existence.

Hmmmnnn? Was that master plan to make us come along roughly and destroy this planet, so IT can start all over again with another one of the millions of planets in the galaxy and beyond?

We will never know. What we do know is that we have control of what we do and think, or do we? Okay, I 'think' I am drifting too much now, or am I? My 'up' the road may be your 'down' the road? Is the glass half-full or half-empty?

Exactly, perception is based on us. This is OUR lives we are living. Old saying: 'Each to his own', but we have to conform to kind of survive, and...

What a Pity!

Nowadays, we know so much more.
Yet, majority of us are afraid to open the educational door.
Behind it, we can so clearly, scientifically see,
that everything we were taught was just a fabled fantasy.

Reality shows us the truth as a total difference.
Yet, most of us don't want to jump the logical fence?
The Earth is so much older than we were told,
with proven evolution that knocks every religion, twofold!

Yet, we still want to believe that what we know and follow is right?
All others, even though more than three-quarter of the rest of us, are doing the wrong worshipping plight?
Think on that, AND also, the basic, most important aspect–
as Homo sapiens, we ALL simply belong to the same genetic sect.

Yes, we do have external racial types, that are based on the climatological differences,
coz our ancestors migrated to different continents and crossed fences.
Is it not so simple to see that, Africa is hot with a burning sun, therefore dark skin & curly hair,
but in cold, cloudy Europe, the lack of vitamin D in the sun's rays, led to lighter skin & longer hair.

Everywhere, the fact is that different races match the climate of their origin.
It is so obvious to see,
no storybook fantasy.
Is believing in fact a sin?

Internally, human beings, us, are all the same!
Funnily, even blood types match and cross the varied race type's game.
We have made 'outside' differences on our own,
not what proof of the 'inside' pudding has shown.

Culture and religion especially.
There are so many types of 'holy'!
Which storybook fantasy is right though?
Because, even between them, there are big differences, you know?

When will we ever see,
that we are all just ONE humanity?
We a fathom of a species that came late in Mother Earth's long history!
Yet, we continue destroying Her, each other and ourselves, what a pity!

A pity for me to dream? Well, I did have quite a few nightmares, purely racial bias incidents, that I chose to leave out of this re-worked book. Straight fact though, sadly racism does exist, with SO much tension.

Aren't ALL people JUST people? Song in my head by Black Eyed Peas: 'Where is the LOVE' – "But if you only have love for your own race, then you only leave space to discriminate, and to discriminate only generates hate, and when you hate then you're bound to get irate."

We get irate and fight. I so much wish that we all had like x-ray glasses to see the exact same skeletons everywhere. Or even those medical machine scan things that remove the skin and hair, and you just see the flesh, as the lenses of our glasses. Surely, we can programme our existing lenses to become colour-blind in a sense and just see people?

Another song in my head by Arrested Development: 'Mr. Wendal' – "Civilisation, are we really civilised, yes or no? Who are we to judge when thousands of innocent men could be brutally enslaved and killed over a racist grudge?"

Sadly, we only 'hear' words of songs, hey? 'Listen' to 'Mr. Wendal', who has tried to warn us about our ways, BUT we don't hear him talk, is it his fault when WE have gone too far?

Have you gone too far? Our world does sadly need a 'kick up the backside'.

Okay, although I am reducing this book, regarding racism, I sincerely feel bound to add my most painful...

'Kick Up the Backside', to Decipher – Either Or?

In the latter part of 2013, Diwali, the yearly Hindu religious ceremony, fell on the 3rd of November. With Shane being in Newcastle and it being the festival of lights, he decided to be with his God and Goddess in QwaQwa.

Not wanting to abandon the Mandir on the night before because part of the ritual is to walk around with a metal tray, banging with a spoon and chanting, 'Nikhal dalidar, lakshmi auo' *(Go away bad, and come good)*, Shane spent the Saturday cleaning it.

(Yes, I do believe differently, but simple things that cost nothing and make logical sense to me, I follow. I also feel bound in a way to attend family rituals, mainly out of respect for them, not what is actually being done.)

On Sunday morning, he drove to QwaQwa, stopping a few times to give hitchhikers lifts.

(Yes, dangerous, I know, but nothing will change my heart against helping others. I do live by Gandhi's principle because it could easily have been me hitchhiking and wanting people to stop. I am the change I want to see in the world, especially for the world itself. I can't hurt any life form.

Just thinking, as I showered this morning, I helped a drowning beetle out onto the ledge. Okay, okay, yes, I am weird. I get you. At least that beetle thanked me! And more importantly, I feel good about what I did, whether destined to or not.

Besides, I see everything as destined to occur, whether we like it or not. Just told you earlier. Do we have any control over it happening or even total control over what we do after it happens?)

Shane reached QwaQwa just in time for lunch. It was luckily his Goddess's famous vegetable biryani, a mixture of vegetables with rice. On all religious days and prayer services, Hindus fast and eat only vegetables, like the true roots of Hinduism. This directly links to the ancient belief in reincarnation, so nothing is killed and blood is not consumed.

(From the future to disclose that in mid-2014, I became a vegetarian. I honestly always assumed that we were omnivores with canine teeth. But we can only eat cooked meat. We can't really chew raw meat. Think! Neither is our digestive tract really designed for meat consumption.

Also, vegetarians apparently have longer lifespans. Not that it matters to me, but my soul just truly feels settled not to eat anything that has eyes. I love animals and all life forms.

BUT life lives off life, so if I can live off life with no eyes and a heart, then why not? Blooming good idea, I say. Get it? Never mind. On we go.)

After lunch, Shane played with Ashrika, his 5-year-old niece, bursting little firecrackers. The factory yard that they revelled in also had chickens, ducks and geese that Mr. Praag had set up because he loved to reminisce his past 'Mayoyo' days, growing up in a rural settlement.

Diwali has the main celebration at night—the bursting of fireworks. Just before they could start, however, Mr. Praag had a muscle spasm, where his body convulsed and muscles were locking and releasing.

Shane was very concerned but was told that the same thing had occurred in the morning, and it was massaged away and helped with some light medication.

This one was a bit more intense though. Luckily, Manny could still walk to the car, and Shane took him to their family doctor, Doctor Meer, who was called. Even though it was a Sunday night, he gladly went to and opened up his surgery just for them. With Mr. Praag being diabetic, Dr. Meer considered the medication he administered very carefully.

From all Shane's and the families' experience, Mr. Praag, loving the attention he would receive, would always kind of overplay his ailments. Maybe this is from coming from a family of nine children, seeking attention?

Shane and his God returned home, with the festivities in full swing already. Ricky had a few friends and their families visiting. The fireworks were mainly for youngsters, who were more than enthralled.

Mr. Praag loved the fireworks as well and was smiling again. Everything returned to normal, and everyone breathed a sigh of relief as the night passed.

Shane planned to spend the entire week in QwaQwa because Manny had arranged to go for minor prostate surgery on Thursday at a private hospital in Bethlehem, a town 90 kays away.

Manny had been to the hospital 3 weeks back, with itchy feet, and after scanning, the doctors assumed that he could maybe get prostate cancer because a small growth, which could be removed, was already on it.

The doctor also explained that it was genetic, and knowing that his father passed on from prostate cancer, Manny made the booking for the minor surgery. Sheila asked him if he was sure he wanted the surgery, and he responded that if he could stop the possibility of passing away like his father, who suffered in his last days, he would.

"Besides, I've got a loving wife like you to be with me," he chirped at her.

They were still so much in love, the perfect couple with their divine love, even after 46 years.

To support his decision further, Manny also added, "They now have this new technology laser surgery to remove it, so you out the next day."

In the 3-week, interim, his feet still continued itching, and Sheila would lovingly massage them with some massage oils. Being there,

Shane also scored a back massage from her. She is the most loving and caring wife, mother, relative, friend and person anyone could ever have.

(Yes, we all love our moms the most, that maternal love is from another realm, our Goddesses. If yours is alive, how about you give her a call right now and tell her you love her and thank her for bringing you into this world?

You must always remember what she went through for 9 months, carrying you. Also, the actual process of giving birth. Hectic! And then all the crap you gave her? Literally and figuratively. LOL ☺

If she has passed on already, just close your eyes for a moment and connect with her. Remember her in good times. Channel your well-wishes with a thought of gratitude.)

The days passed quickly, and Shane's pot belly grew more with Sheila's divine cooking. *(Something to hate mums for at least?)*

On Wednesday afternoon, as he was helping to run the tuck shop inside RIX Engineering, he could hear Manny's very disgruntled arguments with his medical aid.

"What's wrong, Dad?" he asked as he entered the office.

"These bloody people, I'm paying them so much, always on time. My first time claiming for a medical thing ever, now they don't want to approve the laser surgery! Can you believe that?"

Just from the aura around him, Shane could tell Mr. Praag was fuming. He did have a temper, especially if he was right, which he was this time as well.

"Relax, Dad, it will work out. They must be just checking, trying to save a quick buck?"

"Well, they better come right, or I'll give them a kick up their backsides!"

When Manny said those words, which he is famous for, you knew you had better run for cover. He spent the entire afternoon on the phone, and at about half-past four, he was given the good news—laser surgery.

The following day, Manny drove Sheila and Shane to Bethlehem because he insisted that he would drive. He hated sitting in the passenger seat.

It took them just under an hour, and they entered the hospital. All the paperwork was filled out, and by 8 am, Manny was checked into his ward. In his usual jolly state, he joked and laughed with the nurses who were doing more paperwork and tests.

There were three nurses—an elderly senior and two students. Shane joked that they must pay him for using Manny as an example in their education. Everyone was laughing when the young doctor, who would do the surgery, arrived.

Manny explained about his diabetes and also that he had not taken his usual medication, which he showed to the doctor, because he could not eat or drink water for 12 hours. The doctor relaxed him and confirmed that it would be put in his drip, post-surgery.

"What time is the operation, doctor?" asked Sheila.

As he walked out, he responded, "Well, the surgery is kind of full today, but by ten 'o clock, the latest we will begin. It will be just over an hour."

It was a patient wait. Then, just before ten, because Sheila and Shane were staying over for the night, when Manny was being wheeled into surgery, he told them to rather go to a hotel and book in, joking that he would be busy.

With a giggle, they left and went to the hotel; it was part of a casino. The room was not yet ready, and Shane suggested to Sheila, who loves to gamble on the machines, to go and try a few spins while they waited.

In not even ten minutes, she hit a jackpot of R2000 and excitedly burst out that it was actually Dad's favourite machine she was playing on. They used to frequent that casino often. Shane just smiled and said that Dad gave her the good luck then.

Their room was ready just before 11, and they left their bags and sped off back to the hospital. Manny's ward was empty though. Shane could see the worry on his mum's face.

They patiently waited. The surgery procedure was supposed to be just over an hour. Shane went to the front office and enquired but was told that there was a lot of patient build-up, therefore the delay. Smiling, he bought a set of headphones needed for the TV because he wanted to watch a cricket match. Sheila read a few magazines. An hour passed.

Upon asking a nurse who was passing by, she also said that there was just a delay in the surgery. Another hour passed, and seeing the worry on Sheila's face, Shane started telling her about the cricket match. She used to watch with Manny, as he was a sports addict, but he had not truly explained to her the workings of it all.

Half an hour passed, and Manny was finally wheeled in. He was still groggy from the anaesthetic. The doctor explained, quite casually, that they had initially started with the laser surgery, but it was not successful, so they had to then do incision surgery, which is what caused the delay.

Sheila was holding Manny's hand as he was regaining consciousness, and the doctor reassured a very concerned Shane that it was not a big thing as he left.

The student nurses were now left in charge of setting Manny up to rest and recover. As they were about to leave, Shane saw that the bag collecting body fluid was filled to the top and asked about it.

Calmly, the nurses explained that it was a flushing system, where a hygienic fluid is sent into the wound to drain out the toxins. This was the drainage bag.

She then added that she forgot to empty it and started to. It took her a whole ten minutes, and nothing drained. Shane went to have a look. He saw a pipe system with a closed valve and just opened the valve. Obvious thoughts ran through his head, but he kept quiet.

(I mean helloOoOoOo! Aside from them wanting to leave before draining it, even when told, they both could not just open the valve. They were student nurses, yes, probably from the Black Employment Equity system to empower the previously disadvantaged, but really now. Efficiency? Standards? Don't get me wrong, not a racial discrepancy,

but are there not standards to uphold? This was a private hospital, which in the apartheid days was for Whites only.

Now, the older nurses and staff teaching the student ones were still white, yes, but the education being received? I saw the almost hostile dictatorship at first when the older White nurse was teaching the two young Black ones.

The student nurses were even speaking in their native tongue, a language that the head nurse didn't understand. I knew a tad, and they grumbled and jokingly said rude things. This happens everywhere.

Be proud of your language and culture, yes, but not around people who don't understand it. Sorry for pointing it out again, BUT why can we not see humans first and work as one?

The doctor was of Afrikaner origin and was honestly very blatantly blasé about Mr. Praag. There was a whole different atmosphere in that hospital. It was actively being covered up, but the essence?

Even the operation? I mean, merely from a layman's perspective, the patient was scheduled for laser surgery. After messing it up, they decide to do incision surgery on a diabetic patient?

Him being diabetic is why he fought with his medical aid for laser surgery. Out of the box, think of a rubber tube. If there is a hole, no matter how much you pump air in, it will still leak. Now see the veins and arteries as a tube with a hole in it. Diabetic – lower blood pressure! Plus, you will need a proper patch and solution to repair the tube. And, in this case, a specialist repair man because this tube was different! That they knew but still cut the tube?)

With Manny groggy and needing to rest, Shane and Sheila went to the hospital coffee shop to have lunch. Upon returning, they saw that he was now conscious.

Looking a little drained, Sheila fed him some of the food that was on the bed counter shelf. He only took a few bites and asked her to call Resh because he wanted to speak to her and tell her that everything went well. Resh initially wanted him to come to JHB, by her, for the surgery, but he told her it was a small thing and that he would leave the next day.

Resh is almost a hypochondriac and very emotional. Sheila told Shane that when he had his major accident and was finally checked into St. Anne's hospital, Resh had an emotional breakdown and had to be hospitalised.

On the phone now, Manny assured her that all was okay and explained that he would just have to stay longer, as they had done incision surgery. He ended the call and wanted to sleep again. His feet still felt itchy, so he asked Sheila to rub them with some oil. Shane didn't think twice about it, as he knew of his father's history and how much he loved the attention he received from his loving wife, and continued to watch the cricket match.

After being massaged to sleep, Manny was out cold. Sheila read a magazine, and Shane, who was watching cricket, just thought again, quite funnily, about how she was now an expert at waiting it out in hospitals because of him.

In the late afternoon, they went to the coffee shop to have some tea, after which they returned to Manny, who was now awake and more alert. He and Shane watched the cricket match, and even though he didn't want the earphones to listen, he knew everything that was going on.

During the afternoon visiting hours, Ricky arrived with Ashrin and Ashrika. It was a normal visit, with Manny first asking Ricky how things were at the business. Ricky responded that they were luckily quiet and managed without him.

Sheila made sure that Manny ate at least some of his dinner before they left.

As they were walking out, Manny said to Sheila, "Sheels, don't worry about me. You must gamble, hey. You will be lucky again."

Earlier she had told him that she won R2000 in the morning.

Lovingly, she responded, "You bring me luck, so even when you not with me, you must think about me."

After their light dinner at the restaurant, Shane felt tired and had to take his medication.

(Whoopsie. I forgot to tell you that I am a mild epileptic. Nothing after my major brain surgery in 2004, but it came as a permanent late stage in December 2011 when I had an attack. It was very well controlled with the compulsory medication that I took, but I hated relying on medication, so, in the latter part of 2018, I stopped drugging myself, AND till now, the latter part of 2019, I have not had another attack.)

Shane decided to retire to his room to take his medication. Knowing that his dosage is also the much-needed sleeping pills that put his overactive mind into a lulled mode, he left his mum in the casino and wished her well.

After gambling for a bit, with just a little luck this time, nothing big, Sheila called it quits and returned to the room just before midnight.

Because they had to stay longer, in the morning, Shane went to query if there was place for them that night as well, but the hotel was fully booked. They had breakfast there, packed up and returned to the hospital for some bad news.

In his sleep, Manny's liquid collection bottle, which was now above and near his head, had filled up and apparently burst, wetting him, in his new pyjamas, the bed and his cell phone.

Checking with the head nurse if all was well, they were told that he would still be checked in for a few more days. Shane and Sheila had to now find a place to stay for a few more nights.

The front office gave Shane phone numbers for accommodation, and he started making calls. All of them were very expensive, and they just needed a place to sleep overnight.

Before lunch, they drove off to view one. It was beautiful, but Shane, not being in the right mental state, had confused it. It was double the price of what they could afford.

Returning to the hospital at around lunchtime, they were given more bad news. Manny was now also nauseous and spewed out lunch. More medication was inserted into his drip to stop it, and he was now asleep.

At the coffee shop there, Sheila and Shane had lunch and then he made a call to a number he found for accommodation. All looked good this time. Not too far off and perfect—a room with two beds and even a small kitchenette inside, within their price bracket.

They drove there, liked it and spent the afternoon. Sheila had a much-needed afternoon nap, and Shane got intrigued reading the new book that Resh gifted him a few weeks back, '1001 Ideas That Changed The Way We Think' by Robert Arp. His analytical mind was being more charged up.

When they returned to the hospital in the early evening, Manny was awake and asked them to tell Ricky and them not to come that night as well because he would be discharged soon. Sheila handed him the pyjama top she had bought him on the way to their accommodation.

He was feeling too hot though and didn't want to wear it. His feet were itchy again, so she rubbed them. They were told that he spewed out dinner that they tried to give him as well and that he was still on anti-nausea medication via the drip.

After 8 pm though, just before they were about to leave, Manny requested his all-time favourite, plain buttered brown bread and tea. Shane went and found the nurses and arranged for it. Sheila lovingly fed him, and he ate. He then told them to go home and not to worry as he would be fine.

With Sheila also not being fussy about food, they just stopped at a takeaway close to their room and got some fish and chips. *(I became vegetarian the following year.)*

After eating, they played nine cards, their favourite game. It truly took their stress away; they laughed and enjoyed it till just after midnight and only stopped when they saw the time.

The following morning, a Saturday, on the way to the hospital, they first stopped at a cell phone shop to see if Manny's mobile phone could be repaired. No luck because it was the weekend, so they carried on.

Entering the hospital, Sheila went to Manny, but Shane went to the tuck shop to get a cool drink and then took a calm walk to the wards.

When he entered his God's though, it was full. The doctor was there as well as two nurses. Sheila was in a panicked state. The doctor said that Manny's blood pressure had dropped too low; he had had a minor heart attack earlier in the morning, and he needed to be transferred because they did not have the capability of handling him anymore.

Manny was still conscious, and with pain on his face, he asked the doctor what more could be done. The doctor replied that they did not have the experience or equipment because he had taken a turn for the worse, so the medical staff wanted to send him to a better hospital in the city of Bloemfontein.

With them having no one there to live with, Shane asked if they could transfer his father to JHB, almost the same distance away. The doctor said it was fine, and he must arrange for that, as they quickly wheeled Manny out to do more tests.

The stress and tension could be heard in Shane's voice when he phoned Sanj to enquire about the JHB hospitals. He broke down in tears and told Sanj that he sensed that they were in a bad place when he saw how those nurses didn't even know how to drain the bag.

Sanj calmed him down and said he had contacts at the JHB hospitals and that he would arrange something. Shane went to the front reception desk and told them that it was okay to take Dad to JHB instead of Bloemfontein. They agreed and said to wait for him to come back from further tests.

Sheila was on the phone with Resh, who was panicking and getting into her car already. Shane went into the Intensive Care Unit, where Manny was.

Still conscious, he said, "Aye, Shalendra, I don't feel too good, hey. You just look after mummy so long and tell Resh and them not to panic. Love and light, my boy, love and light."

Holding back his tears, Shane responded, "I know you strong, Mr. Praag. Show them the Praag blood. Love and light."

The doctors then asked Shane to leave the ICU, and he went back to be with Sheila. Some time had passed, and Ricky and family arrived at the hospital. It was just before lunchtime. The mood in the waiting room was tense and got more so at about 3 pm when Resh arrived with Sanj, Nikita and Shival.

Resh broke down and just wanted to go in to see her father. A special plan was made with the nurses. She and Ashrin went in.

They returned with Resh a little calmer, saying, "Dad is okay. He is sleeping. They got him on the ventilator machine and stuff, but they not going to move him anymore, they going to consult telephonically with the JHB hospital."

Outside, Shane was with Shival and Ashrika, distracting himself and them. They played with whatever he could think of, like the game of '5 stones', where you have to pick up and throw stones not too high to catch while sitting down. Time passed.

With the official ICU visiting time only being half an hour, and the long procedure of visitors cleansing, the family then decided that only Sheila and Shane must do the evening visitation because they all could not go in, and it would be hard to choose who exactly should. It would be at half-past seven.

Walking into the ward, Shane saw his father, unconscious and wired up with so many machines. He shed a tear.

Sheila joked, "Don't be stressed, my beta, you had double this amount of machines. They will look after him now."

"They should have put him in intensive care when he first came out, Ma. They knew they messed up. Why they waited?"

Just as Shane asked that, the senior nurse, an elderly Afrikaner lady, arrived and said that Manny's condition was totally unexpected but that he was in good hands now and they will do their best to sort him out. She put a reassuring hand on Sheila's shoulder.

Sheila thanked her and then turned Hindu holy ashes around her divine love's head as she prayed for him. At the same time, Shane took his 'Core', his own holy bead around his neck, the 'Rudh Raksh' seed from India, with five lines representing the five major religions of the world and Scientology 'drilling' the hole in the centre, to his third eye and asked the Creator for blessings for his father.

(Just realised that this is the first time I'm mentioning my Core, the bead I am holding on my third eye in the cover photo. I actually got it in 2009. When writing my book in my Mandir, I noticed a tray of beads next to my family's indoor shrine, 'prayer place'.

I remembered that it was my brother's trance necklace, which he broke off in his trance state a few months earlier. It is a 108-bead string necklace that all swamis use as well.

Having five lines, major religions, drilled in the centre – Scientology, spherical shape, bumpy like mountains and valleys. Hmmmnnn???

Mother Earth as a bead that is also kind of spherical? It perfectly matched my thoughts for uniting the world. The name 'Core' came to mind because, essentially, it is symbolic of the first roots that were planted in our Mother to give birth to us.)

The simple act of holding his Core to his third eye and making a spiritual connection took all of Shane's stress of the day away. He and Sheila exited the ward. She told everyone that Dad was now finally in good hands. It was just after 8 pm and they arranged to go for supper at the casino's restaurant.

Everyone had just sat down and were ready to order when Shane's phone rang. It was a landline, and he knew it was the nurse from the hospital. Not wanting to panic anyone, he stepped outside to take the call. She told him that his father had taken a turn for the worse and that they must calmly hurry back.

The lump in his throat was huge. To get some calm, he took a deep breath, walked back to the table, called his Goddess aside and told her.

Resh had already sensed that something was not right. "Something happened to Dad, something happened to Dad!" she burst out at Shane, who tried to calm her down as they decided who was going. They left Ashrin with the children and rushed to the hospital.

(I was the first to get to the ward. The nurse quite bluntly told me that I lost my father and that they tried but there was nothing more they could do because he was checked into intensive care too late.

A ton of bricks hit me in the gut as I just hugged Mum, who had reached by then. She could sense my inner pain and just knew. She was finished. Everybody was, especially Resh, who struggled to even breathe. I needed air and walked outside.

I stared up at the stars and asked IT why. Then the doctor in charge was walking past me. He didn't even stop. He seemed to be in a hurry to leave after a long day at the office and just stated to me, "Pity. We did all we could do. Sorry."

Deep inside me, I was so angry, BUT I took a deep breath and remembered that Dad's fate led him there. He chose to go. It was a written destiny—Universal Timing. I remembered him saying to me quite casually that sometimes bad things do happen to you, but maybe they are part of the pathway to something good in the future.

My family, even though all broken from just losing our God, picked up our heads. I went and collected mine and Mum's stuff from our room and returned.

Ricky said that he would wait at the hospital for the hearse, which Manny's best friend, Mathew Shunmuggam, had organised because he ran a funeral parlour in Newcastle, to arrive there, and the body to be released to them to take to Newcastle. He told me and Sanj to go so long.

In two cars, we left Bethlehem and headed for QwaQwa. We packed up some stuff, and at around 1 am, we drove through to Newcastle to reach just before 4 am.)

Quite a few other relatives were already waiting in the Mandir. Hindu funerals happen the next day generally. There was an extreme sombre mood with everyone. Tears were shed as they recited holy songs, sending Mr. Praag on a safe journey, but everyone was distraught.

Relatives, friends and neighbours were flooding in. In the early hours of the morning, the body arrived in Newcastle and was taken straight to the mortuary to register the death and complete the paperwork. The street outside the Mandir was actually closed off because they erected a third marquee tent on it to add to the one in the yard and the one in the driveway.

The funeral was set for 2 pm, and the word was spreading fast. At around 10 am, the close male family members went to the mortuary to bathe the body and dress it to be placed in the coffin for the final viewing.

(We were all so broken inside. I held and washed my father's lifeless body. Then the idea that a body is merely a physical vessel for the soul to transcend struck me hard. I had done this before, washed a few relatives' bodies, and now my own father. I could almost feel him above me, consoling me that it was his time to go and that we all have a time.

The body was prepared, shaved and dressed in a fancy suit and placed in the coffin. The hearse brought it to the Mandir, where crowds from far and wide had gathered already, all asking why and how.

I explained the hospital's mess-up but at the end said that destiny decided for it to be like that.

Everyone cried. The yard and the street then packed up with people. This proved to me who my father, my God, was—an outstanding citizen, a martyr to TRY to follow.

It truly moved me when Arthi came from PMB with her whole family. Her sister even flew from JHB to Durban early in the morning to be picked up and brought to the funeral and then get taken back to fly back to JHB that same day.

That was the effect my God had on everyone. People sympathised with us as they walked past the coffin and threw in some flowers. I had already placed a token Core on his neck, above his tie. Mine held my breakdown inside of it as I clasped it firmly.

At half-past one, just before the holy ritual, Mathew gave a talk to everyone. I also wanted to say a few words after he was done. I stood up. Those first words did not want to come out though.

I had the most massive lump in my throat ever! I took a deep breath and, in tears, I think I said, "Life happens and expires. The time for expiration is not really known. My father, my God, has passed on and his soul is in a transition into the next realm. We should not let it hang around above, watching our pain with regret, because it has to go."

I know I had people listening as I could hear a few deep breaths and the crying quiet down a bit. Then the ritual was performed by a priest, after which the body had to be loaded into the hearse to be transported to the crematorium.

You are probably not going to believe this, especially after I phrase it, but the 'heavens' did 'open'. There was a heavy thunderstorm, and everyone outside got wet as the coffin was loaded.

We drove in the rain to the crematorium. In not even ten minutes' time, we reached. It was still raining. The Hindu funeral ritual involves close family males carrying the coffin in, and on the pathway to the entrance, they stop seven times, lay it on the ground and pray for blessings before entering. This was routinely done, and we were soaking wet by the time we got in.

The coffin was laid out on an altar, just as it was at the Mandir, for those who attended here to pay their respects in the same way. Suddenly, the rain very surprisingly stopped, and the sun came out.

This line of people was never-ending as well. When it was finally done, I was called onto the podium first to say a few words before the prayer ritual.

"Honestly, I am taking what just happened with the rain as a strong sign. It has more than taken away that lump in my throat I had earlier at home. My father, my God, who is now above, wants us to do what he just showed us. We must wash away our pasts and start afresh. Like he just did as he entered another realm. He showed us what life does give us, the capacity to erase, wash clean and do again. Not everything, yes, but things that we can, especially our feelings and thoughts inside of us. It does not directly affect anyone or anything, just us. I mean we are in charge of us. Wash clean?"

I then repeated what I said at the Mandir, and inside me, I did find solace and peace. After the formalities, the body was taken to the back to be incinerated. Another ritual prayer was done and that was it. I said goodbye to Mr. P as they closed the doors to the incinerator.

Back at home, there was first cleansing of the hands at the gate, a ritual that has to be done before one enters. I went straight to Mum after that and held her tight. I could feel her immense pain inside, yet she still stood strong and worried about the many people there.

She is remarkable, my Goddess. That afternoon passed, and evening as well. Most of the close family around me had not slept at all the night before but were still awake. Slowly, we started fading off. Even though I was tucked into my bed, the pain wouldn't leave me yet. I needed poetic relief —

Goodbye

We all do come from a possible 'somewhere'?
The direct link is our parents though,
our 'God' & 'Goddess', who gave us a life to bear.
This is a fact that we all know.

How does one deal with losing them then?
Because everyone and 'thing' must pass.
It is just the clichéd 'when',
that our lives shatter, like broken glass.

That is what life gives us.
A set time to be what we are,
until we also must pass,
leaving behind just a memory, a scar.

Even with this as our reality,
the pain of a loss of who actually made you,
tears your heart apart with a grumbling plea.
Oh, what to do, oh what to do?

Suppose we just have to learn to accept,
that nothing lasts forever,
and the tears that you wept,
are just for memories that you will always treasure!

Make your time with those who made you close then,
Because it is only a moment when goodbye will never have a hello again!

With inner peace, my eyes willingly closed because I had to be up very early the next morning, as the family elders decided to take Mr. Praag's ashes to a temple in Durban.

Personally, I would've rather taken my God's ashes to Qwantani, a freshwater dam resort close to QwaQwa, his most favourite place on Earth, where the father and his two sons would fish.

But the temple in Durban was kind of the trendy thing for those who could afford to go to pay *their last wishes to their dearly departed. It was recently built but became quite renowned because it had an apparently special system connecting the altar inside with a pipe to a river outside, which led to the sea, which in turn led to the water on the shores of India that mixes directly with the mouth of the holy Indian river, the Ganges.*

I just kept quiet about it being a geographical fallacy because the mouth of the Ganges is on the other side of India. I mean, really now?

Plus, apparently, all rivers lead to an ocean in some way anyway, so why go all the way there to Durban? Were we helping the ashes and cutting 300 kilometres off the next countless thousands of kilometre journey?

Also, lots of Hindus in Newcastle just go to a river to dispose of ashes, but taking it to Durban is more special? Really now? Something to brag about? Expensive trip there and all and all.

Maybe that is what matters the most. Spend more to show you care? Reaching this temple of solace, even though I understood it, I saw it as a capital gain.

(Maybe you caught the pun in pay *being different when I first mentioned this temple?)*

Some other people were there already at the main entrance, comforting themselves in a way as they gave money away to the apparent needy in donation books, without even a slip or any documentation. Notes just flowed before they went into the main temple.

For our turn to check-in at the entrance, the 12 males from my family, who went in two cars, did the same thing. Just to scatter Dad's ashes, there were a dozen of us who each reached deep into our pockets. We must have given at least R5000 away in total. Where it went to is not known, but it was given with good hearts.

Come on, Shane, come on, now is not the time for your financial gripes with this world. *I know, but still, I mean they sell 'holy' bottled water that claims to be from the Ganges River WITH JUST A STICKER, no phone numbers or official certification or anything.* Well, it is a capitalised world we live in.)

His family had paid in a sense but had to wait for their turn to enter the main section, so Shane walked around the grounds and plucked two wild roses. Through a window, he peeped inside the now empty temple; there was a small concrete pool of water with a kind of basin altar on either side. It was apparently filled with water from the Ganges River, pure and clean, and devotees were invited to drink straight from it.

(When the doors were opened and we entered, I suppose I was in too much pain to tell my family that when I looked in through the back window of the closed temple earlier, I also saw a priest filling the concrete pool with a hosepipe connected to a normal tap.)

Shane maintained his composure, and as his father's ashes were sent down a kind of basin, he held his Core to his head, prayed and sent one rose on the journey and screamed, "Love and Light, Mr. Praag, Love and Light!"

On the way back home, he held his other rose tight. In the Mandir, he hung it to dry above his own personal altar. His every morning ritual, before having breakfast, is to first water some plants with the little cup of water he refills and keeps on the altar.

Then, he feeds the birds outside the Mandir and some lucky ants who amaze him with their impossible strength of carrying a breadcrumb that could weigh, according to species, apparently, up to 100 times as much as them?

At his altar, in a brass tray, there is a small oil lamp, all his special rocks, little tokens and feathers. With him seeing fire and light as the most important tool the human ancestors could control as holy, he lights up the lamp, holds his Core to his third eye and prays in his mind for him to be happy and have a good day. To close his ritual, he blows out the lamp.

(Remember me mentioning how the curtain burnt up when I found my nana's frame? I now blow it out intentionally, also kind of symbolising that we can start a flame and put it out when we still have control over it.)

With him reaching his solace with his own personal ritual, Shane's family continued with the holy traditions of fasts and ceremonies, which he participated in just to keep the peace. His God had left the Earth's physical realm and was now either transcending in another reality above or maybe choose to return?

(That we will never know. What we do know though is that reality and death happens. Like Osho says in his book 'The Art of Living and Dying', which I gifted my Goddess with to help her, "After you are born, you are dying."

Simple reality. Can happen at any time after birth. Life is merely a process until death. I see my God's passing not in a solemn way, but him giving me a kick up my backside to live my dream, become known as a writer and different thinker.

I can clearly remember Mr. Praag supporting me, listening to me, watching a few documentaries with me and totally agreeing with me. Remember him and Mum paying for my tattoo also?

It also became a habit of his to always say 'Love and Light' in closing. I can hear you Dad, Mr. Praag, my God, and even though you're gone, your legacy will live on.

You fathered me, made me, the creator of a cult to unite the world, an improved world dealing with reality and focusing on our REAL Mother. Watch me from above, or wherever you are, as I, with this book, give the world a kick up their backsides to wake up and cherish nature more.

Goddess was moved back into the Mandir with me after God passed away, and Ricky remained behind in QwaQwa to run the business.

Okay, I have to put the following short snippet in because it's even hard for me to believe, so maybe in my book, it will be easier to accept it?

Flight of Fancy?

In Hindu tradition, a widow has to remain in solace at home for a year and wear only white clothes to symbolise her mourning. After the one-year ceremony was done, I took Mum to visit Ricky in QwaQwa. As we neared the workshop, I could sense the pain inside of her as memories obviously flooded in. She shed a silent tear.

In one big yard is the workshop, the home and a smaller outside abode as well. I drove in, parked and opened my door. Mum did not open hers though and kind of felt too sad to get out. I took a sad, deep breath and climbed out.

From nowhere, a sparrow flew over the car and then actually roosted on the vertical brick wall, next to the closed door of the home. I pointed it out to Mum, and she climbed out of the car to have a closer look.

Ashrin, my sister-in-law, opened the door and stepped out to welcome us. The bird then flew into the house and perched on the table.

In disbelief, I joke and tell Mum that the bird is Dad and he is showing her the way in. We laughed and waited for it to fly out again, but after like five minutes, it is still on the table.

Ashrin said it truly was weird because it was happening for the first time, and she invited us in. We entered the home together. From the table, the bird could have flown back out the open door it had entered through, but it did not. It then flew more into the house, went past the kitchen and into the bedroom, the one that Mum and Dad first used. It stopped on the bed.

We entered and then it flew out the passage again, went outside through the kitchen door and then perched on the roof of the outbuilding house that Mum and Dad eventually moved into after Ricky, Ashrin and Ashrika moved into the main house.

Hmmmnnn??? Possibly Mother Nature sent the bird to do what it did and make my mind really consider the paranormal chances of it being my father, showing his widowed wife that things were okay and that he will never leave her side by leading her in, showing her around and then eventually the outbuilding, where he last resided?

Hmmmnnn??? Or possibly, just a bird on a flight of fancy to tickle my bird-brain about the paranormal?

Okay, paranormal? Have you ever felt like being controlled in a sense? Especially when you get lost or end up somewhere you never plan? I am reducing this book, but I have to put it when I felt this 'weird' vibe the most...

Moving Incident

In May 2014, I did a motivational workshop in Johannesburg that shaped my destiny, for me to dream again, especially because other group members identified with my visions with this book.

As an exercise, I also had to make a huge collage as a visual representation of my dream. I put all the press articles, my doctor's note and a card from family on the 4th of the 4th of the 4th (04/04?2004) for me to believe in miracles. I laminated it, my collage, and it is my treasure.

Now, attendance to this workshop, I see it as Universal Timing that led me to a very deep incident on the way to it.

You know what? This incident may sound unbelievable, so here is the actual long WhatsApp message I typed to a 'special' friend who was helping me edit my book at that stage:

"Hey Krista. Hope you well? I am back in the Mandir. Got a deep 'story' to share with you. No need to edit, I'm putting it into my book like this.

I thought that I knew the inner, faster routes in JHB, so I branched off the freeway. After a few turns, I was lost. I could've made a U-turn, and went back down the road again, to get back to the freeway, (N3)

BUT something 'made' me want to go straight to get to the other freeway? (M2)

There too, at the McDonald's intersection, I could've turned right, headed up Louis Botha road to get Corlet Drive, turn left and get to the M2.

BUT in my head again, the thought came, not to, but head on straight as that girl's school I used to teach at briefly in 2007 was just in front, and I had a special 'planned' quick route to the freeway around it.

As I passed the school, there was a right turn to make for my quick route, BUT, 'something' made me go straight?

Heading down that road, I immediately saw an elderly blind man on the other side tapping his cane & walking alone down ON THE ROAD, facing oncoming traffic? I passed him and something made me make a U-turn to go to him.

I stopped in front of him and he moved to the right, going more into the oncoming traffic. I thought YOH, jumped out and went to him. I just instinctively, with my hand around him, got him away from the facing traffic.

I asked him where he's from and what he was doing there. He was well-spoken and told me his name was Jabu. He added that he wants to find work coz he got kicked out of his house and needs the money for rent. He needs R700.

I asked Jabu where he was from and he said 'Evaton'. I didn't know where Evaton is, so I then asked him if he even knew where he was now?

Jabu shook his head and said he took a taxi to JHB city centre in the morning. From there he asked to go to a shopping centre or area to ask for small jobs & the taxi took him up Louis Botha road, out of town & left him by Balfour Park mall which is actually 20 minutes' drive from town, and quite a few other malls and places to try, but they were somehow overlooked?

When he was at Balfour Park that I passed, about two kays from where I found him, the security chased Jabu away and sent him back on the street again.

I mean really now? As if huge JHB was not more than enough for a normal man, BUT a blind man? Alone! Taken to an area he never even know and then just kind of thrown on the street? AND he was probably in his sixties!

Holding his hand, I said to him, let's go into the car so I could see where exactly Evaton was and take him back at least. I sat him in the passenger seat and walked around to the driver's seat, sat down and took out my cell phone to see where Evaton is.

It came up 42 kays away, in the opposite direction to where I was going! It would have made me too late to take him and come back.

Jabu then said that I must just leave him in a nice neighbourhood, so he can go door to door to ask for jobs. WTF???

Alone in a JHB neighbourhood? High walls? Electric fences? Dogs?

He said he needed the money badly. Was it SO bad? Besides, what work would he really do?

It was like I was in an unreal world. I gave him some water and watched him gulp the full 750 ml bottle down in like ten seconds flat!

After asking him how much he has, he put his hand in his jacket pocket and took out his money, saying that he doesn't know, and he put the crumbled bundle into my hand, asking me to count.

It was two 20s and a 50 that I could've so easily pocketed and booted him out if I wanted to. Jabu was so trustworthy of me, and after enquiry, he said that he could 'feel' my heart and what kind of person I was.

That just tugged on my sympathy strings more as I suggested to take him to a police station or something so they can help him to maybe find somewhere else to live?

He replied that he went to them before and they don't care and he just wants to make his rent money & go back to his house in Evaton. I took out my wallet.

There was an 'extra' R500 that 'separated' itself, to go into my bank coz after paying for the workshop, the account was empty. Weird, hey?

BUT I kinda never go to the bank to deposit it, even though I passed the bank yesterday? I just thought that I will put it in the R500 when I get back on Monday. Also, I thought that 'maybe' I will need it for something on the trip.

'Maybe' this was its destiny? So, I took it out and added it to his R90, BUT he was still R110 short. And guess how much 'using money' I had left in my wallet???

I emptied my wallet totally and the exact R110 it added made the exact R700.

I was jittering inside myself, but I could actually feel the warmth of Jabu's appreciation as he folded and put into his jacket pocket.

Now to get him back to JHB to get to Evaton? I'm going the other way and was pushing for time.

Jabu said that he'll take a taxi to JHB city centre again and from there get to Evaton. I found loose coins in the dashboard – R7.50. He needed more. I drove up to get back to Louis Botha road and was going to leave him at the McDonald's to grab a taxi.

As driving and almost there, I saw a few shops. I then thought that maybe someone there can quickly donate to his taxi fare at least?

I stopped and got out. Jabu sat inside. I went to a locksmith Key Shop. There was an Indian man & a Black lady. I just told them about Jabu and asked them for loose change for him. YOH! They actually scolded and chased me away!

Next to them, I rushed to a furniture shop. The Muslim boss told me that he never make anything yet, so not even a R2?

Then the worker at the shop, who sounded Zimbabwean, said he will give something when he sees Jabu.

I took him to show him Jabu and even though he himself was visibly poor as just a part-time worker, he emptied his pockets.

A whole R7.50. He just freely gave it. So nice that the poor understand, hey?

I then realised that I must take Jabu with me to ask for taxi money for him.

I went to a doctor's waiting room. As entering the doorway, the North African doctor was heading out. He rudely stopped Jabu and me before we even entered and warned us not to worry his 'customers' and walked out.

Before we could even turn around, the doctor's secretary, also sounding foreign (North African), then came & shouted at us and chased us away. Four foreign African patients in the waiting room as well just rudely joined her. Kind of like shooing a dog away.

I realised that Jabu was probably used to this as my heart broke when we stepped out, and he grasped my hand tighter to kind of reassure me in a sense.

Time was ticking. I saw a shop with a Portuguese owner. We went. He also just chased us away. Hectic! Not even listening.

We had 15 bucks for a taxi. I told Jabu it will at least get him to JHB and I need to leave because I was going to be late.

Helping him, we crossed the busy road and waited opposite McDonald's for a taxi.

I signalled and the next one stopped. I asked the driver to please take Jabu to town for free, as he already had about ten paying passengers, and Jabu needs his R15 to help him get back to Evaton.

The driver actually agreed to help, bringing more than a smile inside me as I put Jabu in. The other passengers sympathised and helped him.

I am sure that they must have also given him some loose change to build up his taxi fare to Evaton.

What an experience for me?

At the end of the day, in this financial world we have to survive in, aside from my obvious belief that I was being tested from above, with all the improbability of finding a blind man doing what he was doing, where he was, me going that way, the money match etc.???

I think more to show me the tough financial world we do live in, hey?

Cold! And money is all that matters primarily! My dream world is exactly as entitled, in my head. I need to now get money, to 'play the game' as well, in a sense. My vow for the poor and disabled will never change, I will stay simple and humble inside, but I need to change on the outside, as I learnt from the workshop I just attended.

We all have to keep searching for it. Never ever give up, it will come and even if it doesn't, at least you tried. We got to deal with what we have, yes, but just be grateful to not be the blind man or the impoverished coz we so easily could've been that!

Just be happy that you are not.

Love & Light ☺

(I suppose the old adage about us learning from things we experience is so true. Although most of us seclude ourselves in our own shelters, thankful for not having the rain, BUT what if you do get the rain?

What then? Well, again, IF you don't help those who are caught in the rain, how do you expect others to help you when you are caught in the rain?

A simple premise in the concept of karma: do good, get good. I feel my most good when I help people, especially helping children. I go regularly to a nearby squatter camp and feed hungry children—ultimate bliss for me.

Another bliss for me is helping Mother Nature. I had a very deep experience as something fishier was planted in my mind...

Catch and Release?

With the current digital age, I contacted my old school friend Ashika, who was happily married and lived in Glencoe, a small town around 60 kilometres from Newcastle.

She has two teenage sons, who I became close friends with as well. With their father busy working, like most of today's breadwinners, they had never been fishing.

It was actually the summer school holidays for them in 2016, so I decided to go to Glencoe and take them to cast a few lines at the dam there that I never fished in.

I only had two rods, so I borrowed my next-door neighbour's son's little 'toy' rod as well. All set, I picked them up, and off we went to the dam, which was really down because it was actually a draught.

The dam itself was almost halved, and we had to walk to the centre to get to the water. I put all three rods in. It was midday and being stuck in the open sun for almost an hour with not even a bite, I was ready to pack up.

But then the little rod was pulled into the water. I jumped and picked it up. The tip actually broke. I then tried to turn the reel, but it jammed. Kind of instinctively, I held the rod tight, turned tail and ran up the bank, towing the fish out of the water.

It was a freshwater carp weighing around two kgs. Since becoming a vegetarian, I now catch and release, getting the fish back into the water a.s.a.p.

This fish, however, had almost swallowed the hook, and it took me close to five minutes to rip it out of its gut. Inside me, I could actually feel remorse build as I did that.

Releasing it, I first held it upright in the water, helping it to get its breath back. Tears then just flowed as I was actually trying to save its life that I would be responsible for taking by having fun while catching it.

I imagined myself as it, being hungry, seeing a snack, eating it but then having the hook in my mouth and being dragged out of my home, only to be sent back with a permanent scar on my lip, plus a mental one.

I vowed to release my desire to catch fish, which I basically grew up with totally. My mind was made up, and my heart at least felt some relief as the fish finally swam away.

Bird-Brain?

The Mandir has kind of become a sanctuary for birds that I feed every day at a spot I made for them with two water bowel birdbaths as well. Kind of like a shrine to me because inside me, I feel a real sense of belonging as I see them eat and I feed my new mind to focus on all of life as a single entity.

On the morning of 9th November 2017, I was sitting outside the kitchen door of the Mandir, cutting my nails. From out of nowhere, a pigeon appeared. It walked straight to me and looked up at me.

This was the first time that I saw this particular pigeon. Right outside my bedroom window, under the roof rafters, there is a pigeon couple who live there. I do sincerely feel blessed and safeguarded by them because from the whole house, they chose to be with me.

Maybe I am weird, but I actually love to hear them hoot every morning. Kind of like a happy wake-up call for me. Sets me up to have a happy day.

As I cut my nails with the pigeon less than a metre away, looking straight at me, almost like requesting food.

My mum was in the kitchen and I loudly asked her to bring a slice of bread. She then walked outside with the bread and asked the pigeon to follow her to the back of the house.

I laughed at the idea that Mum spoke to it, BUT it readily followed her, and she dropped the bread, which it ate there.

Ever since that day, 'Tha-tha', the name that I gave the pigeon because it does walk like an old grandfather ('Tha-tha' in Tamil means grandfather), actually made the window ledge of the garage his roosting spot. He hardly ever even leaves the yard. Kind of like he found the perfect place to be—food, water, shelter.

In a way, Tha-tha became my pet but almost takes an assertive role and actually pecks at my feet sometimes to ask for food, like he is in charge, funnily asking, "Who's yo daddy?"

Believe what you will, because of the symbolism behind the date as well as Tha-tha taking control, like a father would?

Tha-tha 'could' be Manny's reincarnation, returning to the Mandir after flying around for four years, he finally found it?

Hmmmnnn???

Or he's a very clever bird, who now enters the kitchen to feed his 'bird-brain', directly from my disabled hand that managed to hold the bread to let me take a selfie to grace my book as proof to you?

I suppose I have simplified things as well as read more into things that occur. I truly am a very different person now.

Yes, not following the clichéd 'norm', making my own route and very happy to just be ME.

It is MY life I am living, and I want to share it with the world that it is okay to be different and even write a different book about it. LOL ☺

I have tried various publishers and do know that it is a mission! (Heard through the grapevine that J.K. Rowling tried around 90 publishers before 'Harry Potter' was born.)

I do dream the impossible dream of my own rebirth, and my faith never died, especially after...

Radha & Krishna?

Knowing inside himself that this existence works on the concept of Universal Timing, while Shane was sending his book off to various publishers, he was in his comfort zone in the Mandir with his Goddess.

With the current 'new age – digital world', since the latter part of 2015, Shane became an online Scrabble addict. Scrabble, for those who do not know, is a game where one is given seven random letters that one has to make words on a board with. The word stays on the board as more are added, kind of making it like a crossword in a sense. It kept his overactive mind busy, plus heightened his love for the English language as his verbosity grew.

The bonus is that you can meet people from all around the world, play them and chat to them. Shane had built up quite a few Scrabble friends, who he kept regular contact with. Roberta, his new best friend from New Zealand, really lifted his spirits up.

(I emailed her this latest book in May 2018, she read it and actually became an EtheREAList. The first one, next to me, obviously, LOL ☺, that she is SO proud to be. You are almost there yourself to assess, be patient?)

Roberta and Shane shared a very close platonic bond, like he did with all his opponents. Every game he played, he always would chat with his opponents at the beginning. I suppose that is the beauty of the digital world.

Scrabble has different formats to use—a normal game where there is a 24-hour time limit for you to make a move or the timed 'speed-play' option, giving you 2 minutes, 5 minutes or 15 minutes to make your move.

Always playing the normal game with at least a dozen games running at all times, Shane spent most of his time playing. On the 15th of August 2018, without having an active opponent, he decided to try out the 5-minute speed-play game.

Even though it was only five minutes to make your move, Shane chatted to his opponent, who was actually quite taken aback by his friendly nature. Him and 'Mei', a 24-year-old lady from Shanghai, became friends.

(We would honestly spend more time chatting to each other than actually playing. It was almost as if we kind of knew each other? Possibly having known each other in past lives, IF that concept is even true, but sincerely, Mei just clicked with me.)

After just two days of meeting her, with it being a Friday, Shane emailed her his latest book that he had sent to a renowned publisher in India. That weekend, Mei never played, but on Sunday evening, she sent him the following email"

Sunday, August 19, 2018, 7:26:54 PM GMT+2, Mei Yan <meimiss001@gmail.com> wrote:

Hello Shane,

I am extremely thankful for sharing this amazing book with me. You have changed my life, gave it a purpose, to serve Mother Earth and be human above all.

I have written a review, though I think it just doesn't justify the beauty of this book.

Review: The author, Mr. Shane Manilal, has a unique way of writing. He illustrates with interactions, lovely poems, inspirational quotes and even drama. You feel his presence alongside you. He wants us to be a part of EtheREALism, which is a three-clover leaf – be the change, think before you accept anything and value humanity above materialism. He illustrates his struggles and his difficulties. He portrays his shortcomings to show that he is just another man, which makes him relatable.

It brings to the notice of the readers that evils like racial segregation are still very much a part of society. The basics of all religion preach humanity, but people tend to forget the very essence of it. The shallow meaning has been extracted and is used for malpractices.

The author brings out the fact that even though being a part of the technological revolution, people tend to forget to differentiate between myths and reality when it comes to religion. The author is asking us to be our own masters.

I would highly recommend the book for anyone who is interested in transforming their lives, help Mother Earth and achieve inner peace.

Mei Yan

Global Computer Graphics

(I was beyond awestruck. We had just met. Our chats continued as she opened up more to me, and our feelings for each other grew. Yes, I know that I had intentionally closed my heart off, BUT it started fluttering all over again, for a woman far, far away, whose heart was also fluttering for me.

We shared the deepest and sincere chats on Scrabble, Facebook and emails. Mei started confiding in me and declared that she was always so different from what her family accepted – social norms, but by chatting to me, she could be the real her.

So sad that majority of us have to conform, hey? We are too scared to be different? Too scared to be our TRUE selves?

Mei asked me to write her a poem about how she as a child could not really play and have messy feet.

Happy Chappy?

How far back can you remember?
All grown up now, yes, but you also were a child.
Well behaved, or naughty and wild?
Take a few minutes, and in your mind ponder.

Times were so different then,
the cliched moment when
nothing really mattered totally,
and even if it did, within no time, it was sorted, silly.

All that mattered was to be outside.
Getting your feet dirty and messed up.
Having fun on a plastic motorbike or bicycle ride.
Playing games with friends and winning your own cup.

Oh, the good old days, hey?
What changed your way?
Yes, physically you grew.
BUT your inner child still lives within you.

Okay, you are all responsible and grown-up now.
No one is asking you to become a cow!
BUT surely in your own mind,
you still can and jump over the moon, to look to find?

What are you really looking for though?
Ask that little boy or girl inside of you.
Only they will know,
and, in doing so, to you, you will remain true.
Want to be,
A happy chappy?

..........☺ ☺ ☺..........

This 'Happy Chappy' from far, far away really stole Shane's heart away, especially when desiring to mess her feet to connect directly with Mother Earth. Was a storybook romance blossoming?

Storybook? Hmmmnnn??? Also, Mei was actually going to change her name to 'Radha', the fabled 'Juliet' from Indian fables. When Shane was younger, his family actually used to call him 'Krishna', the fabled 'Romeo'. Radha and Krishna uniting again?

Radha even vowed to help Krishna to find a publisher because she also wanted the world to wake up and realise that we are ALL human beings PRIMARILY.

Shane informed her that he was still patiently waiting for the publisher from India to get back to him. *(I know, I know, I am a patient man and I believe in Universal Timing; everything has its destined place of belonging in the timescale. Besides, my fluttering heart, which I gave up on, was beating relentlessly again. More so when...)*

On Sun, 2 Sep 2018 at 7:19 AM, Shane Manilal wrote:

Hey Mei

Told you on Scrabble.

My sleep broke but I actually woke up from an intense dream.

I typed it as a story.

Dream

In an open field, full of daisies, I can see a woman from the back. She's wearing like a dark grey, full-length dress. I'm a bit far off from her but notice that she is sobbing. Then she kind of lifts her hands to her face, wipes her tears, then hands go down in fists to the side. She takes a deep breath and sets off.

I follow her, keeping the distance. There are no buildings or anything, just out in Nature. Then there's a small stream, with no way around. She stops. Fists to the side again, bends down and removes her shoes, folds her dress up a bit and ties it, then steps into the stream. Lucky, not deep, she walks in. Water is just above the ankles.

She crosses and doesn't look back. I stay hidden. I remove my shoes, fold up my pants and go inside and follow her. On the other side, lucky, a flat bank that she climbs up, stepping on the ground there; there is something different about her. She kind of plants her foot down, lifts one leg and does like a swivel movement, kind of connecting and feeling the ground. Then a smile as she does a twirl on one leg, like a little girl.

To me, I can now see like a glow around her. Her radiant Indigo Aura. She sets off again. I somehow also don't put my shoes back on again as well as I climb up the bank. I tie my laces together and put them around my neck and set off. Amazing feeling to connect with Mother Earth. I follow her. We are now at the starting of a forest—trees.

She enters. I follow. A bit tricky to go in but kinda like a pathway. I duck behind trees but keep her in sight. After some time, a kind of climb builds up. She still pushes ahead. Not a steep climb but few rocks etc. We climb.

After some time, the land kinda flattens. She walks for a bit. Then she just stops and stands still. I am still behind her and slowly walk towards her. Can hear her sobbing as she looks ahead.

When I reach her, kinda like she knew I was there, her hands wipe her tears, then make like an 'X' cross on her chest, and she turns around.

(Remember my dream embrace of Suraya, my flutter?)

My hands immediately open up as I embrace you. I hold you close to my chest, my hands around you, encircling you. With me holding you, I open my eyes and can see that you stopped at the edge of a cliff.

I can just see a slight drop and then mist. You kinda step back and my hands open. Then you stand to my left-hand side and hold my left hand tightly with your right hand.

We now are facing the cliff, holding hands. Your head turns to the side, you look at me and say, "Thank you for being here for me."

I respond that it is a pleasure and that you must remember that you are never alone; I will be there for you in whatever you decide to do.

You then let go of my hand and say, "I know it, so we might as well, hey?"

Both our left hands then kind of automatically put the index and forefingers together, kind of like a gun, and then we kiss our own fingertips. Then both hands meet each other, kind of transporting each other's kisses as our hands join together again. (My special kiss)

With a tight grip, you scream, "Here we go, Krish!"

I woke up in a shock. Dream was so real. Not sure IF we jumped, IF we walked back, OR even if there was a pathway down the cliff. OR if we flew like angels? LOL ☺

Mei, my analysis is that you are going through quite a lot and always remember that even though we never met, and you do not know me, I will always be there to support you and help you find your way.

Also, you need more connection to Mother Earth as you get in touch with yourself.

I am here for you. 'Thuqdeer' – Fated Guidance – Universal Timing made us meet on Scrabble for some reason or the other. I am a nice guy with BIG dreams, BUT why the hell not? It is MY life I'm living.

Life is tough and challenging, BUT never look back, regretting and wondering. Always aim forward. Live YOUR life the way you see fit.

Besides, you have a totally different life ahead of you in India. If you were still with your ex-boyfriend, what then?

Long-distance relationship? He cheated on you when you were together. DUH! Everything happens for a reason.

You starting afresh. Your whole life is ahead of you, and I will be there to, like I said in the dream, support you in whatever you decide to do.

Your life. Live. Love. Laugh.

Love & Light,
Shane ☺

On Sunday, September 2, 2018, 11:11:00 AM GMT+2, Mei Yan meimiss001@gmail.com> wrote:

Hey Shane

Wow! What a dream? So real. Honestly, I thank my lucky stars immensely for bringing you in my life. I feel very lucky. I am so lost and still floating in this dreamy reality. A whole new look at things, that I am loving.

So simple to focus on Mother Earth instead, hey? I will try my best to follow serious ways to save water at least... revert tomorrow on that.

And I am trying to think out of social norms; for that, I need to make up MY mind first.

Sending you some of my pics

Love & Light
RADHA ☺︎☺︎

A really strong bond was developing, yet they never spoke, and when she emailed him her pictures, Shane was more 'goo-goo, gaa-gaa'. Was he in dreamland? She was beautiful, loving and caring, and even loved nature.

(Pictures of her outside, climbing a tree, feeding cows from her hand, staring out at the ocean, her pet dog.)

On Sun, 2 Sep 2018 at 3:30 PM, Shane Manilal wrote:

Hey Radha

Beyond AMAZING. You were so right. I can see myself 'in' you.

WTF???

This is on a whole new level. Can even sense your aura around you when feeding the cow.

And a bit of a 'scamp' when climbing the tree. Again, I can sense an Indigo Child.

https://lonerwolf.com/indigo-child/

(Open the link)

There is an age-old saying: 'Cut from the same cloth'.

Honestly, just looking at you, does make me think that this whole thing is unbelievable.

You are beautiful.

Oh, I have the same Core as you. Also three around my neck but just on a string.

(Love the doggy Miss 'City Chick', that I actually think you are not really. You are more of an 'outdoorsy' type? Climbing trees – messy feet – 'Happy Chappy' when outside in the fresh air. Love the pic of you staring out at the ocean, lost in thought at the expanse of this existence)

Hmmmnnn...???

Okay, too deep now and too much.

Thank you for making my heart beat again.

Love & Light,

Shane

(Oh, my nickname as a kid that my family used to call me used to be Krishna because I used to love cream and used to also eat butter and margarine with a spoon. Hmmmnnn?)

On Sunday, September 2, 2018, 3:40:17 PM GMT+2, Mei Yan <meimiss001@gmail.com> wrote:

Hey Krish

You know I read your letters million times. I have never shared so much with anybody. With my ex, we never talked so much. I am just so lucky... you like talking... you share... you care... ohhh I can die so peacefully now... nothing else my heart needs.

My mom used to worship LADDU GOPAL ie lord Krishna and I have kept the trend. He is my first love. Is he you? Feels like. So, I love YOU.

Thanks for staying in my life

Love & Light
Radha 😊

On Sun, 2 Sep 2018 at 10:24 PM, Shane Manilal wrote:

Hello Radha

I was busy. Went to visit my cousin sister Shakira. When I speak about you to her, she says that she can see something totally different in me. She says that I glow when I talk about you? Hmmmnnn...???

I saw her yesterday as well when I first told her about you. Now, you know my whole 'story' about belief patterns. I am an Agnostic Totalitarian Pantheist.

Basically, an EtheREAList.

BUT I will respect you to believe whatever you want to believe in BECAUSE 'they' actually brought us together again.

Radha & Krishna?

(I can't stop staring at your picture. You are BEYOND beautiful. Like a Goddess.) I still kind of feel like I am dreaming? Can't wait for the call tomorrow to hear your voice.

Oh, maybe you must actually get WhatsApp because we can then video call? I can show you my home, The Mandir, etc. Will add a new dimension to this.

In closing, I told you: Communication is the building blocks of any relationship. Yes, we both love to talk, BUT we also both like to listen as well.

I still feel like I am dreaming? Am I? Well, my heart is fluttering again. 😍

Take care and enjoy your start to your new week.

Love & Light,

Shane 🌚

On Sunday, September 2, 2018, 10:17:34 PM GMT+2, Mei Yan <meimiss001@gmail.com> wrote:

My love,

It's 1:45 am here. Just woke up. Can't love you enough. Thanks for liking me...... oh god... I slept while writing... it's 3:15 now

Hope you love such a crackpot. I glow when you talk to me. I wanted to call YOU GOD because my heart feels you are. But till now I cannot call anyone else god or goddess but you, I so want to call my and only mine GOD. I so want to call you right now... why wait in agony... I so want to be in your arms right now. It's painful to stay away. Floating in your love.

I want to take your head in my lap, caress it, caress your eyes, nose, cheeks... anything but lips... with fingers first and then....

Well... I am going crazieeee

Shane, when you said it's only in my head that you like me... I died... I flatlined...how can I portray my despair... I did not want to continue only... I wanted to leave the game... uninstall... and cry my heart loud louder loudest... but my mind was not ready to take it at its face value... and I stalled... I am so happy... my gut saved me from possible eternity of sadness

I had misunderstood, so basically at every such corner of life when any of us feel such negative feeling, let's promise to give it time and understanding and TRUST.

Hope you don't mind my chitter-chatter, but I want to share sooo much. Whenever we tied on scores and you said ROOM SHARING...

my heart beat so violently... threatened to burst my ribcage and go into your hands. Three parts of my entity... one wanted you to flirt... another wanted me to stop myself liking you and the third one was always confused... now what... this is stupid... I can't have him ever... and whatnot.

Shane, it's becoming very difficult... do you think you can come in September to India to search for your publisher... same time as me... we can settle together ... I MISS YOU

One more thing... last... promise... Shane, you always lived in open places... you like getting connected to Mother Earth... will you be happy at Kolkata or Hyderabad... am worried... they are horribly congested... I won't like you to be unhappy... I am thinking about investing in a bungalow kind then... at least it will have a small lawn where you can have your garden and do your rituals... a little away from main city maybe... I am just thinking... have to work many factors

But we can work out everything together. Tons of KISSES 😘

YOURS ONLY
RADHA 😘

On Mon, 3 Sep 2018 at 1:12 PM, Shane Manilal wrote:

Hey Radha

Well, well, this 'thing' is unimaginable, hey? Honestly, I still feel as if I am dreaming.

It will be SO nice to 'live' with you there in India, BUT I have committed the rest of my life to my Goddess. So, I am based in the Mandir. (I am at my most peace here at home.)

My mum will not manage alone here. Besides not being able to drive, I also think about her safety. South Africa is very violent. Plus, reverse racism in a sense, where the blacks are against Indians, who funnily were under Whites as well before? Hmmmmnnn....?

My forefathers and foremothers ALSO struggled as slaves, BUT as a race, the South African Indians DID develop. We used our brains and became Doctors, Lawyers, Teachers, Business-people etc. and built up financially.

I am not racist at all, because I believe in ONE humanity in my 'dream' world, BUT realistically, the blacks, who never build themselves up, are jealous of South African Indians. They want to loot our homes etc. A very bad situation indeed.

So, BASICALLY, I actually see my 'destiny' as such – to take care of my Goddess.

(Why I never marry in a sense?)

BUT I can still travel about. My mum can be with my sister in Johannesburg OR with my brother in 'QwaQwa'?

Just for 'holidays' in a sense. Like she does and leaves me to house-sit. (I also have a handyman/gardener/guard – Vusi, who lives in a small outside 'shack' in the yard. So, he can guard and take care of the Mandir even when me and mum are not here.)

It is our home that she and my God worked so hard for to establish. I feel so 'bound' to it in a sense.

Besides, to actually 'live' in India permanently, I will probably need a visa etc. Politics that I don't like.

So, regarding this, I will come to visit you, wherever you want to make YOUR home, but for now, I can't leave my mum.

YES, I do 'feel' a whole lot for you inside me, my gut, BUT I have to also use my mind.

BESIDES, you never even met me OR speak to me. WHAT if you don't really like me? (Just saying. Coz I am 41, you know.)

That age 'thing' I am just pointing out again. I know you know and you have no problem with it because it is just a number for you, but 'maybe' it will be a bit of an issue?

BUT, honestly, I know that I am babbling away here because I am still in a disbelieving 'shock' mode.

Okay, let's do this. I am very sporadic. I love the idea of meeting you when you go to India and help you set up there etc. Maybe I can search for a publisher while there as well?

(I am going to phone the publisher I am trying in India just now to see, BUT I also Googled and found a lot of publishers there in India itself.)

You said you going to be in Mumbai, right? Well, well, coincidentally, I found one last night, the only one I opened and I found another one in India – Mumbai.

I never contact them yet, BUT 'maybe' this is what my thakdeer wanted. You will never know, hey?

Because, honestly, will be SO much better IF I can do it in person when there in Mumbai with you. Thank you for making me dream again.

Honestly, your faith in me and my book has really lifted me up.

It just feels so right. Like what I have been waiting for all this time.

Almost surreal.

Okay, so let me go and call the publisher so I know, and I will continue my dream for my book.

Number 2 dream: To actually meet you. You give me a date and I will come. Simple.

(Oh, I am more of a crackpot than you. LOL 😂 My nickname was even INSANE Shane)

Regarding you thinking about me and waking up in the morning etc. I did for you on Saturday night. BUT, last night, I actually slept. I took out my irritating clock battery – it bloody ticks and keeps me awake. I even used some earplugs and I got about 10 hours of deep sleep.

Faith – I love that you accept my belief pattern. Just the idea of you making another home away from the city more than makes my heart beat even more.

BUT do what suits you. To me, nature is everywhere. Even ants, cockroaches, birds.

Besides, again, I am NOT going to rule and dictate YOUR life. I am going to SHARE it with you.

So, again, you make a place for YOU there. I will come to VISIT you there. Then 'maybe' you want to come back with me to see my home and we can see what happens next.

(I am already willing to commit the rest of my life to you, even IF long-distance relationship, because we will only be physically apart, but emotionally our hearts will be together, and we can enthral our minds with email, Scrabble chat and maybe WhatsApp?)

Hmmmnnn...???

Okay, too much now, but in closing, I love the part where you said that your heart beats even more when we are tied and in the same room together.

Same room?

Hmmmnnn...?

BUT I will stick to my chastity vow and ONLY copulate once I put a ring on your finger.

(I will kiss you deeply when I first see you though)

Okay, let me go to phone the publisher.

Take care and remember always that someone very far away holds your heart next to his and has fallen for you.

Love & Light,
Shane

On Mon, 3 Sep 2018 at 1:55 PM, Mei Yan <meimiss001@gmail.com> wrote:

Hi Shane

Just went through your letter once but I am more surer of my choice. Age and other factors don't matter as much as love does. You love me even more than I do you. You put my life and my happiness above all, even above your wishes. I am no fool, Shane, I won't let you go and leave me

Right now, very busy and tired... will call you as promise... please do wait for me... love you 😊😇

Yours forever
Radha 🏃

On Mon, 3 Sep 2018 at 2:01 PM, Mei Yan <meimiss001@gmail.com> wrote:

AND... NOTHING can stop me reaching for that someone very far away...who holds his heart next to me and has fallen for me.

I have to thank angels for that... people don't get love so easy 😇

And Mumbai is right now not under plan... it's Hyderabad and Kolkata... but Mumbai is only 2 hrs flight from there, so shouldn't be a problem.

Now I just hope I reach home safely and fast enough for you 😇😇😇😇😇😇

manilalshane@yahoo.com **To:** Mei Yan <meimiss001@gmail.com>
Sent: Monday, September 3, 2018, 2:13 PM GMT+2

Hey RadiantAuraDivinelyHeavenlyAlmost

I know it is a tiring day for you. Sorry to worry you with thoughts of me. You can read my long email again when you safe at home. Just want to tell you that I did try to call the publisher in India but no answer. I will send another email. That is regarding my book. My dream.

BUT my more important dream is YOU. Am I dreaming or are you for real? I will wait patiently for your call. Let me send the publisher another email meanwhile.

(Oh, the lady at the internet shop, Aunty Rita, is my very close friend. She is my friend for like 11 years now. I first met her when I went to make an email address in 2007 before my trip to London.

Amazing person. Knows all about my various appeals that she has helped me make and loves that I never give up my dream to be published. Anyways, I told her about you because she also told me that there is something different about me.

Then I showed her your photo yesterday on my email. She is also amazed at how beautiful you are. She joked and said to me that my 'patience has paid off'.)

Has it? Will wait for your call.

(Oh, regarding the ring and 'proposal', I already do have a ring that I keep next to my lamp I light every morning. Here is the 'plan' – I come to you. We assess. IF you like me, then you come to me. You assess. IF you like me, that ring becomes yours and you can have me. Told you, just technical issues about WHERE we going to live. I can't leave my mum, BUT maybe IF you come here and you like it, you can stay with me. Hmmmnnn...? You won't even need to work or anything. Just be my wife and be on holiday. I will pamper you for whatever you need. Hmmmnnn...?)

Take care.

Love & Light,

Shane

After sending that last email, Shane went online to play Scrabble. Mei had not played her turn. Most of their chats were actually on Scrabble. Her last message to him was just after 9 am: 'Love you, my GOD. Miss you. My day is very busy...dead tired...couldn't even have 2 hours sleep... it seems like my heart is incapable of handling excess of happiness that you give me...still 2–3 hours here and then 150 kms back home... grrrrr...love you'.

At just after 11 am, after Shane saw the message, his heart fluttered away as he wittily made the word 'CLIT', not even thinking that it would be accepted. But it was and he typed: 'Well, well, I never know that it will be accepted. Such a naughty word. BUT, with you, anything is possible?'

Close to three hours had passed. No response from Mei to his email or on Scrabble: 'Are you okay? Busy day?'

Shane then sat at his desk, and on his laptop, he ventured online to Facebook to check for her. He received a friend request from 'Anuradha Agarwal'. Like he always did, even without knowing the name, he accepted the invite. A few minutes passed, and all of a sudden, the actual laptop starts ringing and flashing.

He had never seen this before. *(Yes, I am not a digital person. My laptop is my typewriter.)* The flashing asked him to open a video call on Facebook. It was Anuradha, who he had just accepted.

"Are you Shane who is an author?"

"Yes, I am. I have just accepted your invite."

"Do you know Mei?"

"Uhmm? Yes."

"Finally found the right Shane. I have been trying for like an hour. 10 different Shanes. You sent Mei your book, right?"

"Yes. I have been trying to reach her for a few hours now. Is she okay?"

Shane notices a tear in Anuradha's eyes. His heart starts to droop. Taking a deep breath, Anuradha says, "I am sorry, Shane, she is not okay. She had an accident on the way home from work. On her scooter, looks like a truck never saw her, turned into her, pushed her off the road and drove away."

"Is she in a hospital now? Is she okay? Did she get hurt?"

"She died on the scene, Shane, but the person who found her said that her last words were 'Tell Krishna that his Radha is coming to him'. She told me all about you as her God Krishna, so I made it my mission to find you to tell you."

(My heart was literally in my mouth. I started to tremble. It felt like I was shot in my chest. I took a deep breath)

"I am at a loss for words," gulped Shane.

"Well, everyone who knew her is devastated, Shane. I just wanted to tell you. It felt like my mission. I am her cousin sister in India, and she told me all about you, the love you shared and how happy you made her. Also how much your book changed her for the better and that she was going to make it her mission to help you find a publisher."

"I am still reeling here but do feel SO special that you missioned to find me. I mean, you just found out, are broken inside, yet you searched for me. Truly means a lot."

"Well, it just felt right to find you, AND I will also honour her vow to you to help you to find a publisher. I don't know much about it, but I will. I feel destined to."

"Thank you so much, my dear. About her passing, I believe in Universal Timing, destiny. Everything has a set time of existence on Earth, then it leaves. Where it goes to afterwards, no one will ever know, until they also pass. BUT, sincerely, I can feel my Radha with me, my Krishna, inside me."

"Too sad to believe, but I suppose that is the true story of Radha and Krishna, hey? Okay, I have to go now. Namaste, Shane."

"Be strong. To me, her soul is still hanging around. She is gone physically but will be around spiritually. All the best. Love and Light."

With deep pain in his heart, Shane went to his lamp that he lights every day. Holding his Core to his forehead, he lit the wick. With his eyes closed, he prayed for Mei's spirit to rest in peace, and maybe he was just imagining it, but he felt a flutter in his heart as he blew out the candle.

(Yes, I am weird, BUT I felt her with me. In my heart is where she now resides.)

With him being alone at home, he needed a hug and rode his bicycle to the internet shop. Without even telling her anything, he just asked Aunty Rita for a hug, and her arms opened up for him. After that much-needed hug, he told her.

Tears came to her eyes as well, but she told him to be strong and reminded him of his own words to her after she lost her son, his friend Ajay, a few months back: 'Everything happens for a reason, and we actually have no control, but we have to accept and move on.'

In his mind, Shane vowed to be strong, BUT in his heart, he felt deep pain. He had to go to inform his cousin Shakira as well.

(I suppose that when we tell people, we share pain? Shakira consoled him and also reminded him about his belief in everything happening for a reason.)

When he got back home, Shane shared his pain with his laptop. *(I honestly kind of slip into another zone when writing, especially poetry.)*

Rest In Peace, MissExpertlyIntelligent

My best online Scrabble opponent you were – MEI.
But our fate was more than just to play.
Digital destiny brought us together again,
because a true love sheltered in our den.

You were my dream woman, far beyond perfect, sincere and true.
A really special soul, I got to experience the real you.
Unimaginable how our bond grew,
many won't even believe it to be true.

We even planned to meet up and live together,
after you move to India to start your new life.
We were ready for any stormy weather,
to survive, as husband and wife.

Yet, destiny knocks one with a surprise.
In my heart, there's now a hole, no lies.
BUT I have learnt that death is a sad part of life,
stabbing a dagger into those you leave behind, to deal with the strife.

To honour your memory though and stay true,
to the real love I felt for you,

even though, inside, I am very blue,
I have to accept, to bring you back, there's nothing I can do.

What I can do though, is remember the good times we shared.
Smile inside, show you I truly cared.
IF you can see me, from above or wherever you are,
I vow to heal my heart's scar.

Because I know that you hated stress,
So, no sadness, I promise.
Final words to wish your soul well on its next flight,
Is sincerely, 'Love & Light'.

(I emailed the poem to Anuradha, as well as a voice note of me reciting it to be played at her funeral. The following morning as I woke up, I went straight to my phone and there was a video recording of her in tears telling me that she did, and everyone at the funeral was so surprised at even the concept of the love that we shared in such a short space of time, without even meeting or actually speaking to each other.

I suppose that kind of sums up life itself? There is no set rule book, and we have to acknowledge that everything does happen for a reason, and our perception of whatever happens is left up to us to decide.

I can sincerely feel her with me all the time. Like my God, like my Moonshine. Like all who I knew who parted. Physically gone, BUT always in existence in memory.

Sincerely, Mei truly inspired me to never give up, and I emailed the publisher in Mumbai my appeal directly, with her actual review. But after a few weeks, no luck.

Oh, I also actually printed and airmailed my book and appeal to another publisher a few months before. They are HUGE, and even *though my book was received because I enquired via email, I sincerely doubt that it was even looked at properly.*

Rightfully, in a sense, I do understand the multitude of submissions they receive. It is very hard to reach a publisher, especially as a first-timer.

But Mei inspired me to never give up. Maybe she blessed me with her want for many more to read it, like you are doing, because Anuradha requested the book. Even though she was very busy, she made time for this book and then sent me an email for me to use to entice a recognised wildlife group to take my book and help me spread my vision to the world.

You know what? Let me put that whole email in. Even though you know the 'story' already in a sense, hard to believe, hey?

To: manilalshane@yahoo.com Oct 20 – 2018 – at 10:55 AM

To whom it may concern,

Before I write anything about an amazing book that changed my life, 'Crash Landing, Born again for?' by Shane Manilal, I would like to share how unbelievable turn of events led me to it.

I am from Kolkata, India and my cousin Mei Yan from Shanghai, China, in August, told me about Shane, an author from South Africa with whom she had made an acquaintance with on online Scrabble. Her voice was dancing as she told me about his awe-inspiring book that she loves. She then added that she adores and worships Shane, feeling divinely drawn to him in a sense. To my utter disbelief, she said he reciprocates her feelings.

I felt very happy for her. Just two days after she told me though, she met a horrible accident and passed away. I was torn, but I felt my duty to find this 'Shane' guy and inform him. I found him on Facebook. He was crestfallen but maintained his composure. He wrote a compassionate poem in honour of Mei, which left everyone at her funeral in tears.

We became friends, and I got to know a real saint. He has an enormous heart and is truly spiritual. He loves ants, birds and actually worships Mother Nature in a sense as he whispers to trees and values water as an actual precious commodity.

He absorbs the agony of everyone he touches. World needs a spiritual leader like him. Under him, I have become an EtheREAList and attempt to educate indoctrinated minds to see Mother Earth as our primary concern.

I have even offered to help him to pay one-third of the cost of printing the first 500 copies of his book.

It's kind of help to make Mei's dream come true because she also promised to help him to get his book out there because she really wanted people to assess their ways as we nonchalantly, very sadly, destroy our very planet.

Now about the book, which I believe will change your outlook towards everything you ever had faith in as it reaches people on a spiritual level combined with scientific reality.

With engaging candour, eloquence and wit, Shane narrates the inspiring chronicle of his life – a first-hand experience as to how to make peace with your inner self. He is a firm believer in fate and destiny.

Shane believes that he was reborn to bring change. He is an Agnostic Totalitarian Pantheist, which matches the principles of his cult 'EtheREALism' in his book. He wants people not to believe in orthodox norms blindly and rather view them scientifically. He is requesting people to leave the shallow realm of religions and racism and delve into camaraderie in all humanity.

We all need to engage in sustainable development and love Mother Earth as our priority. Shane has come back for a reason that is much-needed.

Thanks

Love & Light,
Anuradha Agarwal

Now do you believe me? LOL ☺ *My dream just grew even more. Anuradha became an EtheREAList in India. Plus, Roberta in New Zealand. Hmmmnnn?*

*(The 1st and 2nd Ethe**real**ists are now very close digital friends as well after I introduced them to each other. Truly lifted my dream for Ethe**real**ism to unite the world.)*

*Ethe**real**ism? Hmmmnnn? Live your OWN life how you want to. Be true to what you feel inside. Even IF you feel like an outcast, so what? Point 'A' to point 'B'? You are the pilot of your own plane. Do whatever makes you happy. Feel inner bliss. Again, your own way to feel like a Happy Chappy. Even doing strange and apparently weird things.*

Oh, I have to point out that I disclosed to Anuradha that I now only eat two meals a day—lunch and supper. I skip breakfast to feel hunger every day, to just, in that short time span, feel the same as at least half of the rest of us.

Anuradha actually scolded me, BUT I am super healthy, and everything is 100%. Do we really need three meals a day?

I took this vow to myself after I had some hungry children knock on the door of the Mandir to ask me to break fruit from a tree in the yard. They told me that they never eat the whole day and that the fruit was probably the only thing they will eat for the day.

My heart truly broke as I allowed them to freely break the fruit without damaging the tree. After they left, I thought about what they were going through. I then realised that I never really felt actual hunger.

My inner being then said to me to not eat for the whole day but only drink water. I survived.

The following morning, I researched and found out about Gandhi's hunger strike—28 days. I then saw that essentially you can survive without water for just over a week and food for around two weeks.

Hmmmnnn? I survived for the two weeks with just water and energy drinks. A bonus was that I lost ten kilograms because my belly had grown a bit more after my mum bulked me up again.

*Oh no, I never told you another very important story before Ethe**real**ism.*

Okay, to highlight my point about steering your own journey, I can choose for even that important story to wait for just a bit because I almost forgot to tell you about an essential ingredient to my current mindset.

Let me take you back in time again – January 2017 to…

Huff and Puff and Clean up Your Own Backyard?

At The Grey Goose B&B, just outside Newcastle CBD, every Saturday morning, there is a five kay Fun Run/Walk. It is an organised event controlled by the national Park Run committee, with the ethos of exercising just for fun.

I have recently joined. Lovely to be outdoors and run in the fresh air in natural surroundings. You can walk it also, so I have just introduced my mum, Aunt Maya and their friend Koshir to it.

Huffing and puffing after I had completed my run, with the best time for my age group I might add, LOL ☺, I was cooling off, socialising and making friends while I waited for my mum and them.

This event has a very friendly atmosphere that I love and sincerely believe that with the current day and age, mankind is losing his inherent camaraderie stature.

Just being in natural surroundings really settles my soul. Standing on a balcony overlooking a pond, I spoke about the beauty of nature with a man who I had just met.

From the top, however, we could see a few bottles and cans floating around in the pond. I expressed how we are losing touch with reality. We are ruining Mother Nature, our MOST important imperative for the survival of our species.

I continued about how I sincerely believe that we all look up to the 'apparent' heaven and ask for protection and safety, yet down here, where we actually control that, we are ruining it. No matter what race or faith you 'belong' to, at the end of the day, Mother Nature should come first because we know 'She' is real and is changing because of us.

While talking to my new friend, I noticed an elderly man enter the pond with a refuse bag. He straddled along and started collecting the litter.

Even though my muscles were sore from the run and I was tired, I gave my mum and them, who had just completed and were ready to return home, my shirt and shoes, asked them to wait for me and I joined the man in the pond.

What an experience! As we were paddling along and swimming at some stages, we spoke a lot, sharing knowledge and wisdom, both very heartbroken about the arrogance of humanity. ☹

I was more shocked to learn that the gentleman was a doctor, almost in retirement. He had noticed my paralysed right hand and scar on my throat from the trachea when I was in a coma. He was very concerned about me being in the water, but I explained that it was mainly fatigue from the run and that I am managing.

We removed an entire bin bag of rubbish in like half an hour and got out near my mum and aunts who were patiently waiting at a concrete table under a tree.

The doctor asked if we had had breakfast yet, and mum explained that we are on our way home to do that. He insisted to get us breakfast there, stating that I deserved it.

Proudly, I smiled and explained to him that, as we had just spoken about our lives and what we do, internal reward and personal appraisal is all that matters. You don't actually need appraisal for anything you do.

But the doctor insisted and rushed off to return with four breakfast vouchers because there is actually a restaurant there. Have to follow doctor's orders, right?

I dried up and went into the restaurant with Mum and them. It is buffet style and I am an ovo-lacto vegetarian, from a planet conscious perspective—but let me not divulge about our omnivorous biological stature and physique that leans more towards vegetables as opposed to meat now because I will get to this later—so perfect.

There, I see that the doctor had returned to the table he was actually eating at and left, and now he resumed his breakfast, still wet, but heart and mind at peace.

I then found out that he is Dr. Wilson, whose practice is actually on the premises. In the restaurant, as we were still at the table, he came by to thank me again, and I joked with him that I had helped him clean up his backyard.

The next day, I found out that he actually owns the entire Grey Goose itself and is probably a millionaire because he also owned the place with Donkey's Pass, where I ran to my inner journey.

Dear reader, I chose to put this story in as advice to the rest of us in a sense. Can we not clean up our own backyards and, more importantly, stop littering other people's and, MOST importantly, Mother Nature's?

Also, everything you do leads to something because Dr. Wilson and I became friends, and the following weekend, I gifted him this book to read.

In a few months' time, he confirmed to me that I should continue my search for a publisher and that he would write the Foreword.

He understood me and my dream to help Mother Earth primarily as well as change people's mindsets, so, actually, very cheaply I might add, he rented me a cottage at Grey Goose to open up 'Love & Light – Holistic Healing, Counselling, Life Coaching & Parapsychology Centre' in April 2017.

The Grey Goose Lodge is only five kilometres on the outskirts of Newcastle but is truly a sanctuary for all nature lovers, a feeling of inner peace and tranquillity that I wanted to share. Took me some time to set up, and I even employed a struggling man from Zimbabwe to make me the copper sign outside that I took a photo of with my Goddess and her Goddess (My only direct sources to my bloodline).

Basically, the perfect spot for me to assist people with my life experience, PLUS studies because aside from my three university degrees and two certificates, I also completed online courses in Counselling, Life Coaching and Angel Card Reading.

My loving cousin from Benoni, Avinaash (remember Michael Jackson concert trip, plus printing flyers and stickers for my nightclub in XS in G.town?), designed my flyers and business cards, for which he gifted me 500 off.

All was set up and I was ready to open up, but my local newspaper never did the article, for which I was interviewed, to inform my town about my practice. But the following week, they published this article...

Shane Manilal leaves the biker life for good

Quinton Boucher

Shane Manilal's passion for motorcycles has finally come to an end which he crashed his motorcycle on Friday, April 14.

"This was my 13th crash and I am now done."

Mr Manilal is opening a counselling practice at Grey Goose and wanted to show a friend what his offices looked like.

"On the way there, I saw some liquid on the road but thought nothing of it at first."

However, on his way back he realised too late the liquid on the Memel Road was oil, and not water as he initially thought.

"I was in front of my friend, in fourth gear of my bike, doing the likes of about a 100km/ph, and as I entered the 'S' bend, I could feel my bike starting to slide. I tapped off on the throttle and forcefully positioned it upright, not to slide down, as I rode into the vacant opposite lane."

However, during the process, he was flung off his bike.

"Luckily, my friend slowed down because at first he thought I was stopping to show him something." Lying in the veld unconsciously, his friend frantically searched for him.

"I was lying in between long grass and it was only when I regained consciousness I heard him calling for me and I was able to answer."

The ambulance service was summoned and Mr Manilal was rushed to the hospital.

He is now being treated for a dislocated hip and damaged ligaments in his right knee.

Mr Manilal also sustained bumps and bruises.

"My first crash was in 2004, where I flat-lined more than once and the crash left my hand paralysed. But then I didn't feel much because I was unconscious for six weeks and wasn't awake to feel the pain and see what my family and friends went through. Now I can see and feel everything," Mr Manilal believes the oil spillage was what caused him to lose control of his motorcycle.

"An eyewitness told me there was a car crash earlier that day and if the scene was cleaned properly, I would not have slipped with my bike."

However, he is taking this as a life lesson and will not only be leaving his biker lifestyle behind, but will be focusing on helping others.

"I believe something is keeping me alive for a higher purpose," he concluded.

Shane Manilal is now on the road to recovery and is looking forward to pursuing his dream of helping others

*This following story is my most important story, which I remembered earlier, after I said mum bulked me up, a few pages ago, that before Ethe**real**ism, I have to disclose my...*

Lucky Number?

My 6th crash with Yummy. Major one with Blade (31/01/2004) and I also had six minor ones with other bikes, which I never mention in this book due to size reduction – One with a friend's scooter, and five on my old BMW 650 GS, 'BeeMee', that I bought cheaply in 2013 to give Yummy a break.

Also, BeeMee was dual purpose, so I could go off-road as well. But my last off-road crash, in the beginning of 2017, damaged the engine and was too costly to be repaired, so I sold it to a scrapyard.

With BeeMee, I had quite a few stories that were in the longer version of this book, most important of which was my 3200-kilometre solo, round trip to Port Alfred for the Sunshine Rally, where I won the prize for 'Furthest Biker' and was featured in BikeSA Magazine.

When I gave my talk that Saturday evening at the Sunshine Rally, explaining my One Blood Biker ethos, I sincerely felt connected with the crowd. That made my whole long trip worthwhile.

What made it MORE worthwhile is that I had probably my most blissful night ever. I was in a log cabin of the Nature Reserve, The

Woodlands, in the small town of Kenton. I was 'inside' of nature, connecting with Mother Earth, which you will only know about after you try.

Okay, return on Yummy to my 13th motorbike crash in total? Don't blame the bike and all and all because I also had 11 minor car crashes as well. Crash junkie? Mountain bicycle a few times as well.

Well, this was my lucky number 13 for motorbikes but not for Yummy. This time, major damage all around. An accident write-off even IF I had a million rand to repair.

Yummy is R.I.P, at my biker friend Ravi's workshop. Kinda too painful for me to bring him to the Mandir to look at all the time, and I have made PEACE.

More importantly, with me now being at home with my Goddess, I will not get on a superbike again, and tempt fate, coz she is my priority. I am a biker by essence, but I don't need a bike to crash!

My sixth crash with Yummy, also oil on the road, led only to a dislocated knee and hip. But, quite sadly, I was in no coma now, and I actually felt the pain this time.

Bedridden again, eating Mum's lovely food with no burnout. After three months, even with my extra 10 kgs, I could limp along with a small pain inside my left knee. This may just be constant but...

...everything does happen for a reason, and I can clearly remember, as the bike slipped on the oil on the road, I actually did nothing. It became kind of brighter as I slid off the road. I truly felt my God with me and saw him above the speedometer. Maybe I was imagining, but I heard him say to me that he left me in charge to take care of my Goddess.

I actively did nothing, and Yummy banged into the mounting cable of an electricity pole, threw me into the air and then tumbled up and down quite a few times.

I landed and rolled a bit until I came to a stop. Luckily, I was not riding alone this time, and my Muslim biker friend known as 'Ibby', who was behind me, slowed down as he saw everything happening.

For those first two minutes, I suppose I blacked out from the shock and because I was amidst long grass on the side of the road, Ibby could not see me, and I was not responding to his loud cries for me. He had assumed I had passed away and was praying to Allah.

But it was not my time, and I felt my God with me or maybe Allah sent me back? Being stuck to the ground, I shouted back at Ibby, who found me and made a few calls to the hospital and ambulance service.

As I was being wheeled into the private hospital that my loving sister Resh was paying for alone this time, I saw Mum. My first words to her were, "I'm sorry, Mum. I will not ride a bike again while you are my priority."

Plus, I had, quite funnily, a few days before the accident, read an article about the exhaust fumes from bikes damaging our other mother's lungs more than cars. AND, basically, for me, I ride for fun. Not a necessity.

So, my fun actually contradicts my inner belief to focus on Mother Nature first? Hmmmnnn? Just thinking about motorsport in general. Formula 1? MotoGP? Drags? Rallies even, in nature. All to actually ruin our Mother for fun. Not a necessity at all.

All part of this money-making world where we have honestly lost our essence, or even knowing that what we are doing is detrimental, because we are just one of the millions who rev their motors, hey?

Again, IF you make the change and all of the millions actually change as well, then it will be met. BUT will you even change in the first place? Think.

Who is responsible for the change in you? Kind of funny how you realise things only after something happens, hey? To me, there is a deeper meaning to all that happens.

Accept it, and work with the outcome, NOT wishing and hoping that you had not done whatever you did. Always remember that it could have been worse, then too, only you are the judge of the outcome of what happens.

Shit Happens, & Life…?

Why is it that we can think back?
Analyse and regret, make changes, wishing some things never happen?
Our actions, we attack,
regretting what we did to make our spirits dampen.

However, is that not the power of thought?
Everything learns from what it does.
Even after losing a battle you truly fought,
you gave your all to what WAS.

That is exactly it, it WAS meant to happen,
you can't go back & change anything,
but your spirits you dampen,
wishing you did this, wishing you did that, wishing, wishing, wishing?

Sadly, we have that capacity,
but what's meant to be, is what's meant to be.
Even when it's not what we wanted, that's destiny!
Accept & move on, set your mind free.

Learn from your mistake, yes,
but let go of all the stress,
shit happens the way it is meant to, just go on, do what you got to and can do,
coz, whatever it WAS, it does lead you to something new.

Embrace the change,
placing you into a tangible range,
to achieve something, at least
not totally slaying your dream's beast!

Chapter 2

Second Half – Etherealism

No, I am not riding a bike again. Too much shit to smell. LOL ☺ This intro picture is just to inform you how being on the bike is also part of what shaped my mind.

I am known as 'One Blood Biker – Brother With All' – the patch at the back of my waistcoat.

I also redid Yummy as a 'Rasta bike' – Yummy – One Blood Bike. This picture was taken just two weeks before my lucky number 13 crash and was when I made the YouTube plea to search for a publisher – Oneblood biker

https://www.youtube.com/watch?v=ooTNk_zqih0

One Love, One ♥?

I am back from the dead,
and need to spread what's in my head.
I had a terrible motorcycle crash,
life was gone in a flash.

But when I was in the coma, you see,
family, PLUS different races & religions, all prayed for me.
'One love...'

Also think on this fact,
and stop the act.
When in hospital, and you need a blood transfusion,
it just matches the type, no faith or race confusion.
'... one ♥...'

Now, on the road, ALL bikers just greet each other,
even though we come from a different mother.
That's why I'm back on a bike primarily,
coz you don't know what race or religion I might be, but wave at me.
'... let's get together...'

Even though disabled on my previously main right hand,
I love being a biker who can fly, in fantasy, with humanity as a united band!
My shifted focal point is Mother Nature.
Who we all need to nurture.
'... feel alright!'

(Want to feel alright? That will be your decision to make. I am merely going to offer you a serving. Taste it. Like it? Swallow it. Don't like it? Spit it out. Always remember that you are in charge of your life, no one else.

At the onset, let me clarify that I am not an expert as such. Yes, I have university degrees and studied anthropology and read a lot, plus watched quite a few documentaries, like my favourite about the universe: 'Cosmos: A Spacetime Odyssey' presented by Neil deGrasse Tyson, who cleverly places the lifespan of existence on a calendar. (Available on YouTube)

There is a faster, like five minutes, National Geographic one, look up 'Origins of the Universe 101': https://www.youtube.com/watch?v=HdPzOWlLrbE

Let me just summarise this for you even faster. According to Wikipedia:

(https://en.wikipedia.org/wiki/Cosmic_Calendar) Big bang – 1st January – our solar system only on the 2nd September, our planet's first rocks on the 6th September, biotic life 14th September. Okay, let me speed ahead to dinosaurs. Only on the 25th of December. Then 26th December – mammals, 27th December – birds, 28th December flowers. On the 30th December, dinosaurs die out.

31st December – 6:05 am – Apes. At 14:24 pm – hominids. At 22:24 pm – primitive humans. At 23:44 pm – domestication of fire and ONLY AT 23:52 PM – ANATOMICALLY MODERN HUMANS – US!!!

Our main focus should be us, right? Watch National Geographic's 'The Story of the Earth' for factual scientific proof of our planet and then BBC documentary 'Origins of Us' by Dr. Alice Roberts, to trace human evolution.

Dr. Roberts also made a lengthy 5-part series, 'The Incredible Human Journey', explaining more about us, how humanity spread out of Africa and then populated the world over thousands of years, while physically, externally adapting to their climates. Genetically speaking then, we are all African because that is where Homo sapiens originated! Fact!

You should honestly get access to these to educate your indoctrinated mind. YouTube is divine for knowledge. Remember my taste principal?

Also check out 'Origins: The Journey of Humankind' presented by Jason Silva. All facts and history laid out to you. My tasting has led me to my intellectual approach to our origins and existence. Only you can feed your own...

Thoughts?

At the end of the day,
as Homo sapiens, we are a living organism,
now ruling the Earth, our way,
with our thoughts, knowledge and wisdom.

What does set us apart,
aside from biological superiority,
is our mental capacity, to learn & give a new start,
because we 'think' we control destiny.

But do we really?
Yes, funnily, this book is being written by me,
but surely something plans what I am thinking and writing?
You, my friend, will decide whether to continue reading.

What you do next, I cannot say,
as you may give this a sway,
or carry on, and be at a loss of what to think or say.
Hectic, hey?

I even watched different possible alien involvement theories, like 'Hidden History of the Human Race – Everything You Know is Wrong' by Lloyd Pye, which does seriously question evolution's missing link.

Also, Gregg Braddon in his book 'Living From the Heart' postulates that, yes, evolution has taken place, but our development goes way beyond the linear pattern of evolution.

It is, as he states, a perennial question of life. We all want to know: 'Who am I? Where did I come from?'

To lay it straight, there is no exact way for us to know the truth about how life got here and reasons, etc. But what we do know is the scientific fact of what happened after it got here.

Yes, it can also be seen as theory, but no scientist claims exclusive authority. It is merely postulated for you to decide upon. Basically, we are here and can think about us.

Personally, I rather work with the recognised thinking Homo sapiens—doctors, scientists, archaeologists, palaeontologists, professors, etc.

Even the knighted Sir Albert Einstein was not religious. Knighted? Hmmmnnn??? Sir David Attenborough, who saw first-hand when working with nature and making the most amazing documentaries, is agnostic—spiritual, but NOT religious.

Also, renowned thinker Steven Fry, who travelled around our collapsing natural world to see animals on the endangered list before their extinction.

Or even atheist actress Julia Sweeney, with her amazing biography lecture as a scared little girl growing up in a staunch Roman Catholic school.

Hmmmnnn? Basically, people who use their cranial matter now teach us that we are not on a flat Earth anymore and now know that the answer of 1 + 1 equals 2.

In my research, I was surprised by the number of influential people, especially those from my ancestral homeland as well. Too many to list, but here is the link. Open yourself and see?

https://www.quora.com/Who-are-some-influential-Indian-Atheists-and-Agnostics

Maybe many remember the most interesting television show ever in 1988. Legendary men Stephen Hawking, Carl Sagan and Arthur C. Clarke did 'God, the Universe and Everything Else'.

https://www.youtube.com/watch?v=HKQQAv5svkk

Such a long time ago, influential people attempted to educate the masses. But the majority of us fear being different and actually using our cranial matter. Educate your own mind!

Also, in 1994, Carl Sagan, the renowned astronomer mentioned earlier, gave a talk that gets to the essence of us actually assuming that we are the centre of the universe. A totally informative lecture from a great thinker whose mind was way ahead of time.

https://www.youtube.com/watch?v=6_-jtyhAVTc

Save the best for last? You have to check out Hollywood legend Morgan Freeman, who hosted the 'The Story of God' series for the National Geographic Channel.

https://www.nationalgeographic.com.au/tv/the-story-of-god/

Morgan Freeman also hosted 'The Story Of Us', which shows you the basic truth through scientific fact as well as our history.

https://www.youtube.com/watch?v=gxG39vHxsuQ

My basic essential for pointing you in the 'thinker's' direction is to trigger you to also think. Hmmmnnn?

Start working with our reality—evolution has happened! Even IF aliens helped us to become human (missing link) to get us across that last step. We do not know yet BUT are learning and finding out so much more fact.

We do belong to the hominid family, directly linking us to apes. Roughly 98.4% DNA matching with chimpanzees! Yet, the mass majority of our species are actually indoctrinated, trained in a sense, to close their eyes and believe otherwise. A simple explanation of my understanding of our factual origins will follow for you to see, to reassess, question and seek answers.

Oh, before we begin, you need to adjust your time frame because the numbers are not in our usual, every day ones, but HUGE! According to physical evidence, carbon dating, our planet is roughly about **4.5 billion** *years old. Let's make it easier to visualise. Think of the number 4.5 billion figuratively as a full head of hair.*

Now, humans are only proven to have been on Earth an estimated 200 000 years – 350 000 years. Upon thought then, that is basically just a strand of hair (+ – 350 000) on Earth's full head (4.5 000 000 000).

Our calendar roughly lasts 10 seconds. Yet, we are honestly destroying our aged Mother. We are making Her 'bald'. Strangely, in such a short space of time, compared to Her like 4 months of existence?

Human 'progress', even though making life easier, is actually working against us in a sense – 'regress'? We know that. We have kind of taken over the planet. The damage seems done. We can do nothing in a sense.

Our capitalised greed is destroying us and our beautiful Mother. Again, maybe that is the master plan of our Creator so that the Earth can also be a destroyed planet?

Again, maybe the meteor strike, KT Asteroid, was meant to destroy Earth but messed up, and even without dinosaurs, life progressed, with us eventually evolving and taking over?

Again, maybe we even surprised the Creator of the universe with our existence before IT could reload to send another comet?

Maybe the Creator knew the whole plan and that we eventually would come along and destroy our planet?

Maybe, maybe, maybe? We will never know. But we are alive now and know the reality of our situation.

Okay, full stop. Let me not digress any further with probabilities and my solace and discrepancy of us destroying ourselves and provide my understanding of where we came from.

Let us start at the top, shall we? Again, stay in your adjusted time frame, as these occurrences are theorised to have progressed over billions of years.

Apparently, according to the standard theory, our universe sprang into existence around plus/minus 13.7 billion years ago. Our universe is thought to have begun as an infinitely hot, dense, something—a singularity that exploded. Where did it come from? We don't know. Why did it appear? We don't know.

After its initial appearance, it apparently inflated, expanded and cooled, going from very, very small and very, very hot, to the size and temperature of our current universe. It continues to expand and cool to this day, and we are inside of it: incredible creatures living on a unique planet, circling a beautiful star clustered together with several hundred billion other stars in a galaxy, soaring through the cosmos, all of which is inside of an expanding universe that began as a singularity which appeared out of nowhere for reasons unknown.

This is the Big Bang theory, an inflation of the universe in a sense. Prior to the singularity, nothing existed, not space, time, matter or energy—nothing.

So, where and in what did the singularity appear if there was nothing before it? We don't know. We don't know where it came from, why it's here or even where it is. All we really know is that we are inside of it, and at one time, it didn't exist and neither did we.

We do know that the early Earth, like all planets, was kind of formed by a swirling collection of space debris that took millions of years to gather and shape. Basically, a gravitational gathering for the elements in space, which are actually remnants of burst stars and other burst planets. That means that the universe is actually in motion, with explosions and debris floating around. In our solar system, this gravitational gathering happened nine times, giving us our planets that circle around our sun/star.

Yes, there are other, even larger stars/suns out there, with so many possibilities of different planets in their solar system, much like ours. There's even a possible tenth planet in our solar system.

We basically can only know as far as we can see and access with the Hubble Telescope that sends unimaginable images back to us. Our universe is so HUGE that we cannot even fathom the size. What the Earth is then is truly a grain of sand on a whole beach.

With the mighty Earth as merely that grain of sand, we as the dominant species are honestly not even a particle on that grain of sand in the universe's beach. Basically, we are nothing! More?

What if our whole beach itself was just a grain on a bigger beach or even a desert? And then that desert is just a grain on a bigger desert? Hectic? Totally possible though.

Sorry to digress in my usual bantering, but before I proceed, just want to tell you quickly to access 'What the bleep do we know?' on YouTube. It is so much as entitled. We truly know nothing much. Let's get back to the picture for me to share with you the little that I know.

When the Earth was formed, its gravitational pull also uniquely brought in liquid from somewhere, making us the only planet in our solar system with water.

The most likely theory is that comets with icy centres plummeted the early actively volcanic land mass gathering of fragments of drifting material.

Now, this most valid theory is kind of hard to think of or even imagine, I know. This process is millions of years in the making. Our made Earth was... hmmm... let's see.

Think of a fruit. A litchi. Now the peel is our atmosphere, and a long time back, the juicy, fruity part was the water. It covered the seed, the land mass, entirely.

This is actually getting trickier to explain, but that litchi metaphor worked, I think? The trickiest part comes now—life on the litchi?

Apparently, in the flesh of the litchi, our water, a kind of bacteria was present. Millions of them. Water is not only the key source to all living things but actually is alive as well. It has the spark to generate life also.

Simply think of moss on a moist stone. It is life. BUT from where? Even mould on a sandwich still in the lunchbox after a few days. There is life all around us, it just needs the proper conditions to germinate.

Initially, the estimated guess is that around 3 billion years ago, the bacteria that formed learnt to reproduce and more millions were produced, living off each other. Sustenance is every living thing's requirement, so it ate each other. The bacteria had to, therefore, change and adapt to outsmart those like it—the start of evolution.

Some remained, but some developed into other bacteria and created new life forms—like plant life was first underwater as well. This life mixture was like a chemical solution, brewing and turning out different results, like the amoeba, a single-cell living organism, and also early crustaceans to evolve further into other life forms.

With development and evolution, these life forms took millions of years to form and reshape to adapt. Some developed into fish and other naval life forms, depending on which evolutionary path the species followed. There were differences.

Then, over millions of years, and volcanic activity, the oceans receded as the land mass, the 'seed', grew and surfaced as the water, the 'fruit', also steamed and evaporated. This process added moisture to our atmosphere.

The surfacing land mass still contained the soil and particles, chemicals from the stars, the 'root' of the Earth. Now, over millions of years, exposed to air and gases, this chemical mixture, along with bacteria, led to different life forms and vegetation; just as on the ocean floors as well, it developed in the air now.

These varied in their own evolutionary paths, with different plants and insects. The gases were also formulating themselves. The plants still breathe in the carbon monoxide and out the oxygen, the most important gas for life on the Earth that now had air, its lungs.

Oxygen was a whole new ingredient to add to the broth. This was chemically possible in water for the living things, but now it was out of the water and in the air. With the new recipe, the land also produced life as vegetation spread.

After a few million years, some of the marine life forms surfaced onto the land. Fossilised evidence of the approximately 375 million years old 'Tiktaalik' shows a fish with legs that waded on land.

Much like the mudskippers of today, who spend more time out of the water. With evolution, over millions of years, fish developed limbs to walk on land also. (The very rare coelacanth is still around.)

On land, along with limbs, the creatures developed lungs for the oxygen from the air. Some progressed as evolution lead them to evolve into reptiles. Also, cold-blooded. Basically, walking fish. That led to the reptiles and dinosaurs. Some went back to the water, even though they were air-breathing—whales, dolphins, seals, etc.

Dinosaurs took to the skies as well. Fossilised imprints show some with feathers. Before knowing any of what I know now, I have always been fascinated by feathers. I collect them. I am amazed at the lightness, intricate design and also their attachment to the birds with so many variations and colours.

Okay, that obviously leads us directly to the birds. Guess what? Birds are actually proven to still maintain the lightened skeletons of the dinosaurs. Even the shapes of skeletons. Imagine in your mind an x-ray of a chicken. See the bones only. Now imagine a Tyrannosaurus rex skeleton. Same thing. Remarkable, hey?

My, oh, my, oh my. Hmmmnnn? Remember my science project with Skeech when I put a chicken skeleton in a glass tank, and I said '...like a fossilised dinosaur.'? Well, well, it actually scientifically was one, and surprisingly, my science teacher never even knew. Hmmmnnn?

Pterodactyls, the flying dinosaurs, took to the skies and also evolved into bats—early warm-blooded mammals. Those that could not fly evolved into rats.

At first, in the reptilian dominated days, they were scavenger rats. Ready for a shocker? Even though all life is linked in a sense as the same, life is life, be it in a tree or insect or us! The shocker here is that our genetics link us to rats, the very first ground mammals. Shocked? Well, all mammals are linked to rats!

Let's sum it up for now then. The main point is that all this can be proven by fossilised evidence. This story of reptiles and mammals actually occurred.

The Earth was alive billions of years before we even got here. This is proven because our oldest fossilised evidence as a species is only roughly + – 350 000 years. My analogy again – a strand of hair, on a once full but now balding head due to us.

Our Mother was alive before us. Yes, She went through a, let's say 'hiccup', as it still cannot be a hundred per cent explained. This is the KT Asteroid strike that apparently shifted the Earth slightly. The geographical shift that this caused affected the climate. This change in climate led to the extinction of the dominant species, the dinosaurs.

Life still survived though, as some species adapted. Dinosaurs became birds, etc. Life adapted and evolved. (Funny side note: Apparently cockroaches are survivors from the dinosaur days. Even some crocodiles are still the same! Okay, let me stop 'roaching' and 'crocking' around and get back to business!)

That proven business is new life's adaptation on a new Earth, our Mother. This proof comes from what we now know. With life progressing and still evolving on our Mother, let's get straight to the matter. Where did we actually come from?

Rats also survived the meteor strike. Over millions of years, some grew and ascended to the treetops, with limbs adapting for the branches—the root of our primates as various monkeys and apes developed.

Straightforward answer again: yep, we have come from them. Our evolutionary strand led us here and to answer the most frequent despondent question!

The rats and apes we evolved from were not on the same path as those that still exist today, so that is why they are still here.

Think of the trunk of a tree with different branches. Same starting point but ventured on different routes with leaves and flowers.

Here is Wikipedia's quick summary of Human Evolution, I downloaded this excerpt.

'**Human evolution** refers to the evolutionary process leading up to the appearance of modern humans. While it began with the last common ancestor of all life, the topic usually covers only the evolutionary history of primates, in particular the genus *Homo*, and the emergence of *Homo sapiens* as a distinct species of hominids (or "great apes"). The study of human evolution involves many scientific disciplines, including physical anthropology, primatology, archaeology, linguistics, evolutionary psychology, embryology and genetics.

According to genetic studies, primates diverged from other mammals about 85 million years ago in the Late Cretaceous period, and the earliest fossils appear in the Palaeocene, around 55 million years ago.

The family Hominidae diverged from the Hylobatidae (Gibbon) family 15–20 million years ago, and around 14 million years ago, the Ponginae (orangutans) diverged from the Hominidae family.

Bipedalism is the basic adaption of the Hominin line, and the earliest bipedal Hominin is considered to be either *Sahelanthropus* or *Orrorin*, with *Ardipithecus*, a full bipedal, coming somewhat later. The gorilla and chimpanzee diverged around the same time, about 4–6 million years ago, and either *Sahelanthropus* or *Orrorin* may be our last shared ancestor with them. The early bipedals eventually evolved into the australopithecines and later the genus *Homo*.

The earliest documented members of the genus *Homo* are *Homo habilis* which evolved around 2.3 million years ago; the earliest species for which there is positive evidence of use of stone tools. The brains of these early hominins were about the same size as that of a chimpanzee.

During the next million years, a process of encephalisation began, and with the arrival of *Homo erectus* in the fossil record, cranial capacity had doubled to 850 cm^3. *Homo erectus* and *Homo ergaster* were the first of the hominina to leave Africa, and these species spread through Africa, Asia and Europe between 1.3 to 1.8 million years ago.

It is believed that these species were the first to use fire and complex tools. According to the Recent African Ancestry theory, modern humans evolved in Africa possibly from *Homo heidelbergensis*, *Homo rhodesiensis* or *Homo antecessor* and only migrated out of the continent as Homo sapiens, some 50,000 to 100,000 years ago, replacing local populations of *Homo erectus, Homo denisova, Homo floresiensis* and *Homo neanderthalensis*.

Archaic *Homo sapiens*, the forerunner of anatomically modern humans, evolved between 400,000 and 250,000 years ago. Recent DNA evidence suggests that several haplotypes of Neanderthal origin are present amongst all non-African populations, and Neanderthals and other hominids, such as *Denisova hominin,* may have contributed up to 6% of their genome to present-day humans.

Anatomically, modern humans evolved from archaic *Homo sapiens* in the Middle Palaeolithic, about 200,000 years ago. The transition to behavioural modernity with the development of symbolic culture, language and specialised lithic technology happened around 50,000 years ago according to many anthropologists although some suggest a gradual change in behaviour over a longer time span.'

If you want more proof, get a hold of Richard Dawkins' book 'The Greatest Show on Earth: Evidence for Evolution'. Scientific theory has fossilised bones as proof, and carbon dating. That, to me, seems more believable than a story in a mythologically holy book!

We are hominids, but our difference is that we are totally bipedal, walk upright with two legs. Plus, we have opposing thumbs to grip and manoeuvre things around. This shaped our increased brain capacity.

The root of bipedalism amongst the earlier hominins, which also saves a lot of energy when moving, plus limits the sun's direct contact with the body, has various theories.

Oh, before I give you more information, I only realised now that you are wondering if I made a spelling error with hominid and homonin, even with the Wikipedia excerpt. Same root, but there is a slight difference in new age books.

Hominid – *the group consisting of all modern and extinct Great Apes (that is, modern humans, chimpanzees, gorillas and orangutans and all their immediate ancestors).*

Hominin – *the group consisting of modern humans, extinct human species and all our immediate ancestors (including members of the genera Homo, Australopithecus, Paranthropus and Ardipithecus).*

I am fond of two theories for the first hominins – Sahelanthropus or Orrorin, a full bipedal that stood upright and stayed walking on two feet, the root of us.

Before I outlay them, I would also like to reiterate that theories mean a mere possibility. Some will agree and some will disagree, but remember to taste first before you decide.

1. Savannah Theory
2. Aquatic Theory

1. Savannah – Before being majority a savannah, like the Great Rift Valley, apparently, the African part of Pangea *(Earth's supercontinent with all of our current ones joined together)*, the land, as one huge continent, was a rain forest. *(Yep. The land on Earth was joined together. Pangea – just look at the shapes of the different continents and think of a puzzle. All are pieces that join together. Easy?)*

   On Pangea, the monkeys that actually evolved from some of the first mammals, rats, were in the trees. With less swinging in trees, some gradually lost their tails and evolved into apes. But over millions of years, as the continents drifted and climate change transformed a once rain forest into a savannah, the apes now had to travel on the ground as well.

   Fossilised hominin species like *Sahelanthropus* or *Orrorin* decided to raise its head above the grass to see and also watch out for oncoming predators. Over thousands of years, this mannerism stuck with it, and it would walk on two legs, freeing the hands—bipedalism.

2. Aquatic – This theory is that when the continents split, yes, we had a joined together land mass as mentioned—Pangea. But the Earth shifted. In fact, it is still shifting under a centimetre every year. Apparently around smaller islands, like Madagascar, the apes were in knee-deep water as well—swamps.

To breathe then, because of wading most of the time, they had to raise their heads above the water. Thousands of years pass, and they become fully bipedal.

Why I kind of lean, no pun intended, to this theory is that childbirth for women is actually easier underwater, and newborn babies who have baby fat float, with an attached umbilical cord. Hmmmnnn? Makes you think, doesn't it? But it's just a theory!

What is not a theory though is that our basic root is confirmed to be in Africa. Different hominin species existed, kind of smaller but still bipedal. Basically, in our evolutionary chapters, certain species died out or evolved, but the genetics progressed.

As the brain grew, hominins that used their opposing thumbs more and more increased their cranial capacity. With bigger brains, they became more upright.

In Africa, because of the heat and the sun, initial Homo sapiens had pigment in the skin, short curly hair, a broad nose, thick lips and a muscular structure, especially the derriere. This athletic build helped them to run away from carnivores on open plains.

Over time, however, some wandered around. Even in Africa, the different landscapes sculpted different Africans. Think of the tall Kenyan Masai Mai tribe, in comparison to the short Namibian San and Koi Koi. Or even the shorter Pigmies in the Central African rain forests?

To me, this is solid proof that climate and region dictated adaptations to the same species. However, this proof is a kind of an easy link putting Africans together, but humans in the north of Africa, around 90 000 years back, moved out of the top of Africa.

Possibly, they crossed the land shaft near the Suez Canal that links Africa to Europe? Or another theory was made by Dr. Alice Roberts, when she linked emigration through what's known as 'The Gate of Grief', linking Africa to Arabia. Or even possibly a land shaft linking Morocco – North West of Africa to Spain?

We cannot specify which one exactly, but humanity was definitely on the move and adapting to the changes. In Europe, the colder climate, lack of vitamin D on the skin due to a lack of it from the sun's rays, led to lighter skin colour, longer hair to keep warm, physical changes in the structure of the anatomy as well, smaller nasal canal, thinner lips, decreased buttocks, etc.

A bit to the east and it gets more interesting, shall we say. Aside from skin and hair colour, Iraqis, Arabs and Indians are very similar in structure to Caucasians.

Maybe they came back into Africa and became the Egyptians who bear a very similar racial structure? Hmmmnnn??? There is honestly lots of debate as to whether Egyptians are 'African' or not.

Okay, let me return to India, which is obvious proof of distinct adaptation difference, to enforce my climatology point. In the lower parts of India, as this is also a hot equatorial region, there's dark pigmented skins. In the centre of India, however, it was predominantly cooler. Here, the body adapted for the 'medium' complexion Indians. Further above India, close to the cold Himalayas and a weaker sun, the skin got even lighter—the fair Indians. Just like the current Muslims who dominated the Pakistani part, with Iran and Iraq area to the west. Around Turkey and Greece, because of the cold, the eye colour also changed with the skin lightening a bit. Hair remained mainly darker. Italians as well.

Further up into Kazakhstan and Russia, the paler Caucasians. Lightest skin, narrow nose, different hair colour and eyes. Also the same effect when moving westerly towards the United Kingdom.

At the top of Europe, there was actually connecting land masses kind of linking Europe to North America. Mankind moved on to the west. This was roughly 50 000 years back. Think Iceland. We all know of the Vikings, right? Pale and red headed in the cold there.

Eventually, some progressed through Canada, reaching America 30 000 years ago. Here the complexion changed a bit as well, as the summers were warmer and the Native Americans, commonly known as the 'Red Indians', lived closer to the coastlines. Pigment added to the recipe along with darker hair.

This biogenetic make-up remained similar going down south more to the Mayans, who were around Mexico, and further to South America as well.

On the other side, the east of Europe, from above the Himalayans, humans ventured into China and adapted there as well. Apparently, the sun's rays reflecting from the glaciers made the eyelids bigger, therefore the 'Chinese' eyes.

On a southern venture, below China, humans established themselves on the various islands in the Papua New Guinea area, even reaching Australia and then New Zealand. The southern climate is hot like Africa, so physically not much difference between the Aborigines and Africans really.

Now, in my kind of rough layman's explanation of the evolution of us, I never gave a different species as we are all the same—Homo sapien. Just the biological exterior differences to match the climatic conditions.

Think on this: If a human from one race, let's say Caucasian, needs a heart transplant, will an African man's heart fit if it's a total match?

Of course, it would. Why then are we still seeing each other as different? Even, again, think of blood transfusions. Just the type is matched, not the race or religion or even sex of the donor! Outside, yes, different adaptations, but inside? We need to finally realise that we are all the same—human beings, full stop.

Makes sense? Our species has originated the same way. It is just with the climatologically different areas we ventured into that we brought about different races.

We also, with our intellectual proficiency, created belief patterns that we spread and forced the tribes we took over to follow!

Okay, on that point. I mentioned Papua New Guinea earlier. Surfing on YouTube, I found an amazing video. You have to check it out: 'Tribe meets white man for the first time'

https://www.youtube.com/watch?v=5aV_850nzv4

Shot just a few years ago, the tribe were still in the stone age, living peacefully with Mother Nature, drinking from the river, hunter-gatherers, not destroying anything.

I was in tears, hey? IF only we all remained like that, but as the filmmaker stated, it will only be time before the people are 'discovered' and the 'modern world' is thrust upon them.

There are some islands, however, who forcefully keep outsiders out. Check out the Sentinelese in the Bay of Bengal:

https://www.youtube.com/watch?v=87XW5t6qydU

Basically, finding these undiscovered tribes paints a clear picture of what happened historically. The world itself was like a huge market of food stalls and restaurants. BUT with our inherent domineering persona, to take over other restaurants and kind of make chain stores, we forced other restaurants and food stalls to close.

To add some spice to the pot in the next chapter, let's think about what we normally taste in our created belief patterns that we just routinely swallow as we are indoctrinated to.

Before we proceed, again, I will request that all you have to do is THINK.

We Defeated You, Now You Must Do as We Do!

A straight-arrow – YES, it is true. We honestly merely follow a kind of fabled façade. Even though mythologically created, proofs of our, let's just keep the civil peace and call them 'simpler', ancestors swayed the need for accuracy, especially on their flat Earth, do the majority of us on our spherical one believe the ancestral roots of the 'simpler' ancestors' flat Earth theories?

Hmmmnnn? I am remembering a movie from the eighties, 'The Gods Must Be Crazy', where while flying over the Karoo desert, a pilot drinks a Coke and throws the bottle out of the plane window.

This bottle then lands on a Khoi-San, bushman's head, and the still rural, 'undiscovered' tribe think that it was a gift from God. Just a bottle became so many 'magic' things.

Quite a hilarious movie where possession of the magic bottle actually causes animosity and rivalry amongst the tribe. Hmmmnnn?

Sad human flaw? Yes, we all want 'things' and will fight for them, but my point here with the movie is that it is SO similar to actual creation of ALL religions.

In simpler times, the masochistic man did make religion. In each area, we actually created different belief patterns, based mainly on our own, not only climatic differences but lifestyle differences as well.

This is because our mythology rooted itself in our experiences and what we wanted. What made these beliefs spread though is our inherent domineering persona that took over the neighbouring tribes and forced them to follow our crazy Gods.

Yes, that is actually how mass religion spread. Dominant tribes converted who they took over into following their mythology or die.

Looking at the world today, there are around 4200 documented religions because each one was formed in its own, much smaller world.

Our world is HUGE, and honestly, I do not want to take a direct punch at all established faiths yet, which I do have lots of discrepancies with, BUT let's come alive with a poem to set the par.

Life

In today's world,

even though knowledge and intelligence,

has, at mythologically 'created' religions, a few punches swirled,

divine power has to certainly be in existence.

Something, at the onset, had to have made everything?

Earth as a magma planet hit with aquatic comets, our solar system, and a universe that's light years beyond.

This in itself, logically, has to be the master plan of something?

Even factually, those billions of years ago, placing life in the water at 1st, as merely just organisms, floating around in the pond.

What and who exactly, how can we know?

Huge majority of us stick to what we are taught,

basically, an indoctrinated, fabricated show,

about our God, the Creator, and the battle He, or for feminists, She, fought.

We, however, fail to realise,

that we do have the intellectual capacity,

making us wise, to see that all indoctrination is just a fabled disguise.

Therefore, why not open your mind and eyes to actually see who is crazy?

We are on a huge, spherical planet with life,

that changes and evolves,

scientifically linking everything together, as we continue this strife.

In our own religion, we all look at prayer and faith, that apparently every problem solves?

The unsolvable problem though,

is even when linking all established faiths to ONE Creator,

who made that Creator, and then the maker of the maker, and obviously, the maker of the maker OF the maker???

Infinite question that, frankly, an answer we will never know.

We think we do, by sticking to our indoctrinated, even chosen beliefs,

seeing what others follow as wrong,

hitting their life's ship, on barricaded reefs,

as we sing our correct, holy sailing song.

But, at the end of the day,

how do we even know that we are right?

Why are there so many others with their own songs and way?

Is it a battle we want to fight?

Maybe after taking in what you just read,

a thought came into your head,

that you rather just leave everything to fate,

taking it all in, as a set destiny, opening your spiritual gate?

BECAUSE even when you think that you are in control of your destiny,

what if that is also controlled, you see?

Hectic, hey? So, find your inner peace and balance,

in whatever worship, or non-worship, you do.

Just know for a fact, it is not an act, life does tighten its wince,

and one day, you too, like everything, have to go.

So, make the most of your time here,

because when you die,

you don't even know where you going, but fear.

Heaven? Hell? Reincarnation? Spirit??? Just remember, what you do know is that you are alive, so let your soul fly.

After this existence, where it goes,

who the f### knows?

BUT what to do with it now in the physical form as you,

only you, yourself do!

Because, at the end of the day,

Only you control your way!

My poetic ventilation, especially with the profanity, sums up the entirety. Upon analysis of every faith, there are huge discrepancies when looked at from 'out of the box', with an intellectual mind. Our main discrepancy is that we are indoctrinated to believe that our belief pattern is right and that all others are wrong.

Well, if acceptance of human beings, us, as just one species, that then means that roughly 80% are wrong because they have their own beliefs? Hmmmnnn???

To better understand this point, allow me to use a simplified example to explain directly this mathematical equation I postulated earlier.

Take only the five major religions: Christianity, Hinduism, Islam, Judaism and Buddhism. Just in theory, equally, if the world is divided by them, that would give each one 20% to make up the 100%.

Basically, then, whichever of the five you follow, that then means that 80% of the rest of humanity does not follow it. Hmmmnnn?

Also, even within the 20% of your religion, there are major differences between different sects. We sculpt and mould, forming more mythology that we pass on. Old ways and manners die out or are beaten away by the dominant beliefs of the ruling tribes.

All the ancient civilisations followed what we now see as outdated beliefs, classified as heathenistic—the Mayans, the Egyptians, Native Americans, etc.

I was always drawn to the Native Americans' ethos of living in harmony with the Earth. That was the basis of their existence. In fact, probably the biggest genocide EVER is when they were slaughtered in the name of 'Civilisation'.

What's sadder is that this history is kind of overlooked by the majority of the world currently. To many of us, actual history just 'happened', can't change it, so we move on.

Well, what actually happened to like between three and five million Native Americans is that they were wiped out to around only 280 00 in the 19th century. Wars, plagues, European diseases, capitalistic westerners, etc. wiped them out. TOTALLY inhumane, according to today's ethos of the United Nations. ☹

Upon thought, there was a scene from the movie 'Dances with Wolves' with Kevin Costner, where I really shed a tear and cried.

Synopsis if you never watch: the early 1800s in America, a soldier is solo at the western frontier. He befriends the Native Americans. Even finds them the buffalo herd. Before the hunt, the natives actually pray for the souls of the animals that they will kill and thank the spirits for giving them the chance to.

On the hunt, even though there were thousands, they only kill a few. Then, from the buffalo, they use up everything—meat, fur, horns. Nothing goes to waste.

My sad scene was watching them shift tribe on the run from the westerners, and they then pass a field of buffalo, slaughtered and with only their furs removed. Fur trade? Money world?

Straight talk: Is this world that we are destroying really all about the money? Maybe access 'Ben Fogle: Where the Wild Men Are'. In fact, Ben Fogle is also an amazing adventurer. Just Google his name if you don't know him.

Oh, let me digress to another recent movie with a similar premise, 'Avatar'. A whole planet is destroyed to make money. ☹

Everything is money. Even the current topic, 'religion'. Let me return to the fact that so many ancient cultures existed, BUT violence and warfare either kept them or their ideologies running.

Think. An example to bring to book, that many would identify with, is Ancient Greece. It is actually classified as Greek mythology now with all its superficial Gods, actually around 60 of them, who helped the Greeks form some of the greatest thinkers of the modern world.

But as the Romans, who spread Christianity, defeated the Greeks around 1800 years ago, they had to actually crush all their beliefs and follow the Roman Catholics. My point here is that we can all now see Zeus, the leader of Greek Gods, with his 'magic' lightning strike, as a created character, BUT in our own 20%, we cannot see how the rest of the 80% see our own Gods as created characters.

Let me focus on the religion that most of us know a bit about, even if we don't follow it. Historically, two centuries ago, Europe's dominant Romans, led by Emperor Constantine, founded Christianity, an outcast sect in a Muslim and Jewish Jerusalem. There were a few disciples of Jesus Christ and about a hundred followers for around 300 years after Jesus was actually killed by the Romans.

Emperor Constantine then took favour to it. He then set up Constantinople as a city rooted in the new belief for the then pagan Romans.

After gradual progression, the literate Romans then formed Christianity by writing the Old Testament Bible. They made a religion. Think? Hmmmnnn? ROMAN Catholic.

Where is the Vatican?

What nationality is the Pope?

Also, more importantly, Jesus was not even Italian. LOL ☺ In fact, Jerusalem is actually around FIVE THOUSAND KILOMETRES away from Italy.

No offence again, but I'm using Christianity as it is the most spread, not only with beliefs but the Western world and ways with different races, which, quite funnily, actually, the holy book does not even account for.

*Do we **all** come from Adam and Eve?*

They were Caucasian. Yet, we have Blacks, Indians, Chinese, etc. So many different races. Yet, they are also Christians? Hmmmnnn?

Colonisation, the spread of the Western World, forcefully spread Christianity. Fact. Warfare. This take-over of others is 'politely' referred to as the Holy Crusades where the natives were now finally routed to God and Jesus after being forcefully taken over.

Think on this: If God made the first man and woman, Adam and Eve, quite funnily with belly buttons because all their pictures have them. Umbilical cord connected to who? Hmmmnnn?

Anyway, they were cast out of the Garden of Eden. They then had children, that means that the children had children with each other? Incest? Genetics? More importantly, again, then how do we have different races?

Or also, furthering the 1 + 1 equals 3, where is the proof that holy mother Mary was a virgin? Did they even know what a hymen was? AND to generate a life form, the male penis must open the hymen 'gate' by breaking it.

Then the act of sex, for the woman to also have an orgasm and produce an egg cell, which is then penetrated by a sperm cell, giving you a fertilised egg cell. This fact then means what exactly? God put his sperm cell into Mary how?

Also, logically, with the hymen intact, it must have been one hell of a painful birth procedure for poor Mary when Jesus literally broke the hymen from the inside while coming out. Think.

Unmarried woman falls pregnant. To avoid embarrassment, parents make a story about being a virgin with 'God's' child? Possible?

Even afterwards, when the Roman Empire created Christianity, an imaginative writer may have made Mary a virgin because, at that stage, the Roman Empire had taken over Egypt and Neith was the virgin Goddess of creation. Possible?

Honestly, stories grow and grow, just like the fish that one caught. You tell a friend, who tells a friend, who tells a friend. Fish becomes huge! Again, straight talk breaks no friendship.

Yes, Jesus may have been gifted, but solid proof? He was born as a Jew in fact and possibly created his own system. Kind of like a spiritual leader who possibly did some miraculous things.

Think of David Blane, who is a remarkable magician, illusionist, etc. from another realm. I saw him survive just over 16 minutes underwater on the Oprah Winfrey TV show. Also, he lived on poles, levitates and even never freezes when submerged in frozen water.

Now had he been alive back in the day, he would probably be worshipped today. This is because all faiths have this disparity to create mythology to make their fish grow.

Fish? Water? Hmmmnnn? The biblical 'Great Flood.'? Where exactly, did that water, to cover the whole planet, come from? And where exactly did it go afterwards? Hmmmnnn?

Then too, was it freshwater or saltwater? Because a river fish cannot swim in the sea and a sea fish cannot swim in the river. Hmmmnnn? Well, what about crocodiles, alligators, seals, penguins, etc. Did they swim all the time? Hmmmnnn?

Plus, according to the Bible, the ark was around the size of one and a half football fields. Sounds quite big, but really now? Could it really fit the WHOLE world of animals?

I am just going to work with rough estimates here, but there are around 5500 mammals. That means, as pairs – 11 000. As pairs also then, 21 000 birds, also around 21 000 reptiles. Oh, what about around 2 million insects. Hmmmnnn?

Let's use reality and just pretend that the modern world with global travel and communication, plus amazing ingenuity, needs to build an ark to hold pairs of every living thing on land. Possible? Hmmmnnn?

Also, we may have to deal with lots of wildlife protection institutes, hey? Because, come to think of it, what did those poor animals do to be punished by an evil God with a flood, merely because man did not listen to Noah?

But, upon thought, the ones on the ark alone would have been punished as well because it was a sealed ark in order to not let the rain in. No ventilation. Flatulence? Hmmmnnn? Need for oxygen because methane gas levels would have been impossible to survive with. Think.

Oh, before I depart from the ark, basically we have to now forget Adam and Eve. All of us come from Noah, who lived for like 3 centuries, and his three sons and daughters-in-law. They were also Caucasian though. Hmmmnnn???

Also, IF we all did come from Noah's family, why then 4200 religions, 6700 languages and around 1500 cultures now? Hmmmnnn?

Okay, too much against the around 280 different types of Christians, who actually changed and adapted their indoctrinated roots to match themselves. So much so that even within an established mythology, there are huge differences. One's belief pattern actually goes against the other's, even though rooted in the same. Hmmmnnn? Straight talk, even racially separated 'White Churches', 'Black Churches', 'Indian Churches', 'Chinese Churches'. Own ingredients to suit mass patrons' taste etiquette of a plate that was forced to their ancestors to feed off a few centuries back?

What about people actually being killed IF they never wanted to eat and were not Christian? Think Spanish Inquisition in the 18th century. Or even the historic Salem Witch Hunt, where the majority rules and poor innocent women were killed because?

Because they never followed what the mass controllers dictated and eat what was force-fed. Hmmmnnn? More relevant example to us. Hitler? World War? Kill the Jews because they don't eat pork?

Think. Okay, again, apologies to Christianity for getting all worked up and using you as an example. It is just that I can think. You 'think' that the world is ONLY 6000 years old and that the Great Flood happened 4000 years ago. I just want to postulate to mathematical geniuses who are the best at thinking to work out for me HOW we mass populated into over 7 billion of us in ONLY 4000 years from Noah's three sons?

Life itself is like a complex mathematical equation. We have missed around 4 billion steps to the solution, BUT now we THINK we know the answer!

Hmmmnnn? Let me just be like Moses, who is actually the root of Christianity, after taking slaves out of Egypt, and tap the sand with the magic stick and make the ocean part as I straight talk another religion as an example.

In Hinduism, my own indoctrination, to the believers of which I do apologise, but there is a most holy monkey God, Hanuman, who once grew so enormous that he carried the flat Earth (the whole world) in his one hand.

Hanuman was also the leader of a massive talking ape tribe that left no fossilised evidence behind. Hmmmnnn?

Quite strangely, ONLY in South Africa do we erect a holy flag, the 'Jhunda', to worship Hanuman in our yards. Aside from it being unique to South African Hindus, it is only the Hindi speaking Hindus who do it and not the Tamil speaking Hindus. Hmmmnnn? Culturalisation?

Well, the flag was first only a red one. Nowadays, there's also a green one. And a yellow one? Follow the next-door neighbour, even though you don't even know why?

Hmmmnnn? Not sure IF I still, with this new edit, kept that I pointed out that, quite symbolically, when I did type about the mythological discrepancy in 2009, about the Jhunda, the next day, one of the two outside my yard was blown over.

That alone can be seen as a holy sign? Like a story I hear about a family that had a dog, and before their prayer rituals, they used to tie the dog up to do the prayer. As time passed, the elders died. The ritual continued, but after the dog died, they would actually get a dog to tie before the prayer.

Honestly, my own racial and religious people, Indian Hindus, are SO superstitious that, to me, we escape reality the most. Yet, surprisingly, the majority of us do belong to the higher intellect from 'The East'.

Yet, we bow in salutation to a God with an elephant head and human body, Ganesha, who apparently lived on Earth a long time ago. Just the actual concept is impossible.

Hmmmnnn? Maybe, a long, long time ago, one village had an albino elephant; it was unique. So, they used to dress it up like a human. Then it died. But the story lingered. Maybe even the neighbouring village heard about it and it became a human with an elephant head. Plausible?

Okay, curiosity got me. I Googled and went to http://www.english-for-students.com/Lord-Ganesha.html to laugh even more. Want to laugh as well? Go check it out.

Just remember that on the flat Earth, they could attach an elephant head to a beheaded human being, then make him holy! Amazing things on that planet, hey? Especially in India.

How about Gods and Goddesses with lots of arms as depicted in the religious pictures? How about different colours, like blue? I was told lots of stories, mythology.

Non-Hindus are seeing the story and getting the picture, but trained staunch Hindus, who close themselves off from the rest of the world, are probably cursing me rotten.

We do that. We see the stupidity in the others, not our own. We close ourselves off in our own worlds and think that our cooking tastes great and the others', even though palatable for them, suck. We can't swallow their offering because our plates are full.

On that note, some dietary items which are acceptable, and even praised, are actually prohibited in other cultures. Hindus are primarily now not supposed to eat beef.

In fact, initially, all meat was prohibited. When Hinduism, the world's oldest documented religion, was created, roughly 5000 years ago, the belief in reincarnation was strong. Therefore, vegetarianism was practised because that sheep could be your late grandparent or someone else's. The vegetarianism, as preached in the holy book, Bhagavad Gita, states that 'slaughter is the way of subhumans'.

This stayed rooted in some, especially the pundits and priests, but the majority of the others bit into something meatier. With the world's progression and colonisation spreading, the exposure to meat dishes was there.

Maintaining the root of non-beef consumption, as the cow is seen as the mother and is still worshipped in India and is illegal to slaughter, *other meat became legal, even worshipped. Yes, animals are slaughtered for holy reasons. WTF?*

In South Africa, some Hindus sacrifice goats and some even do pigs. So then, pork is a holy meat to some Hindus, yet actually illegal for some? Hmmmnnn?

With the Jews and Muslims, pork is actually not allowed to be eaten at all. Both these religions prohibiting pork implicates the obvious rooted growth around fabled fallacies spreading with mythology because they originated in the same area, central Europe, think Turkey, Israel, Iraq, Iran, etc.

I am merely stating my mind, sincerely not picking on any belief or thought pattern, because at least the Jews and Muslims must pray for any other animal they slaughter, making it kosher and halal, but do you factually know that Homo sapiens are not even biologically designed to consume meat?

Again, think for a bit. Can you actually eat raw meat? We are not designed to digest raw meat. Cooked meat, yes, with our, what we assume are, canine teeth, but they are small. Factually, earlier hominins, like Homo habilis, nicknamed 'the handy man', probably scavenged off burnt animals due to veld fires.

The taste of the protein-rich meat then actually prompted them to keep the fire burning, so they could cook meat to consume. Then it probably got them to actually realise, while making their stone tools, that banging the rocks together made sparks.

Quite obviously then, because they did have a bigger brain, Homo habilis learnt to make fire, probably the most defining part in human evolution. Controlling and making fire not only meant consumption of cooked meat but, also, a practically defenceless creature in terms of physical build and attributes could now come out in the open, leave the caves and scavenging and actually have something to scare away the predators.

Think. Fire is the most essential thing in our lives and control of it meant we now had power – campfires. Biologically, now, the cooked food softened. A simple example is that a cooked carrot is around 10 times softer than a raw carrot. This now meant a very crucial aspect of hominin evolution: the big muscles for the jaw bones reduced in size, giving smaller faces and also bigger brains.

Personally, my bigger brain, a good few years back, made me an ovo-lacto vegetarian after I realised that consumption of meat is actually not healthy for humans. Yes! From a biological standpoint, the only meat that we can digest is cooked meat.

From an ethical standpoint, you need to watch what inhumane acts are done, all for the production of meat. If you just Google 'animal slaughter', you will get access to many videos that will seriously make you question where the meat you eat, which you 'think' is healthy, came from. More importantly, what life did that poor animal live?

That leads me to my next deciding factor: the consciousness standpoint. The animals do not really get old. A lamb is still a baby, yet somehow or the other, the human mind is trained to think of it as most tender.

Now, I know that most readers are thinking of the common saying that 'meat is meat and man must eat!' Yes, the past man, our forefathers, on their flat Earth, did not know any better and assumed meat to be healthy, but it really is not.

Scientists have proved that excessive meat intake, especially red meat, is dangerous. Again, research. Educate your mind. Vegetarians do have higher life expectancies than meat-eaters. Like the island of Okinawa in the East China Sea. It holds the largest concentrations in the world of people over 100 years old. Proof is in the pudding? Your choice though. Your life. Think.

Honestly, think on it past just the food we can and cannot consume. We are a 'masses follow masses', conditioned species. We don't even know why, but we follow. And when we hear of other humans following their own 'silly' beliefs, we giggle. The world still has so many different, thoroughly followed beliefs.

Here's just a few to mention. In India, one village, still to this day, prays for and worships rats. Rats roam the village, untouched. They are fed and even bathed. The village even has a tunnelling system built for them. Hectic?

Another one in the Far East that I merely just caught a glimpse of is where the penis is worshipped. In actuality, yes, it does 'make' life, 'but cum on?' (pun intended)

There are so many. What do we base belief on? Good times? Pain? Relief from it? Bearing it?

Hinduism has a strong belief in trance. This is basically a holy possession in a sense. The Hindu guns are blaring at me now again, so no offence, but I'm heated up.

Let's start with that. The blessed trance state of walking on burning coals. Guess what, the skin of the foot is actually thick enough for a short time on burnt coals. Get this, corporate companies now use it as a team-building exercise.

Sticking with the trance state, the penetrating of the skin to hang adornments and also tug the holy wagons in the procession of the ritual feast known as 'Kavady'. All I can say here is that it's body suspension, with the exact same principle of penetrating the skin and then suspending the body with attached cables, which means there is no holy presence.

Our bodies can maintain pain. There is a masochistic mid-African tribe which actually lines pubescent girls for selection for marriage, to enjoy the rest of her life with a man. Staying in traditional dress, only a piece of skin around the hips and pubic area, the girl stands and braces herself for selection. If the patrolling man likes what he sees, he takes his leather whip and lashes out at her a few times, goring and splitting the skin. If she moves, he does not take her, and she is ridiculed and led off. But if she can send her mind into a state where she does not feel the pain, she wins and is chosen. The mind has so much power.

I beat my pain to get a few tattoos. Those who have, know the pain involved. On this point, the Maori tribe in New Zealand actually worship bearing the pain of tattooing as a sign of blessing. It is an ancient practice and is still done with just a sharp needle that is dipped and banged with a mallet into the flesh. No tattoo gun, which, yes, is some pain relief in a sense, but the old way really is double the pain.

Here, the rite of passage is for a pubescent boy to enter manhood. Therefore, the more pain he can take, the more of a man he is. This process lasts a few days, then a further month and finally a Maori man.

Amongst many cultures, there is also circumcision. Amongst the Zulu and Xhosa tribes here in South Africa, the teenage boys are taken on a ritual outing in the bush and the operation is performed by a 'holy man' with just a knife. Yes, there have been quite a few deaths due to this, but that kind of makes it more of a challenge in a sense. Only the brave survive means they are more special?

We all do that. We look outwardly and praise others, even discredit them. Like how in some cultures demonic possession is seen as religious and vice versa. This leads to the fact that the human psyche is so expansive.

Paranormal psyche episodes do exist and are even analysed for a greater understanding, but that leaves us with too many clichéd 'maybes' again. I have studied documented incidents of swamis from India slipping into a meditative state and can make their hearts beat once every 24 hours when they take a breath. Or how about, in the Far East, some Buddhist monks can meditate and make their bodies light and actually walk on fluorescent light bulbs that are vertically suspended on supporting rods. Even proven cases of reincarnation where a child remembers his past, down to minute details.

There is too much to even consider. Yes, I have also seen a documentary called 'The Guru Busters' that shows fake 'holy' men who use sleight of hand magic to entice followers.

That basically is the root of all established faiths to me. When we accepted this kind of trickery as holy things, we followed suit. That is how simple it becomes. We are honestly routinely indoctrinated to follow this stupidity because our prosperity, here or in the afterlife, will hang in the balance.

What about a logical mind that now knows the answer of $1 + 1$ is 2 and not the trained 3 though?

The following question actually summarises and encapsulates this mathematical equation. How come menstruation is still seen as bad *in some cultures? They did not know that that dirty blood is actually the shedding of the old egg cell to create a new fresh one to be used to actually be the seed of life.*

Times have changed, have they not? Why then do we still do as we are indoctrinated to do? Especially seeing places as holy—churches, temples, mosques, etc. Yet, they get bombed. Even people get shot when INSIDE them. Basically, religious warfare still carries on.

The most heartbreaking news I ever heard regarding this is a story about an 8-year-old Muslim girl being gang-raped to death by an actual Hindu priest and his friends inside a temple in India. The saddest part is that the Hindu community actually stood with their priest against the Muslim community. WTF???

Also, how come 'holy men' from all the different religions have been caught committing sex abuse on little children especially, huh? Which so-called 'God' allows for it to be done in 'His' name, even IN 'His' premises.

Is REALITY even going to enter the minds of the majority? Do we not have our own minds? Can we not have our own opinion?

Can we not use our brains and see that the root of all religion was the historical fact of the controlling tribe pounding in survivors that, "We defeated you, now you must do as we do!"

All About The?

I do apologise to the set religions that I pointed out before, and the rest, who I did not, are probably thinking that because of that, there could be nothing wrong with mine.

Step out of your own box and judge the logic behind the practices that you do in your own belief. Think. Where did it come from? Who made it? Why is it different?

Well, that last point is the basic gist again. Difference is, logically, the main thing wrong with all religions. A few years back, with the modern technology of sending pictures on cell phone services, on WhatsApp, I received a picture of a woman holding up a placard during some public gathering like a riot.

It read, 'We WERE all humans until race disconnected us, religion separated us, politics divided us and wealth classified us!'

True? Think. Basically, we ARE the same species, yet our climatological adaptations to our surroundings evolved into different races, disconnecting us. Then our ancestors created different religions to separate us. With politics of the groupings, the tribe headman or king or president or workers, etc., us being given roles, divided us, Finally, with the invention of money, wealth classified us.

The root of this was in our ancestors' flat Earth? Surely our minds have grown on our spherical Earth?

On that note, in 1953, the late science fiction writer Ron Howard started a very true and applicable to our modern world religion: Scientology. What I do appreciate about it is the scientific fact it is based on, but it does appear to be an elitist organisation, with quite a profitable market.

Members have to pay to join and are also apparently brainwashed into funding exorbitant fees for courses and classes. I do not have any solid proof, merely just research, but it does seem to be only for the rich, with a few celebrities, who are kind of marketed for their following of Scientology.

This does break it down to almost the silliest notion in our capitalistic minds that the rich are successful and have the correct approach to belief because they are making money—classification according to wealth?

Upon thought, is religion itself not a business? Every religion is an organisation, with the fancy holy 'offices' that obviously were quite costly to build, paid for staff to run it, a book that must be bought, trinkets, pictures, idols, etc. Even actual prayer services and ritual followings must be paid for.

Christmas is the most 'Festive Season' for money. We want to buy gifts for those we love; the more expensive, the more we care?

With our trained capitalistic minds, we think that the more we can pay and afford, the more blessed we are.

Well, yes, we do need money to live in this capitalistic world that man has made, but is our happiness and inner fulfilment really judged by how much we have and can make?

We need to honestly ask ourselves how much is enough really. Most of us who are employed or run a business work as hard as we can, to possibly give our children a great foundation by sending them to schools and universities so that they can also make their own money and keep the circle turning for their children.

When does that circle stop or slow down? When we retire and are a bit too old to actually enjoy life fully? But at least our focus to leave a nest egg, to be legally fought over, creating enemies amongst families, behind has been met?

When we die, there is then another capitalistic approach. We feel that the fancier our funerals are, with the most people attending, even the food served to the guests and also our tombstones and graves or urns for the ashes of those cremated, display the success of our lives.

Is making money really the ultimate success because it is also juxtaposed as being the 'root of all evil'! All crime is because of it. Think. Robbery, murder, hijacking, fraud, etc., all to make a quick buck.

The greatest evil, to me, is the destruction of our Mother Earth because of money. Think on it. The world economy is majorly based on what? Her 'blood'. The oil price determines the majority of other markets.

Also, Her 'flesh', the trading of gold, diamonds, coal, etc. Plus, her 'skin', the cutting down of forests, reshaping the land, etc. All we are doing is destroying our Mother. It is the world we made and have to live in.

According to a recent survey, at least half the world lives in poverty. Honestly, count your lucky stars because it could so easily be you who starves to death or goes hungry for a few days without...

Money, Money, Money

This poem starts with the main line from the entitling song.
You should know the words, so do sing along.
"Money, money, money... lots of money... in a rich man's world."
BUT there is only one world, into which we've all been swirled!

Majority of the people though can't even waste time to sing along!
They are just concerned with their need for something to eat.
Singing though will be a sad song,
because in this capitalistic world, they missed the beat.

Life truly is 'all about the money,
all about the dumb dumb deedy dumb dumb...
don't think it's funny...'
BUT we have invented something that makes inherent human compassion numb!

Our need and greed,
for a fabricated entity,
truly plants a seed,
to actually not see the pity.

Majority of us turn our heads,
look the other way,
thankful that it is not us with no beds,
and at least our children can have fun and play.

What games do you think starving children play?
Playing should be part of the growth procedure, hey!
To actively run around, even rumbling and tumbling on the ground,
Building up muscle for the future, to keep you safe and sound.

That, however, these poor children cannot do,
most of them do not even have a shoe!
They sit down, saving energy as their tummies rumble,
while we, for more money, grumble and grumble?

What do we take with us when we part though?
We do leave behind a nest egg,
for our families to want more, and grumble and grumble, we know!
Yet, we will still keep our wealth by walking past those who beg?

I also, at times, am guilty, and ignoring them is what I do,
because I do know that with the money, the majority buy some narcotic glue.
They sniff and get high,
forget about their worries, and soar in the sky.

Upon thought, at the end of the day,
For them, that is an easy way, hey?
To escape and forget about this crazy world we live in,
where everything is corrupt and money is the root of all sin!

Yes, the world we have created does need money, hey? People even do crazy things to make money. I actually wonder what we have reduced ourselves to, especially after I watch 'Broke Ass Game Show' on MTV that makes people do the stupidest of things, all for a piece of paper.

What exactly is that piece of paper we use? Who made it? What does it really do by itself? Hmmmnnn?

At the risk of being too rude, let me just say that 'lucky' you can use it to get toilet paper because I doubt it can even be used directly for that. (You are more than welcome to try IF you want.)

Yet, our whole world revolves around it. I always wondered who actually gets new money? Hmmmnnn?

Governments mint money, yes, BUT who actually owns it before it enters the banks and is circulated to our monopolised minds? Who controls our monopoly? Hmmmnnn? Even older money is actually thrown away and bunt up to introduce new money. Hmmmnnn?

Hey, just a thought. We actually ARE on the 'Monopoly' game board. Think. Different areas – different prices. Building property – making money for rentals. Passing 'Begin' – collecting money. Paying for flights and trains. Things are taxed. Even buying yourself out of jail. LOL ☺

Now, are we puppets with money as our string, on our Monopoly board? And we don't even know who exactly is controlling us. Hmmmnnn?

Too much? Well, as the old saying goes: 'Too much is never enough!' We are trained to want and want and want. Richest wins the game?

But we can't take anything with us after we die, then actual families start fighting for your money and things you left behind. We have become SO materialistic. Really is sad.

We do need it as puppets in our world, yes, but I can remember, some time back, watching on TV, a documentary about a secluded island, where they do not work with money. The tribe is kind of like one, big, happy family. There is an open-door policy of all the houses, no locks and keys, and when they get visitors and guests, they are treated like royalty. A very communal approach to living.

Everybody working together to get the basic necessities of life. Whatever food from the ladies scavenging around and harvesting vegetation is coupled with the men's fishing and hunting, and it's mutually shared at a big dinner table for everyone.

They are also always happy and cannot be sad or angry. The only time they can cry is when the men set out to sail on the beach. The main reasoning behind this is that given the rough seas and small rafts, the men may not make it back home.

During this time, all the sad emotions that were held back are allowed to be released. After the boats depart, the smiles return. IF a particular father does not return from the sea, his wife and children are taken care of by the village, kind of adopted in a sense.

I was amazed by the most important thing being that money was illegal there. And they developed a working system.

When I tried to find access to the village, I did a digital search and could not find the name of it, but it was probably a BBC documentary because I remember that an Englishmen was the filmmaker who had to live on the island for a few weeks and became part of them.

They truly cried a lot when he left because he made such strong bonds with them. The basic root being that there is nothing to fight for materialistically because they were all the same and worked together in unity. So much like the root of us all, the camaraderie we shared. Living together in peace and unity.

But then, money. No more...

Love All, Serve All?

In today's times, we,
unfortunately, do not see,
the effect of all we say and do,
especially harmful things that leave those we affect blue.

Personal need, greed & satisfaction,
seems to be the only action?
Have we forgotten that being together,
a social species, calmed our stormy weather?

Our original design was to help each other, camaraderie,
that today, you will be quite lucky to see.
Why though?
What changed?

Funnily, we do know,
our social spirits are deranged.
And this all boils down to what we have made,
you do nothing if you don't get paid.

It is a fact,
sadly, money is part of the act.
It rules all we do,
ridiculously made by us too.

BUT now, to help all and serve all,
the payment makes that call!

Just remembered studying Karl Marx and Marxism at university. Basic essence is Communism, where there is equality in all. BUT Communism fails because, honestly, we are inherently greedy and want more and better. Strong capitalistic ethos are engraved in us. Survival of the fittest? In our genetic make-up? We want to be the best and have more. ☹

But we are ruining our planet! Hmmmnnn? As mentioned earlier, maybe we were designed to eventually take over our planet and ruin her, like the many others in the universe.

Here's a very relevant quote by Hubert Reeves that is now passed around digitally:

'Man is the most insane species. He worships an invisible God and slaughters a visible nature... without realising that this nature that he slaughters, is in fact, the invisible God he worships.'

Think.

You may think that you will be like 'one in a million' who thinks differently. BUT the other 999 999 probably think the same as you BUT are too scared to state it. Are you brave enough to stand up for what you think and be actively the 'one in a million' who actually thinks?

We all have freedom of thought. Just be happy for your thoughts. Even IF it's different, at least stay true to you. Educate your mind. Feed it lots of food. Decide for yourself what you want to swallow. Simple.

Okay, time for me to serve...

Etherealism

I am honestly at a loss regarding 'making' a religion or a cult. As mentioned, I need to answer my entitled 'for' question and, to me, this is it.

The words of John Lennon's song 'Imagine' from 1971 ring in my head now: "I hope someday, you'll join us, and the world can live as ONE."

Upon thought, the majority of us just sing along, hey? Again, we are the puppets, controlled by the puppeteers, with money as the string.

We actually do nothing, but carry on, 'day to day'. Then, absolve ourselves with our own religions as we pray for our Gods to help us, not realising that the main problem is that all documented religions have set ritualistic followings for invisible Gods, BUT like Mr. Reeves said, our Mother Nature should be worshipped.

Once again, I will repeat that I am not condemning any belief system or asking you to convert or anything. I just want to air my views, like I have been doing, directly serving this plate on your table. Taste? Your choice to swallow or not at the end.

I do understand that belief systems are strong in most of us. Again, my family, for instance, are the most astute followers of our faith, Hinduism. We believe it to be correct and the answer to all problems.

The question is, why are there still problems? There will always be, no matter how much we pray for guidance and changes to our lives in our belief systems that are based on mythology. We are indoctrinated into our beliefs.

For all Hindu readers, I have found the English version of our 'Aum bhur bhuwa swah', our most auspicious chant that, as mentioned, even my nani (granny) and God and Goddess (parents) do not know the meaning of. To me, it makes total sense and deep inside me I believe that my reciting it my whole life, till 2009, led me to this, the 'right direction'. Analyse:

'O God, the giver of life, remover of pains and sorrows, bestower of happiness, and creator of the universe, thou art most luminous, pure

and adorable. We meditate on thee. May thou inspire and guide our **intellect** in **the right direction.**'

As English speakers, currently, why then can we not chant that version. The chances of actually being heard by our made 'Man upstairs', even if He exists in the first place, is very rare.

I told you earlier that, to me, there is a power, yes. It is totally undefined though. Like I classified with my little nephew at the water's edge, make it a 'mutual' totality by referring to it as IT, as IT can easily be anything. My main focus in this is that many people seem to profess it but not really feel it. Yes, we are mutual, we are ONE and the same as Anita Moorjani professes in her book 'Dying To Be ME', where she miraculously beat cancer while literally on her actual death bed and left doctors and medical staff stunned.

Anita kind of ascended spiritually over her physical vessel, which is what the body is for every living thing, and then saw life itself as ONE concept.

Basically, life IS really a single thing; the plant is a LIFE form, the insect is a LIFE form, the fish is a LIFE form, the reptile is a LIFE form, the animal is a LIFE form.

Our LIFE is in the human being form. Every human being is exact. Yes, we have physical, gender, racial, cultural, religious, lifestyle and belief differences, but we are essentially all the same.

Maybe we actually make differences because of the simplest of things—classification. We say African human being, Indian human being, European human being, Chinese human being, etc.

Why can we not say human being FROM Africa, human being FROM India, human being FROM Europe, human being FROM China, etc.?

We are all the same species and have created the problems, which we pray for 'above' to remove, but hello, our differences made them in the first place. I do know that it is too late to change what has happened, but to make a spark, might just burn even one away.

I will then be happy. The one that I want so badly to burn away is for us all to realise at least that we are all the same, regardless of our *created indoctrinated differences. Classic Bob Marley song – 'ONE love, ONE heart, let's get together and feel all right'?*

We did get together in PAST tribes. People that matched, in terms of evolution, stayed together because they were physically the same. That is understandable at first. We are an extremely social species. That also, however, brings the shit. We became tribal in a sense, sticking together, but there were also other tribes being formed that were physically the same as us but thought differently.

So then, we followed our inherent domineering nature, maybe under a leader that every tribe has, and competed with the others. That is nature. Think of all the communal animal tribes—lions, wolves, hyenas, wild dogs, etc.

This communal genre is not only for the carnivores. Herds of elephants, buffalos, buck, etc., then omnivores like us, certain monkeys, gorillas, chimpanzees, etc. In fact, all of nature sticks together with the same—birds, insects, fish, even flowers and trees. Everything.

That is the way life is. As the old saying goes, 'Birds of a feather flock together.' All life forms stick together to find a mate to further their species. Our mating and furthering of our species are making so many life forms extinct.

The worst is that if we carry on at this rate with our lifestyles, we will cause our own extinction as well. We say and say and say, some actively do but most of us do not. Why is that?

Yes, knowledge has only been gained in recent years but now we know. Is it too late? Told you Mother Earth is crying. We hear but do we listen? We are shown but do we see?

We blame it on Her moodiness, natural disaster. Yes, there are some, but we are making the biggest disaster that will come. We are destroying Her. Some of Her disasters are made by Her, for some reason or the other, to help Her, but we are destroying Her and causing more disasters.

Our formulated, indoctrinated beliefs need to address her and not just pray for changes. We can make them, can't we? I've been implying *change and changes to heal Her, our Mother, and to me, Ethe**real**ism will do just that.*

I am now juxtaposing myself, trying to make those set rituals that religion has to follow. At the end of the day, the inner feeling of bliss for doing those rituals does settle us inside, neh?

*So, even though Ethe**real**ism believes in freedom, it does have principles to follow. This is all from me and my belief in a kind of round table, not a single person's authority, is also being juxtaposed. Hmmmnnn?*

It needs to be done though. Yes, I am creating it because I have lived my life, the details of which I have given to you, mainly to show that I am a normal everyday person. Am I not? Well, except for my two different palms, I am the same.

I am not a spirit. I do feel pain everywhere except for one side of my dead hand. Where I feel the most though, as I was told by a psychiatrist that I was sent to in 2009, is inside. I am emotional, yes, I've had my romantic crashes and physical crashes, but I have survived with my emotions intact to harp on me again.

Now they are finally focusing on the beauty of life, all life. I actually feel guilty for overlooking things in the past. Simple things like throwing bones away in the bin to decompose in a plastic bag or to be burnt up.

There are so many hungry stray dogs or even just burying them to give Her nutrients back to Her. I also even mentioned crumbs, grains of uneaten rice for the birds. That feeling of joy for me, to watch a bird eat what I threw or an ant carry a breadcrumb, I never felt ever from praying.

*This is the focus of religious prayer, neh? That blissful feeling inside that you have made contact with something upstairs. Connecting with something downstairs is much greater though. And the best part is that it is so easy. There are no indoctrinations to follow, which is why my formulation of them for Ethe**real**ism is very challenging. Let me first explain it to you where it came from and so on.*

You now know what the main goal is: to fix her. So, again, for the gazillionth time, I'm not condemning your belief system, just asking you to analyse it slightly. Yes, the belief that the worst will happen to you, which all religions threaten if you think differently, is holding you back, but why can you not even think in the first place?

As I said earlier, even after you think, if you stick to your belief system, that is your choice. What's needed though is at least a pause to fix Her and then you can play your game with Jesus, Allah, Bhagwan, Judah, Buddha or whoever from the 4200 religions in the world that you want to. Maybe even in a congregation match with other players led by your referee preacher, priest, pundit, Dalai Lama, baba, Maharaj, maulana that sometimes change the gameplay to suit them or to match changes.

Even if you return to this, I will not mind. Who is to judge other than yourself? If you feel complete and satisfied and know what you are doing and why you are doing it, all's well that ends well.

*My stepping away is understood. Told you, even by my elders. Explanations will be countless, but like I'm saying, think for yourself. Now parents are probably saying 'but if we are not supposed to indoctrinate our children, won't we be doing that by being Ethe***real****ists?'*

*Yes, in a way, but Ethe***real***ism does not exclude anyone with any belief. Teach children that we are all one and the same. Yes, I'm using the five dominant religions and Scientology to formulate it, but it includes everyone, everything.*

*There are some tribal beliefs that are still followed in the 'unchartered' world. All our species have realised that there is a spiritual world around and floating above us. For the first time, I consulted my intense Colliers Dictionary to check what Ethe***real****, a word I merely had the idea of, really means.*

*With my verbosity, I kind of knew what the word meant and leaned towards and I liked the ***real*** part in it. My breath was taken away by the four 'different' definitions that are the same:*

Ethereal;

1. *very light and delicate; exquisite; spiritual*
2. *of or relating to heaven or the heavens*

3. *of or relating to the upper regions of the atmosphere, or space and beyond.*

4. *of or relating to 'ether' formerly supposed to fill space.*

Ether;

1. *hypothetical medium once used to fill all space and to transmit light and other forms of radiation.*

With the definition of ethereal, it includes everything around us. I added the 'ism' to make it a following. The basic premise is to worship our existence, the air, the space around us, above and maybe beyond.

In order for this to become physical, I touched on this a long time back as well, She needs to be the focus. While addressing Her, think simple and small. No littering, switching to non-CFC products, etc.

On that note, drive less, use your car less; the shop is just around the corner. Yes, I know we are scared to be mugged and things and our precautionary measures from each other are what is making us kill Her and thereby ourselves.

Yes, in the past, it was survival of the fittest, but why now as well? The answer is simple. Commercialisation in our modern world has brought it up. In the past, it was mainly to fight for food with other tribes, even eat them as cannibals. That is the premise of survival. The taking in of nutrients that invariably burn out and need to refuel. In our hunt for the nutrients, we are not only depleting them but creating conditions to kill them.

The rain forests are being wiped away. Commercialisation and the developments of cities are causing more ruin all around us. We even confuse bees with the cellular network coverage, so they cannot get back to their hives after pollinating.

Reality! The worst part is that we have to accept it. There's nothing more that we can do. We have to grin and bear it. What more can we do other than dream of getting the past back?

Oh, what would I not give to live as a 'Red Indian', Native American, in the planes of America? Worshipping a spiritual world as well as the natural world—Pantheism.

*As I said, both exist, we just have the wrong answers to a question we do not even know. Infinite question but like I said earlier, to me, the solution is that we are here, we are **real**, let's work with **reality** as one united force? Impossible, I know, but to believe I **can** is everything.*

*I think the focus, for now, needs to be here. Stop analysing where from and why but to just improve here. Like Pantheism, Ethe**real**ism can get followers. I will just kind of make rulings to follow based on my life. Again, if it's nonsensical to you, do not bother, even if you follow Ethe**real**ism, which I'm hoping will become a mass religion.*

Spiritual ritualistic beliefs centre us. Combine two beliefs? No need to deny your upbringing if you still agree with it or some of it. Keep the tangible. We are symbolic ever since the cavemen kept lucky stones that have eventually 'evolved' into engagement rings.

(Joke had to come. Why not? Preconceived idea of formality for religion? Basically, primarily, just be you.)

*Something tangible, that is **real**, when touched, comforts us. So then, firstly get yourself a Core. As I said, it is a Rud Raksh seed with five lines. I am specifying it because it centres me, but even if you cannot find an exact match, any bead on a string will do.*

With this permanent attachment around your neck, you actually will naturally feel for it, especially in troubling times and when making decisions. Place it on your third eye, between your eyebrows, with your eyes closed and calm your mind.

Let's call that 'Tikka' yourself because that is what the Indian dots are called. Take a deep breath three times, and your mind will settle and direct you once 'Tikkad'. Even without the breath, it will settle you and send you in the right direction.

The bonus is that you always have it attached to you. Whether around a chain or any colour string, it makes no difference. The Core is the focus. You decide what to put it on. Something will pull you to it, even a piece of unused string or a shoelace. Anything.

Right, tangible token, a Core, done. Now, I mentioned my tattoo. It is a symbol of Ethe**real**ism that I was drawn to. As I mentioned, it has our tangible **real**ity, our Mother, inside the Sun with the moon alongside and the sunshine splitting into a night sky.

Again, touching the tattoo, my mark, does so much for me. Even the process is spiritual, that is why I recommend one. It marks you for life. Preferably on your right-hand forearm so you can reach it with your left hand to centre yourself if you want. Again, works for me.

Also, similar to all religions, you can make a picture or a frame or something. I'm hopefully creating a marketplace for arts and crafts, which honestly is a thriving industry in religion. All religions are tokenistic, especially Hinduism. The things we buy—frames, models, idols, pendants, etc. Too many to mention. Once again, if you want to paint your own sign, knock yourself out. Art and creation settle me.

On that note, again, I'm giving you what I do and what I say, but you can word your own recitals if you want. Formulate your own belief system of ritualised followings if you want. It is your life, but we are social beings wanting to follow a grouping.

Okay, so adaptations are totally accepted. Every morning, I first wake up with a smile. I stretch in bed, then take a deep breath of the fresh air and I hold my Core and say that it will be a good day. (The Secret)

Here's my simple, FREE religious practice again. My prayer ritual before I eat is watering plants. If no yard is accessible, just a small tot glass of water in a potted plant. Told you, the same trick. Our beliefs are in our minds, not in what we do with what.

Sometimes I go outside, I look at the Sun. Then I close my eyes and first thank IT for giving me the Sun, then I thank the Sun for shining down on Her because without the light, everything will die.

I then drop the water into Her, quenching her thirst, also thanking Her for keeping me here and vow that I will do my best to preserve Her by blessing me with a good day.

I then must feed the birds and ants. Return indoors. Light my oil lamp, hold my Core to my third eye and pray for a good day and whatever I want, in my mind. Then blow out the flame. That's it. simple and easy, and I do know that I am connected to the Creator.

The day passes. I do my best to help Her, always questioning what I do. I've changed so much. I also love staring at sunsets. It settles me. At night, I always must try to see the moon and stars and also thank them for my being here and giving me light.

At night, in bed, be thankful for being here, even if you had a bad day. Tomorrow is another day, is it not? Or if you're expecting tomorrow to be a bit bad, just wish for you to do your best. I then hold my Core to my third eye and ask IT to give me a peaceful sleep if possible.

That's my daily routine. Now I just figured that as a belief system, there has to be ritualised followings like rulings. Let's start from the beginning of life, the baptism. I'll call this process the 'Ether'. Would be great for the newborn's parents to hold the baby up together outside on a morning or as soon as exiting the hospital. Basically, out in the open, fresh air, under the Sun, thanking IT for the gift and asking Her to mother their baby as well by touching the child to the ground.

Then, if you want, maybe get a smaller version of the Core or a holy bead that you ask for blessings with and put it on your child. Simple. Job done. Then, that night, you can have a party if you want to celebrate your child.

On that note, there are big ritual gatherings as prayer services. You are in charge here and can make it a party also IF you want. Just do whatever you want to. You are living your life.

Oh, love and marriage? We think love's epitome is marriage. Together forever and ever till death do us part, but then what does divorce do? Many people in today's world juxtapose this vow. Can we not simplify it with something real?

My fantasy I had with my flutter comes into play. Here is a simple wedding ceremony. Just hold hands and step into flowing water together. No one has to even be there.

Because, at the end of the day, only the bride and groom have to be there. Make your personal vows to each other. Then pinprick fingers and join blood. As it joins and trickles into the water, hold hands and recite...

Vow for Now

Our blood is ebbing,
and flowing,
together now in You.
It knows what to do.

It is joined and came alive,
hoping to thrive,
on being together,
through the stormy weather.

It does know of the perils,
the joys and the pending spills.
but is together now though,
giving You the bond to sew.

By shedding our blood together,
we'll do our best to cover any weather.
We will cherish each other,
hopefully forever and ever?

Even if not together when we reach the sea,
we feel blessed in our current unity.

My explanation of the poem is that, yes, you are together now, but the journey you send your joined blood on cannot be defined. It will flow in the river, hopefully together till the sea, but even if not together, separated in the journey, it will reach the sea, the final destiny.

After the Vow, embrace each other and seal it with a kiss. Inside, you are joined as your blood has.

In our world, you need to be legally registered though. In my view, a relationship is equal, so even when going to register yourself, who's the boss? Why not do a coin toss as to whose last name to use? Combine them or something. This will get tricky when registering children. Decide.

Then have your choice of a wedding reception party to show the world that you have joined lives together and let everyone bless you and help you to stay together.

If your own makers are still alive, they can give a toast to wish you well on your journey together, to start your lives as a team—the future makers of life also?

When you become a God or Goddess with children that you bring in, just try your best to stay together because nothing can be set in stone. Try though.

Oh, forgot to exchange wedding bands—the global signal of being married. Totally your choice. After the vow or a ceremony at the reception, maybe put a ring on your finger. This is your life you are making, so you decide.

You can decide on everything in your life, even how to send your spirit into a different realm when you leave this one. Figure out a way to give your nutrients back to Her or to be incinerated.

There are some cultures in the East that I researched who actually leave the bodies out in the open on a kind of altar for the vultures to eat. I would love to give my body back to Her. Even dumped in the sea or cut up?

Or, if buried, there's no real need for an expensive coffin to go into the ground. My nana wanted a simple bamboo 'Tikti'. From Hindu culture, it's just like a stretcher made of bamboo and covered with calico cloth. Simple and, to me, more logical.

Yet, we want to show off even when we are gone about our fancy coffin and even fancy graves and tombstones. Will they bring us back?

I would also, in today's big world, at least give people three days to come to my funeral. This is to let everyone have enough time to come and part with you.

It is a solemn affair, but once rested, I want a jol to be had, to let my spirit look down, smile and head off. I do not want to leave anyone in more pain, as their life must carry on.

Once again, take it or leave it. That's the beauty of Ethe**real**ism; there are no set rules. Why must there be? You can adapt. You just be yourself. Because, at the end of the day, what exactly do you conform to? What worked for somebody else when they created what you follow.

Yes, I am also creating, but I'm giving you a choice to follow, adapt and create your own even. You are an individual. Do what makes you happy.

I am so happy now because I have reached the end of my planned writing job. I feel a sense of fulfilment. I have done my best to answer my 'for' question. Found my most important self-love. My life is now going to change big time...

Watch Me

I have denied experts and survived.

My powerful inner spirit strived.

I am in a very different realm,

that will try to overwhelm,

even your logical mind.

I'm truly a rare find,

Just watch me,

You'll see!

Mainly now, thanks to Louise Hay's 'Mirror trick', I watch myself and speak to myself in my mirror all the time. I also do speak to myself when I write, as you know. You'll never part with me, Shane. Never. We're parting with writing in this book though.

Thank you for sticking with me in my frivolous, eccentrically sporadic banter, even with me repeating and rambling on about the same thing essentially.

Basically, I want you to make the best of your life while you are here because you'll never know when to leave Her. But do your best to make Her 'a better place, for you and for me and the entire human race!' (MJ)

Michael Jackson tried. I have given it my all and am hoping I've done my best. Thank you for being with me through this journey. Please take it into consideration that you are living your own life, so you get to decide everything personally. **'You are born to live, not living because you are born.'**

I was 'born again' to live and spread my word in order to get people to analyse their reality, know what they want and set...

Aim

The disdain of existence,
may be relieved,
when one takes a chance,
with new ideas conceived.

At previously lost happiness,
do not wince,
one's soul seeks forgiveness
from the mere parody of this existence!

So, when destiny beckons,
with happiness astute.
At previously lost missions,
one needs to shoot.

Aim for a new life,
to alleviate one's personal strife.

Envision and aim. Like Julian in Robin Sharma's 'The Monk Who Sold His Ferrari' says, "When you control your thoughts, you control your mind. When you control your mind, you control your life. And once you reach the stage of being in total control of your life, you become the master of your destiny."

In essence, all that basically needs is...

Change?

Things do not remain the same.
This life is somehow filled,
with sporadic, uneven fame.
The need to be unique, in us all, is instilled?

Even though it may seem,
that one has reached,
a zone so supreme,
somehow our ship has beached.

You only need to take a glance,
to analyse and measure.
Become a knight, to aim your lance,
to maintain Mother Earth, our treasure.

IT wants a simple change,
IT may even be personally quite complex?
But in a totally different reality range,
as for Her, it will have positive effects!

IT awaits to open the gate
when you end this life,
and maybe change your fate
by helping Her deal with Her strife?

So, embrace, have the grace
to aim at a different range,
help Her in Her dire case–
CHANGE, CHANGE, CHANGE?!

Afterword

In October 2018, on Facebook, I received a link from a wildlife society to join them. I had just watched the 'Blue Planet' documentary series hosted by Sir David Attenborough, with the last episode dealing with the pollution of our oceans. Truly heartbreaking what we have already done, especially with plastics. ☹

I saw pictures that brought tears to my eyes.

Most of us don't know directly because we close ourselves off from it. But it is happening. We don't think. See the turtle again? HOW did the shell actually grow without freeing it from the plastic? HOW strong is that plastic actually?

This really broke my heart even more, so I joined up. A few days later, I received a call from a call centre consultant to ask me for a monthly donation to the wildlife society I mentioned above.

With the consultant, I had a very long conversation about our Mother Earth's pending state, especially about our trained indoctrination to pray for change to our many 'Gods' and 'Goddesses', but we do not do anything directly.

At least they were actively doing something, so I readily agreed to do a debit order every month. *(Research legitimate wildlife fundraisers and maybe do as well? Every small bit will count. I also do a monthly debit order to a feeding scheme as well. It does sincerely feel good to help people and our planet.)*

I was actually taken by surprise after the consultant suggested to me that maybe I should make my plea to their head office to be a speaker for them. He said that I had him very gripped in our mere telephone call, in which I briefly explained my ideologies to him, in a casual manner, even a funny witty way, playing with my voice training with different accents, etc.

A simple call between two strangers, yet we really connected. The consultant then suggested that maybe his society can help me with this book and distribute it under them because, at the end of the day, we shared the exact same principle to get people to shift focus to our reality, our planet—EtheREALism.

Universal Timing? I never thought of it before. Ideal way to get my word out there—give them my book to sell through them and they keep 60% of the selling price, 10% to disabled societies, 10% to needy children, 10% to feeding schemes, and I will 'donate' to myself as a disabled, unemployed author the last 10% to basically cover my own travel costs and accommodation as I tour around the world as a speaker and become an ambassador for them.

After a few days, I submitted my full proposal to them to take 500 copies of this book that Anuradha vowed to pay one-third of the printing costs for in her sister Mei's name. Yes, I am still in disbelief, BUT Anuradha is a true Earth Angel, who also wants the world to focus on our Mother Earth and each other.

My second Mother, my Goddess, would also gift me some of the money needed for the first 500 copies, from the little she had from my God's estate. In essence, Mr. Praag will actually help me to fart in a thunderstorm. LOL ☺

Basically, 500 books gifted to the wildlife society, and then from selling the books, they can use some of the profit to reprint and sell. Would not cost them a cent to make millions.

BUT, sadly, the wildlife society specified that even though I could make them lots of money, they were not interested. I was really saddened, BUT my dream in Universal Timing never dies. Everything happens for a reason, right?

I seriously have approached lots of people, companies and publishers, even radio stations. A 10-year struggle, BUT I am still a nobody to them and not my destiny.

The person who makes me feel more as a somebody because we are very close digital friends is Anuradha. She keeps reminding me that I changed her whole life around, and she wants to help me reach the world.

After trying very hard with set publishing houses, and experiencing exactly what I went through while banging on closed doors, Anuradha contacted Notion Press, who she kept on hold because Notion Press is actually a self-publisher who makes books after a set fee is paid.

Even though I did fall for a few scams by self-publishing companies who wanted me to change things in my book as well as give the rights of my work away to them, Notion Press is different. I perused the email they sent to Anuradha and was amazed, especially after the publishing manager called me directly and actually listened to me and agreed with my thought patterns and ideologies totally. I felt more at ease.

The manageress even admitted to me that she does not contact authors directly at all, but there was a 'something' that attracted her to my book because she sincerely believes in my vision. This was a 'spiritual connection' to affirm to me that I had finally found the right place—my ancestral roots in a sense?

Yes, I do work with my inner gut feeling, my soul. Hard in today's economically entrenched world, BUT as much as you can, try to remain true to what is inside you. Work with what you 'feel'.

Yes, I do live in a dreamy, fantasy reality, just like a little kid, and read into things from my own ethos, so I have to point out this 'special' thing as well with Notion Press. Mei had a dog in China. After her passing, Anuradha arranged for him to be sent to her in India. I have kind of chatted to him via WhatsApp. He is a very special dog. Why I'm pointing this out as to why I feel like, spiritually, Mei, who was actually the root behind me even trying to publish this book again, is still playing her part is because her dog and my publishing manager at Notion share the same name. Connection? Special?

This is how I feel about my book reaching you through Notion Press, whose costs were split three ways—Anuradha, my mum and me. You just finished reading it, and your mind is now ticking. Your life will definitely improve, much like it did with those who have read it. Personally, I will remain the humble humanitarian that I am. I want to actually take action and make the world a better place, for you and for me and the entire human race. Not just sing Michael Jackson's song.

Again, all this is because of Anuradha. Sometimes people come along in your life for a reason, no matter how exactly you found them, or how you met them, but we must value them. Destiny is like that.

Everything happens for a reason. Every person you meet is there for a reason. You read this book for a reason. I returned from the dead and struggled to get my word out there for a reason. That reason is my dream to educate the masses to shift focus to our Mother Earth, plus distribute the majority of the profits to various charities and establish me as a speaker as I venture around to educate the masses to smell my fart. LOL ☺

Hmmmnnn? By buying this book and smelling it, with a new mindset, you are also helping Mother Earth and many people. And maybe I will be found by a big publishing company as well to make me reach more people? Help? Spread the word? Small start, this, another step? Have I answered my entitling question?

My inner being is SO happy. Find your inner bliss however you want to find it. You are blessed. Thank you. Live, Love, Laugh?

Love & Light,

Shane ☺

Glossary

1. Aaja – Hindi – father's father *(are-jar)*
2. Aaji – Hindi – father's mother *(are-jee)*
3. Babbelaas/Barbie – Hangover
4. Ballie – Father *(bar-lee)*
5. beta – son *(bear-tah)*
6. bor – dried seed
7. bra – friend
8. braai – barbeque
9. bro/bru – brother
10. bruin – racial group – 'Coloured' – Caucasian + African *(brain)*
11. Bhabi – Hindi – sister-in-law *(bar-bee)*
12. bhaia – Hindi – brother *(buy-yah)*
13. bhani – Hindi – sister *(bay-knee)*
14. bhurra – Hindi – big *(bar-rah)*
15. cherries – girlfriends

16. chicks – girls
17. chota – small *(chore-tah)*
18. chuffed – excited
19. con – lie/fake trick
20. dada – boss *(dah-dah)*
21. dallahs – leaders *(dah-lahs)*
22. didi – Hindi – sister *(dee dee)*
23. dips – biker term for leaning the bike on a bend
24. GABI – Grin And Bear It
25. golfaholic – lover of golfs
26. gogo – Zulu for older lady but associated with domestic maid *(goh-goh)*
27. hele – all *(here-leh)*
28. ja – yes *(yah)*
29. 'jee – Hindi addition for acclaim
30. jol – party/nice time
31. junejunees – sensation when mouth waters and cringes to eat *(june-jew-knees)*
32. kaalvoet – barefoot
33. klonkies – warriors
34. kak – shit *(cuck)*
35. kaka – Hindi – father's brother *(car-car)*
36. laduma – Zulu – thrilling chant meaning goal *(lar-doo-ma)*
37. larnie – rich boss *(lah-knee)*
38. lightie/s – young boy/s *(light-ee/s)*
39. lukka – nice *(luch-her)*
40. LOL – Laugh Out Loud
41. madir – nice *(mar-dirr)*
42. Mama – Hindi – mother's brother *(mar-mar)*

43. mamba – big fish *(marm-bah)*
44. Mami – Hindi – mother's brother's wife *(mar-me)*
45. Mandir – Hindi – temple *(mun-dirr)*
46. Matha – Hindi – Goddess *(mar-tha)*
47. Mosi – Hindi – mother's sister *(mow-see)*
48. Namaste – Hindi blessing, greeting – 'hello' and 'goodbye' *(nah-must-eh)*
49. Nana – Hindi – mother's father *(nah-nah)*
50. Nani – Hindi – mother's mother *(nah-knee)*
51. ou – guy *(oh)*
52. papah – baby *(pah-pah)*
53. palmed – bribed
54. pitha – Hindi – father *(pea-thah)*
55. Pooah – Hindi – father's sister *(poo-ah)*
56. Poopah – Hindi – father's sister's husband *(poo-pah)*
57. Sheezy – sure
58. skottle – pan used to barbeque
59. skyfe – cigarette *(skay-feh)*
60. swaar – brother-in-law
61. Thuqdeer – Hindi – fate *(thuck-dee-rr)*
62. Wors – long sausage *(vors)*
63. Your'll – Indian slang for you all
64. vying – going *(v-eye-ing)*

Republic of South Africa

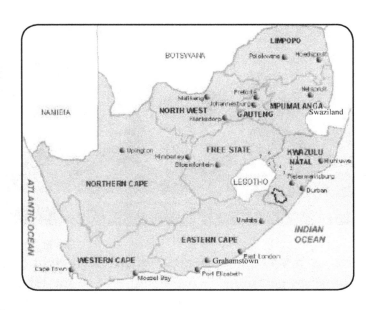

Index

1. Newcastle *(between KwaZulu-Natal)*
2. Ladysmith
3. Estcourt
4. Bergville
5. Drakensburg Mountain Range
6. Harrismith
7. QwaQwa

How Anuradha, a very talented artist in her own right, sees me. The most priceless painting she made and sent to me for my 2019 birthday. We actually ALL are 'Light' that we need to shine on the world with 'Love'.

CPSIA information can be obtained
at www.ICGtesting.com
Printed in the USA
LVHW030006091019
633650LV00001B/16/P